WJEC
Physics
for A2 Level

**Gareth Kelly
and Nigel Wood**

Illuminate
Publishing

Published in 2016 by Illuminate Publishing Ltd, P.O Box 1160, Cheltenham, Gloucestershire GL50 9RW

Orders: Please visit www.illuminatepublishing.com
or email sales@illuminatepublishing.com

British Library Cataloguing in Publication Data

A catalogue record for this book is available from the British Library

ISBN 978-1-908682-59-8

Printed by Severn, Gloucester

10.20

The publisher's policy is to use papers that are natural, renewable and recyclable products made from wood grown in sustainable forests. The logging and manufacturing processes are expected to conform to the environmental regulations of the country of origin.

Every effort has been made to contact copyright holders of material reproduced in this book. If notified, the publishers will be pleased to rectify any errors or omissions at the earliest opportunity.

This material has been endorsed by WJEC and offers high quality support for the delivery of WJEC qualifications. While this material has been through a WJEC quality assurance process, all responsibility for the content remains with the publisher.

WJEC examination questions are reproduced by permission from WJEC

Editor: Geoff Tuttle
Design: Nigel Harriss
Layout: EMC Design Ltd, Bedford

Acknowledgements

The authors would like to thank Keith Jones, for his eagle eyes in spotting mistakes and for his many insightful suggestions. One of the authors (GK) also thanks Penglais school, Aberystwyth, for allowing him the use of a physics laboratory.

Major thanks to the long-suffering wives of both authors who had been labouring under the delusion that their partners had retired!

Cover Image: © Pavel L Photo and Video/Shutterstock

Image credits:
p. 10 (top) Christian Bertrand/Shutterstock; **p. 10** (center) Martyn Goddard/Getty; **p. 10** (bottom) FAECIASP/NASA/Conicet of Argentina/Getty; **p. 14** NASA/Hubble; **p. 15** (top) Mim Friday/Alamy, (bottom) deviyanthi79/Shutterstock; **p. 17** Tristan3D/Shutterstock; **p. 32** (top) Keystone/Getty, (bottom) PlusONE/Shutterstock; **p. 39** Edwin R.Jones, University of South Carolina; **p. 51** (right) northallertonman/Shutterstock, (left) W Bradley, Gooder Lane Ironworks, Brighouse/Creative Commons, (center) Alan Levenson/Getty; **p. 56** Adam Hart-Davis/Science Photo Library; **p. 66** (top) Science Photo Library, (bottom) Cavendish Laboratory, University of Cambridge; **p. 83** Yury Dmitrienko/Shutterstock; **p. 98** Eric Schrader/Creative Commons; **p. 107** miha de/Shutterstock; **p. 112** Nigel Wood; **p. 115** Emilio Segre Visual Archives/American Institute Of Physics/Science Photo Library; **p. 118** Carolina Biological Supply; **p. 123** denisk0/iStock; **p. 131** Boris15/Shutterstock; **p. 136** RGB Ventures/SuperStock/Alamy; **p. 137** NASA/CXC/Caltech/M.Muno et al/Creative Commons; **p. 138** (top) Stocktrek Images/Alamy, (center) John Vickery and Jim Matthes/Adam Block/NOAO/AURA/NSF, (bottom) DR. Rudolph Schild/Science Photo Library; **p. 139** NASA/CXC/Penn State/G. Chartas et al/NASA; **p. 144** Javier Lorenzo, University of Alicante; **p. 150** Nigel Wood; **p. 152** USA Lilbrary Of Congress/Science Photo Library; **p. 154** (top) Klinger Educational Products, (bottom) Shin Okamoto/Shutterstock; **p. 156** Jean Collombet/Science Photo Library; **p. 159** Nigel Wood; **p. 163** Nigel Wood; **p. 168** Nigel Wood; **p. 176** 3B Scientific Limited; **p. 187** Giphotostock/Science Photo Library; **p. 206** Daniel W. Rickey/Creative Commons; **p. 207** (top right) Suttha Burawonk/Shutterstock, (center right) thailoei92/Shutterstock, (left) thailoei92/Shutterstock, (bottom right) EPSTOCK/Shutterstock; **p. 208** Hank Frentz/Shutterstock; **p. 210** Sovereign, ISM/Science Photo Library; **p. 211** (right) Pr Michel Brauner/Science Photo Library, (left) Science Photo Library; **p. 213** (top right) Michel Brauner/ISM/Science Photo Library, (left) SMK4pix/Shutterstock, (right) Michel Brauner/ISM/Science Photo Library; **p. 216** Philippe Plailly/Science Photo Library; **p. 217** Brian Bell/Science Photo Library; **p. 220** Richard W. Thorpe; **p. 221** (top) wavebreakmedia/Shutterstock, (center) stihii/Shutterstock, (bottom) studioloco/Shutterstock; **p. 228** epa european pressphoto agency b.v./Alamy; **p. 230** DR. Gary Settles/Science Photo Library; **p. 231** hiloi/Shutterstock; **p. 232** (top) ARENA Creative/Shutterstock, (bottom) Martin Cloutier/Shutterstock; **p. 236** 1059SHU/Shutterstock; **p. 241** Ken Edwards/Alamy; **p. 242** Paul J Martin/Shutterstock; **p. 244** Us Department of Energy/Science Photo Library; **p. 247** Besto Instruments; **p. 273** Falcon Acoustics.

Ilustrations by: Laszlo Veres/Beehive Illustration: pp. 149 (right), 153 (top), 199 (top), 227 (bottom); Mark Duffin: p. 220 (bottom left); EMC: pp. 154 (center left), 158, 217 (top).

Contents

Unit 3: Oscillations and nuclei

Unit 4: Fields and options

Mathematical and data-handling skills

Answers

How to use this book

This book has been written to support the second half of the WJEC A level Physics specification. The layout of the book in the first two sections matches that of Units 3 and 4 respectively of the A level Physics specification. It provides you with information which covers the content requirements of the course as well as plenty of practice questions to allow you to keep track of your progress and to prepare successfully for your A level examinations.

This book covers all three of the Assessment Objectives (AOs) required for your WJEC course. The AOs are:

- **AO1, Knowledge and understanding** of physics ideas and practice. This comprises 35% of the A level examination including the specified practical activities.

- **AO2, Applying knowledge and understanding** of physics ideas and practice, which comprises 45% of both the AS and A level examinations.

- **AO3, Analysing, interpreting and evaluating scientific information, ideas and evidence** which is 25% of the A level examination.

The main sections in the book are the A level Units 3 and 4.

- Unit 3 covers Oscillations and nuclei
- Unit 4 covers Fields and options

There is an additional section on

- Mathematical and data-handling skills

The Units 3 and 4 sections also include much practical and mathematical material in context. This additional section covers aspects of these skills which are more appropriately dealt with separately. It should be read alongside the Practical skills and Mathematical skills sections of the AS book.

Level of coverage

This book contains material which is examined only in the A level units of the course. The level of coverage in both the main body of the text and the practice questions includes concepts and treatment designed to stretch and challenge students.

Practice questions

As well as the self-test questions in the margin of the main text, each section in Units 3 and 4 ends with an exercise of practice questions. The mathematical and data-handling skills chapter also includes both self-test questions and a terminal exercise, which are not tied to specific subject content but are designed to allow you to practise these essential skills.

As well as containing material relating to the content of the sections, the exercises in Units 3 and 4 contain data-analysis questions around the specified practical work for the unit.

Some questions also relate to the content of more than one unit: the A level examination requires candidates to answer such synoptic questions, which bring together a range of ideas from the physics course. The solutions to these exercises as well as the self-test questions are to be found at the end of the book. Unless the self-test and exercise question asks for the reasoning or working to be shown, the solutions to mathematical questions are generally limited to a final answer rather than showing the way of reaching the answer.

Examination style questions

Following the section on mathematical skills you will find a set of examination-style questions. As the style and the level of demand of the new A level examinations have changed significantly, these questions are not past-paper questions but have been specially written to reflect the changed demands. Similarly to the exercises at the ends of the sections, parts of these questions have also been written with the synoptic requirements in mind and draw together material from different sections of the specification. The solutions to these questions take the form of model answers which, in many cases, provide more than the minimum answer to achieve full marks. In some cases alternative, equally valid, answers are given.

Section B of Unit 3 of the A level examination consists of a structured set of questions based upon a passage, i.e. a comprehension question. Unit 5 of the examination consists of a practical examination. Whilst this book gives important background for your preparation for these units, they are not discussed in examination terms. Some pointers for them are included in the A2 / year 13 Student Resource Guide.

Margin features

The text is supplemented by a number of margin features:

Terms & definitions

These are physics terms and laws that you need to be able to quote without further information.

For example:

┌── *Terms & definitions* ──┐

In the context of laboratory investigations into radioactive sources, **background radiation** is the nuclear radiation which is detected in the absence of the source under investigation.

Physics examination papers always contain a few marks for defining terms or stating a physical law.

Study point

Some ideas from the main text are further developed in Study points. These are used for material which is important for you to understand but is tangential to the flow of the text. Some Study point boxes include material designed to extend your understanding beyond the requirements of the specification.

For example:

▼ **Study point**

ω always has a positive value so we don't need to write $\omega = \pm 20$ s^{-1}.

Self-test

Questions to enable you to check your understanding of the subject at that point in the text. They are often short calculations (though some are longer and some require short verbal analysis), which require you to apply the concepts developed in the main text. The answers to these Self-test questions are given at the end of the book.

Material developed in these is designed to make you think more deeply about the subject. They usually include questions for you to consider, discuss and answer. They will be most useful for students intending to progress to higher study in physics or engineering. Unlike the self-test marginalia, the answers to these questions are not provided.

MATHS TIP

These refer to particular techniques and often direct you to the Mathematical skills section for a fuller treatment.

Matching the specification content to the sections in this book

Units 3 and 4 in this book are generally arranged to follow the WJEC specification. For example:

Section 3.1, Circular motion on pages 70–79 deals with Unit 3 Topic 1 of the WJEC specification, with the same name.

However, some areas of physics are related closely to material in more than one unit. The understanding of the nature of electric fields, for example, relates to the content of Unit 4, topics 1 (Capacitors) and 2 (Electrostatic and gravitational fields of force). To facilitate the use of this book, a grid is provided on page 300 to match specification areas to book sections.

An extract is shown below.

Book			Specification	
page	section	section title		
10	3.1.1	The kinematics of 'uniform' circular motion	3.1	a,c,d
12	3.1.2	The existence of a resultant force for an orbiting object	3.1	e,f
13	3.1.3	Centripetal acceleration	3.1	e,f,g
14	3.1.4	Centripetal force	3.1	e,f,g
21	3.2.1	Introduction to simple harmonic motion	3.2	a, c
22	3.2.2	Equations of simple harmonic motion	3.2	b, d
23	3.2.3	Characteristics of simple harmonic motion	3.2	e, f
25	3.2.4	Examples of simple harmonic motion	3.2	i, j
28	3.2.5	Energy of oscillation	3.2	k
29	3.2.6	Damped oscillations	3.2	l, m
29	3.2.7	Forced oscillations	3.2	o, p, q
40	3.3.2	Expressing the amount of substance	3.3	d, e
42	3.3.3	The ideal gas equation	3.3	a
43	3.3.4	The behaviour of molecules in gases	3.3	b
44	3.3.5	Molecular motion and gas pressure	3.3	b, c
47	3.3.6	Molecular energy	3.3	f

The A level examination

WJEC A level Physics aims to encourage students to:

- develop essential knowledge and understanding of different areas of the subject and how they relate to each other
- develop and demonstrate a deep appreciation of the skills, knowledge and understanding of scientific methods
- develop competence and confidence in a variety of practical, mathematical and problem-solving skills
- develop their interest in and enthusiasm for the subject, including developing an interest in further study and careers associated with the subject
- understand how society makes decisions about scientific issues and how the sciences contribute to the success of the economy and society.

Examination questions are written to reflect the assessment objectives (AOs) as laid out in the specification. Learners must meet the following AOs in the context of content detailed in the specification.

Assessment objective 1

Learners must: Demonstrate knowledge and understanding of scientific ideas, processes, techniques and procedures

30% of the marks for questions set on the examination papers are for AO1. As well as pure recall, such as stating laws and definitions, this includes knowing which equations to use, substituting into equations and describing experimental techniques.

Assessment objective 2

Learners must: Apply knowledge and understanding of scientific ideas, processes, techniques and procedures:

- in a theoretical context
- in a practical context
- when handling qualitative data
- when handling quantitative data.

45% of the marks for questions set on the examination papers are for AO2. Bringing together ideas to explain phenomena, solving mathematical problems and performing calculations using experimental results and graphs are categorised as AO2. Application involves using the skills you have acquired in situations you have not previously encountered, e.g. in synoptic questions.

Assessment objective 3

Learners must: Analyse, interpret and evaluate scientific information, ideas and evidence, including in relation to issues, to:

- make judgements and reach conclusions
- develop and refine practical design and procedures.

25% of the marks for questions set on the examination papers are for AO3. These marks include determining quantities using experimental results and also responding to data to draw conclusions.

A level Physics written papers

Unit 3 (2 hours 15 minutes)

The paper has two sections:

Section A consists of approximately 7 structured questions, each of which contains several parts, with a total of 80 marks. The questions comprise a mixture of short and extended answer parts.

Section B consists of one 20-mark structured comprehension question.

Unit 4 (2 hours)

The paper has two sections:

Section A consists of approximately 7 structured questions, each of which contains several parts, with a total of 80 marks. The questions comprise a mixture of short and extended answer parts.

Section B consists of four structured 20-mark questions, one on each of four optional units.

Quality of extended response (QER)

Some questions assess how well you can present a detailed argument. These are called Quality of Extended Response questions and are indicated by QER next to the mark allocation. These questions are worth 6 marks and the examiner will assess how well you communicate your physics as well as the standard of the physics itself.

Synoptic questions

To quote from the specification, 'Learners' understanding of the connections between the different elements of the subject and their holistic understanding of the subject is a requirement of all A level specifications. In practice, this means that some questions set in A2 units will require learners to demonstrate their ability to draw together different areas of knowledge and understanding from across the whole course of study.' This is known as synoptic assessment. So, in answering questions in Units 3 and 4, A level students will be expected to be familiar with AS material.

Questions in one A level unit will not focus specifically on content from other units but some questions in each exam will draw upon skills and knowledge acquired in studying for other units. For example, the knowledge of conservation of momentum gained in Unit 1 could be required when discussing α and β decay in a Unit 3 examination.

Mark allocations

The mark allocation for each question part is given in square brackets, e.g.

Explain the purpose of damping in a car suspension system. [3]

The mark allocation gives a good clue to the detail required in your answer. The [3] is a clue that the examiner will expect you to make three valid points in the answer.

Questions involving calculations

Depending on their complexity, questions which require the calculation of a result will generally be allocated more than one mark. Unless the question specifically asks for the working to be shown, full marks will normally be awarded for just a correct answer, consisting of a number, a unit and (in the case of vector quantities) a direction. However, incorrect answers will only be given credit for correct stages in the calculation, so you are advised always to show your working. Normally the number of marks available without the correct answer is one fewer than the total allocation.

Error carried forward (ecf)

This is also referred to as *consequential marking*. The principle is that the result of a calculation, in one part of a structured question, is treated as correct if it is used in a subsequent part of the same question. It is often not applied within a question part. See the Study and Revision Guide for further discussion of the application of *ecf*.

Command words used in WJEC exam questions

Examination 'questions' are not usually phrased as questions. They are usually instructions to do some work. Examiners choose the command words carefully so that you understand the sort of answer required. This is a list of the most common command words used.

State

Give a brief, concise answer with no explanation. For example: **State the value of the intercept**.

For questions which require the value of a quantity to be stated, it is expected that no calculation (or at most a trivial one) is needed.

Describe

Write a short account with no explanation. For example: **Describe the relationship shown in the graph**.

Experimental methods can also be asked for using this command word.

Explain

This requires reasons to be given. Depending on the question a description might also be required.

For example:
Explain which of these samples has the greater activity.

In this example the sample with the greater activity has first to be identified (which might not itself be allocated a mark). The explanation might need to include a calculation.

Suggest

This command word often occurs in the last part of a structured question. There might not be a definite answer but you are expected to put forward a sensible suggestion based upon your physics knowledge.

For example:
Suggest how you could investigate whether the assumption was correct.

Calculate

Use one or more equations to find the value of an unknown quantity.

For example:
Calculate the amplitude of the oscillation.

Determine

This command word is often used instead of calculate. Give a numerical answer by manipulating data you have been given. There is no absolute difference between the words but determine tends to be used in situations where an additional process is required beyond a calculation.

For example:

Determine the decay constant of the radioactive sample.

In this case you might first need to draw a suitable graph.

Compare

For example:
Compare the heat transfer from the two systems.

Make sure that you make a comparison rather than two separate descriptions, e.g. 'System A has a greater heat transfer than system B.' Depending on the context, this type of question might require a numerical answer, e.g. 'The heat transfer from system A is 2.5 × that from system B.'

Discuss

This is often used for questions on the practical application of scientific ideas or technological developments.

For example:
Discuss whether the expenditure on space research is justified.

As with suggest there is no correct answer to this type of question. The examiner is looking for reasoned arguments.

Overview:
Unit 3 Oscillations and nuclei

3.1 Circular motion p10

- The period of rotation, frequency, speed and angular velocity of rotation.
- Angles expressed in radians.
- Centripetal acceleration and force.
- The use of the following equations relating to circular motion.

$$v = \omega r \quad a = \omega^2 r \quad a = \frac{v^2}{r} \quad F = \frac{mv^2}{r} \quad F = m\omega^2 r$$

3.2 Vibrations p21

- The definition and characteristics of simple harmonic motion (shm).
- $a = -\omega^2 x$ and its solutions $x = A\cos(\omega t + \varepsilon)$ and $v = -A\omega\sin(\omega t + \varepsilon)$.
- The graphical representation of shm – the variation of kinetic and potential energy.
- Shm in spring systems and the simple pendulum.
- Damped free oscillations including critical damping and the effect of damping in real systems.
- Forced oscillations and resonance – practical examples.

PRACTICAL WORK
- Measurement of g with a pendulum.
- Investigation of the damping of a spring.

3.3 Kinetic theory p38

- The molar gas constant and the Boltzmann constant.
- The equation of state for an ideal gas, $pV = nRT$.
- The assumptions of the kinetic theory of gases;

$$p = \frac{1}{3}\rho\overline{c^2} = \frac{1}{3}\frac{N}{V}m\overline{c^2}.$$

- The Avogadro constant, the mole, molar mass; $k = \frac{R}{N_A}$.
- The mean translational kinetic energy of the molecules of a gas, $\frac{3}{2}kT$.

3.4 Thermal physics p51

- Thermodynamic systems and their boundaries.
- Internal energy, U and the first law of thermodynamics, $\Delta U = Q - W$.
- Calculations using $W = p\Delta V$, $W =$ the area under a p–V graph and $U = \frac{3}{2}nRT$.
- In practical terms $\Delta U = Q$ for solids and liquids.
- Thermal equilibrium as the definition of temperature equality; absolute zero as the minimum energy state.
- Temperature differences determining heat flow, $Q = mc\Delta\theta$.

PRACTICAL WORK
- Estimation of absolute zero by use of the gas laws.
- Measurement of the specific heat capacity for a solid.

3.5 Nuclear decay p66

- The spontaneous nature of nuclear decay; the nature of α, β and γ radiation, nuclear equations; $^A_Z X$.
- Nature, range and penetrating power of α, β and γ radiation.
- Background radiation.
- Activity, A, the decay constant, λ and half-life $T_{\frac{1}{2}}$.
- The radioactive decay equation, $A = \lambda N$ and the solutions:

$$N = N_0 e^{-\lambda t}, A = A_0 e^{-\lambda t}, N = \frac{N_0}{2^x}, \text{ and } A = \frac{A_0}{2^x}.$$

- The derivation and use of $\lambda = \frac{\ln 2}{T_{\frac{1}{2}}}$.

PRACTICAL WORK
- Investigation of radioactive decay using a dice analogy.
- Investigation of the variation of intensity of gamma radiation with distance.

3.6 Nuclear energy p83

- The meaning and use of $E = mc^2$.
- The unified atomic mass unit (u); calculations of nuclear binding energy and binding energy per nucleon.
- The conservation of mass / energy to particle interactions including fission and fusion.
- The binding energy per nucleon curve.
- The relevance of the binding energy per nucleon to nuclear fission and fusion.

Oscillations and nuclei

This A2 unit has three basic themes:

- **Circular motion** and **vibrations** might initially appear disparate topics but the mathematics of the two is very similar. This is because the projection of a circular motion is a vibration. Many terms are common: period, frequency, angular velocity. They both require the high level use of trigonometry, with angles expressed in radians, and of vectors. The understanding of these topics is key to tackling the later topic of orbits.

- **Kinetic theory** and **thermal physics** both deal with the energy of systems. The understanding of these areas of physics was the key to the development of machines for transferring energy. The former concentrates on molecular behaviour and the latter on the bulk behaviour of systems but there is a great deal of overlap of the concepts.

- **Nuclear decay** and **nuclear energy** complete this unit. These topics are first encountered in pre-A level courses but the mathematical development of them, especially in the description of decays in terms of exponentials, makes this one of the more demanding A level areas. The quantities such as the mole and molar mass introduced in the kinetic theory topic are once again required – emphasising the unity of description developed in the A level Physics course.

Content

3.1 Circular motion

3.2 Vibrations

3.3 Kinetic theory

3.4 Thermal physics

3.5 Nuclear decay

3.6 Nuclear energy

Practical work

Practical work is integral to any physics course. Unit 3 provides a wealth of opportunities for students to hone their practical skills as well as to develop their understanding of the contents.

3.1 Circular motion

Fig. 3.1.1 Circular motion in Barcelona

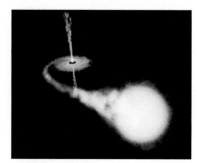

Fig. 3.1.2 Wall of death rider

Every six seconds or so, the excited riders on the swing carousel in Barcelona travel along a horizontal circle about 20 m in diameter. Nor are they alone in this behaviour as the wall of death rider shows. We all, along with other denizens of the Earth, execute a multiplicity of motions that, if not exactly, are close to circular:

• Daily rotation around the axis of the Earth

• Annual path of the Earth around the Sun

• 240 million-year orbit of the Solar System around the 8 kiloparsec distant galactic centre.

As this section is being written, astronomers are celebrating the detection of the gravitational waves which were emitted when two black holes, previously in mutual orbit, coalesced into a single body. This one observation simultaneously confirmed the outstanding prediction of General Relativity theory and provided the first direct observational evidence for black holes themselves (rather than the matter which spirals into them).

To get back to Earth, this section introduces the mathematics needed to describe and understand the behaviour of circling objects – their motion and the forces which produce it. The mathematics is not difficult but the physics concepts contain some traps for the unwary. You have been warned!

3.1.1 The kinematics of 'uniform' circular motion[1]

For the most part we shall limit our study to objects which are moving at a constant speed in circles. This is not unduly restrictive – we shall still allow the swing carousel to spin up and down but only consider the forces and motion when it is at a steady speed. Option C, The physics of sport, considers angular accelerations and decelerations.

Fig. 3.1.3 Angular position

(a) Angular position and speed

Consider an object, **P**, which is moving at a constant speed, v, in a circular path of radius r about a central point, **O**. We measure its linear motion along the circumference from the point **X**. The angular position, θ, as shown is measured from the line **OX**.

We define the **angular speed,** ω, as the rate of change of θ, i.e. the angle swept out by the radius per unit time.

Mathematically: $\omega = \dfrac{\Delta\theta}{\Delta t}$

Units

Scientists often express angles in degrees (°) for reasons of convenience and the familiarity of the audience with these units. Astronomers often subdivide degrees into minutes (') and seconds (") of arc. However, for rotational and vibrational purposes we shall always use radians. The definition of radian measure is covered in the Year 1 / AS textbook, Section 4.2.4. Hence:

Unit of angular position, θ : rad.

Unit of angular speed, ω : rad s^{-1}.

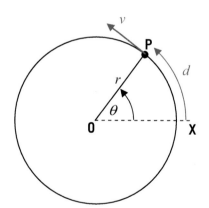

Fig. 3.1.4 Multiple orbits in a star / black hole system

---

1 The quotation marks around 'uniform' are because an object which is travelling in a circle has not got a uniform velocity because the direction is changing – something to be aware of.

Example

A vinyl record disc spins at 33 revolutions per minute. Calculate the angular speed in rad s^{-1}.

Answer

In 60 s, the angle swept out $= 33 \times 2\pi = 66\pi$ rad

$\therefore \omega = \dfrac{\Delta\theta}{\Delta t} = \dfrac{66\pi \text{ rad}}{60 \text{ s}} = 3.46$ rad s^{-1}.

(b) Relationships between angular and linear speed

The linear quantities in circular motion are indicated in red in Fig. 3.1.3; we can use this diagram to relate them to the rotational quantities.

By definition $\theta = \dfrac{d}{r}$

If **P** is moving with speed v, the increase in distance, Δd in time Δt is given by $\Delta d = v\Delta t$

So the angular position increase, $\Delta\theta = \dfrac{v\Delta t}{r}$

$\therefore$ Dividing by Δt and using $\omega = \dfrac{\Delta\theta}{\Delta t}$ gives $\omega = \dfrac{v}{r}$

Example

Calculate the speed of the Earth's motion around the Sun   [1 AU $= 1.50 \times 10^{11}$ m]

Answer

The angular velocity $\omega = \dfrac{\Delta\theta}{\Delta t} = \dfrac{2\pi}{(86400 \times 365.25) \text{ s}} = 1.99 \times 10^{-7}$ rad s^{-1}.

$\therefore$ The speed, $v = r\omega = 1.50 \times 10^{11}$ m $\times 1.99 \times 10^{-7}$ rad s^{-1}
$= 29.9$ km s^{-1}.

Angular speed or angular velocity?

Motion in a circle is directional so it has vector properties. However, the direction of the linear velocity is constantly changing. The one constant aspect is the direction of the axis of rotation, so this is chosen as the direction of the angular velocity vector (using the right hand grip rule, see Fig. 3.1.5). We shall not be treating the vector aspects of angular motion in this course, though it is touched upon in Option C, The physics of sports. In spite of treating only the scalar aspects of rotational motion, it is common to refer to angular speed as angular velocity and we shall use the terms interchangeably in this section of the book, in the same way that people often talk about the *velocity* of light.

Self-test 3.1.1

The magnetic tape in an old reel-to-reel tape player is fed through the play-back heads at a constant 9.5 cm s^{-1}. Calculate the angular velocity of the reel when the diameter of the tape spool is 15 cm.

Self-test 3.1.2

The sound waves on a 33 rpm vinyl record are produced using wavy grooves.

(a) Calculate the wavelength of the grooves for a 1 kHz note at the outside of the 30 cm diameter disc.

(b) How would a 1 kHz groove be different near the centre of the disc?

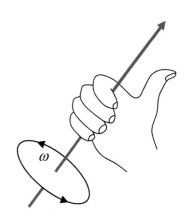

Fig. 3.1.5 Direction of angular velocity vector

(c) Period, T, and frequency, f, of circular motion

The definitions of **period** and **frequency** in the context of circular motion are essentially the same as for waves or any other periodic phenomena. The relationship between them is:

$$\text{frequency} = \frac{1}{\text{period}} \qquad f = \frac{1}{T}$$

In one revolution, the angular displacement is 2π. We can use this to relate the period and frequency to the angular velocity, ω.

$$\omega = \frac{2\pi}{T} \quad \text{and} \quad \omega = 2\pi f$$

These relationships will occur in the same form for vibrational motion. We shall explore the connection at the end of this section.

3.1.2 The existence of a resultant force on an orbiting object

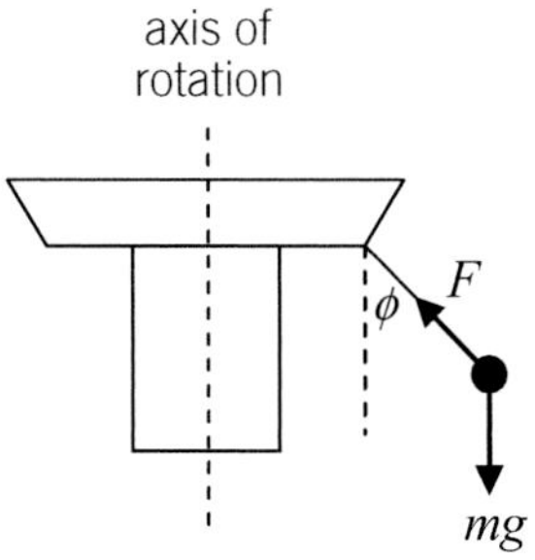

axis of rotation

Fig. 3.1.6 Forces on a swing carousel rider

Consider the forces on one of the fairground riders in Fig. 3.1.1. You may find it easier to use Fig. 3.1.6. Once the carousel has settled down to a steady angular velocity, the riders move in horizontal circles so they have no vertical motion. The vertical components of the forces on the riders therefore sum to zero. There are two forces, that of gravity (mg) and the tension force (F) in the support

So $\qquad F\cos\phi = mg \quad$ i.e. $\quad F = \dfrac{mg}{\cos\phi}$

However, there is a horizontal component to F, so there is a resultant force on the rider:

$$F_{res} = F\sin\phi$$

Substituting for F from above:

$$F_{res} = \frac{mg\sin\phi}{\cos\phi}$$

But $\qquad \dfrac{\sin\phi}{\cos\phi} = \tan\phi \; \therefore \quad F_{res} = mg\tan\phi \qquad$ horizontally to the left

After a half revolution, the rider will be on the left and the resultant force will be horizontally to the right. Hence there will always be a resultant force pointing to the centre of the circle. In the other examples too:

- The wall of death exerts an inward force on the driver as well as a frictional force to hold him up against gravity.

- The black hole and star exert attractive gravitational forces on each other.

We'll now proceed to examine the necessity for such a resultant force and how it depends upon the characteristics of the motion.

3.1.3 Centripetal acceleration

Consider the object in Fig. 3.1.7. It is moving at a constant speed of 20 m s^{-1} with in a circle of radius 5 m. We'll calculate the mean acceleration between points **A** and **B**, i.e. $\frac{\pi}{4}$ either side of the vertical line **OY**.

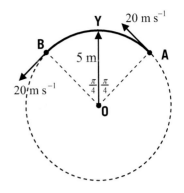

Fig. 3.1.7 What's the acceleration?

The horizontal component of the velocity is unchanged between **A** and **B** at $20\cos\frac{\pi}{4}$ m s^{-1} to the left.

The vertical component of velocity at **A** $= 20\sin\frac{\pi}{4}$ m s^{-1} in direction **OY**.

The vertical component of velocity at **B** $= 20\sin\frac{\pi}{4}$ m s^{-1} in direction **YO**.

$$= -20\sin\frac{\pi}{4}\text{ m s}^{-1}\text{ in direction }\mathbf{OY}$$

$$\therefore \Delta v = v_{\mathrm{B}} - v_{\mathrm{A}} = -20\sin\frac{\pi}{4} - -20\sin\frac{\pi}{4} = -40\sin\frac{\pi}{4}\text{ m s}^{-1}\text{ in direction }\mathbf{OY}.$$

To calculate the mean acceleration we need to know the time interval between **A** and **B**. From the result of Self-test 3.1.4, we can now write:

So the mean acceleration is given by:

$$\langle a\rangle \;=\; \frac{\Delta v}{\Delta t} \;=\; \frac{-40\sin\frac{\pi}{4}}{\frac{\pi}{8}} \;=\; -72\text{ m s}^{-1}(2\text{ s.f.})\text{ in the direction }\mathbf{OY}$$

You should check this result. The figure of 72 m s^{-1} is not so important but the direction is: the mean acceleration is towards the centre of the circle. If you do Self-test 3.1.5, you will find a result with a very similar magnitude (slightly greater) but in the same direction: there is always a mean acceleration around the point Y in the direction YO.

We'll now build on this result and calculate the instantaneous acceleration at point **Y** itself. To do this we'll need to use the result

$$\lim_{\theta\to 0}\frac{\sin\theta}{\theta} = 1,$$

which follows from the definitions of θ (in **rad**) and $\sin\theta$. Fig. 3.1.8 is just a redrawn Fig. 3.1.7 with general values of the variables.

We could use components again but, for flexibility, we'll use the vector diagram, Fig. 3.1.9, to calculate Δv.

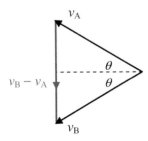

Fig. 3.1.9 $v_{\mathrm{B}} - v_{\mathrm{A}}$

The magnitudes of v_{A} and v_{B} are equal and as they make the same angles to the horizontal.

$\therefore$ by symmetry, $v_{\mathrm{B}} - v_{\mathrm{A}}$ is vertical

and $\Delta v = v_{\mathrm{B}} - v_{\mathrm{A}} = 2v\sin\theta$

The arc length AB is given by: arc $AB = 2r\theta$

$\therefore$ The time Δt, between A and B is given by:

$$\Delta t = \frac{\text{Arc AB}}{v} = \frac{2r\theta}{v}$$

Hence the mean acceleration between **A** and **B** is given by:

$$\langle a\rangle = \frac{\Delta v}{\Delta t} = 2v\sin\theta \times \frac{v}{2r\theta} = \frac{v^2}{r} \times \frac{\sin\theta}{\theta}$$

and the direction is downwards in Fig. 3.1.8, i.e. always towards the centre of the circle.

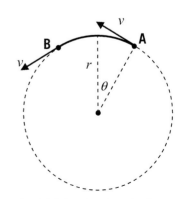

Fig. 3.1.8 Centripetal acceleration

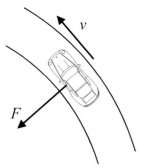

Fig. 3.1.10
M51 – the Whirlpool galaxy

Fig. 3.1.11 Car on a bend

We can use this result and $\lim_{\theta \to 0} \frac{\sin\theta}{\theta} = 1$ to calculate the instantaneous acceleration at the top of the arc, i.e. when $\theta = 0$. The answer is the very simple equation $a = \frac{v^2}{r}$ which, applying $v = r\omega$ we can write as $a = r\omega^2$.

Because this acceleration is directed towards the centre of the circle it is referred to as the **centripetal acceleration** (from the Latin, to seek the centre).

Example

The international space station (ISS) orbits at a mean height, h, of 412 km with an orbital period, T, of 92.69 minutes. Calculate its acceleration.

[Mean radius of the Earth, $R_E = 6371$ km.]

Answer

The angular velocity of the ISS $= \frac{2\pi}{T}$ $\therefore$ $a = r\left(\frac{2\pi}{T}\right)^2 = \frac{4\pi^2}{T^2}(R_E + h)$

$$= \frac{4\pi^2}{(96.29 \times 60\text{ s})^2}(6371 + 412) \times 10^3\text{ m}$$

$$= 8.02\text{ m s}^{-2}. \text{ (See Study point)}$$

3.1.4 Centripetal force

We can now use Newton's laws to move from kinematics to dynamics; we can explore what makes a body move in a circle at a steady speed. N1 and N2 tell us that, if an object is accelerating, the resultant of all the forces on it cannot be zero; there must be a resultant force in the direction of the acceleration, i.e. towards the centre of the circular path. We can calculate the magnitude of this **centripetal (resultant) force** using the N2 equation, $F = ma$.

From above: The centripetal force, $F = \frac{mv^2}{r}$ or $F = mr\omega^2$.

Some examples of centripetal forces

1. Every one of the HI regions (the pink clouds in NGC1566 – the spiral galaxy in Fig. 3.1.10) is acted on by the gravitational fields of all the other objects in the galaxy. The resultant force on each HI region (as well as on the stars) is towards the centre of the galaxy.

2. The resultant of the tension and gravitational forces on the riders on the swing carousel is towards the axis of the carousel.

3. The sideways grip of a car's tyres on a horizontal road as it negotiates a bend (Fig. 3.1.11) is towards the centre of the bend.

It is worth examining that last one. Tyres have a maximum grip, F_{max}, which is normally expressed as a fraction, μ, of the normal contact force, C, between the car and the road. On a horizontal road, C is equal (and opposite) to the weight, mg, of the car.

i.e. $F_{max} = \mu C$

Typically $\mu \sim 1$ on a dry road and 0.2 on a wet road. Using this information we can estimate the maximum safe driving speed around a bend.

Example

Estimate the maximum safe speed for a car to drive around a tight bend on a country lane, which has a radius of curvature of 20 m, if the road is dry.

Answer

Assuming that $\mu = 1$.

The only horizontal force on the car with a component along the radius of the bend is provided by the grip on the tyres. So this provides the necessary centripetal force.

The maximum grip $F_{max} = \mu C = 1.0mg$

With a speed of v, the centripetal force $= \dfrac{mv^2}{r} = \dfrac{mv^2}{20 \text{ m}}$

$\therefore$ At the maximum speed $\dfrac{mv^2}{20 \text{ m}} = 1.0m \times 9.81 \text{ m s}^{-2}$

$\therefore$ Cancelling and re-arranging: $v_{max} = \sqrt{20 \text{ m} \times 9.81 \text{ m s}^{-2}} = 14 \text{ m s}^{-1}$

Self-test 3.1.7

Repeat the calculation in the example for a wet road.

Self-test 3.1.8

Why do the answers to the example and Self-test 3.1.7 not depend upon the mass of the car?

3.1.5 Two common misconceptions

(a) Centrifugal force

When a car goes quickly round a bend, passengers 'feel' a force throwing them to the outside of the bend. Similarly in the fairground ride, the 'cage', the punters 'feel' a force pushing them onto the wall. This sensation is interpreted as indicating the existence of an outward acting force, hence *centrifugal* [Lat: fleeing the centre]. This is an understandable interpretation but it is false – there is no outward force, but there **is** an inward force, as we have seen, which somehow the passengers are ignoring. Suppose (for the moment) that there were an outwards pushing force: what external body is exerting this force on the people and what is its N3 partner force?

This misinterpretation is not confined to rotations: Fig. 3.1.13 is of a (cartoon) airline passenger, but it could equally be a car passenger. The aeroplane is accelerating to the left for takeoff. For twenty seconds or so the seat exerts a forward force on the passenger which accelerates him to take-off speed. But what does the passenger 'feel'? He 'feels' pressed backwards into the seat.

The whole problem is trying to apply Newton's laws of motion in an accelerated frame of reference (accelerating plane or rotating car). It is *possible* to do so by inventing additional, 'fictitious' forces.[2] However at A level, we'll stick to inertial (non-accelerating) frames of reference.

There is a similar effect with gravity and weight. There is no sensation of weight if gravity is the only force on us, as in the ISS or during a dive from a high board. We know that gravity is exerting its ineluctable tentacles on us but we are only aware of it when we are stopped from falling freely by the upward force of the floor or a chair. So the *downward* sensation of weight arises from the compression of our tissues between the normal contact force (which is upwards) and the sensationless downward force of gravity.

So the apparently outwards force on the cage passengers is a psychological misinterpretation.

Fig. 3.1.12 The 'cage'

Forward acceleration

Fig. 3.1.13 Force on an accelerating passenger

---

2 Another fictitious force, beloved of meteorologists, is the Coriolis force, which they use to explain the circular motion of air around low and high pressure regions.

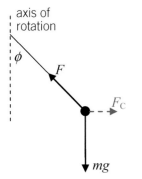

Fig. 3.1.14 Swing carousel rider with fictitious force

Unfortunately, adding this fictitious force can lead to correct results even if the physics is wrong. The red dotted force in Fig. 3.1.14 is this 'centrifugal force'. If a student argues that the three forces, F, mg and F_C are in equilibrium, the triangle of forces can be used to calculate F_C. By a sleight of hand this then becomes the centripetal force which can be equated to $mr\omega^2$: two errors leading to a 'correct' answer.

(b) An additional centripetal force

When solving physics problems, e.g. the swing carousel, students are often tempted to include an *additional* centripetal force over and above the actual forces. This would complicate the carousel force diagram even more. The important thing to remember is that there is no separate centripetal force: it is the resultant of all the forces on the rotating object.

▼ **Study point**

In simple questions, the loop-the-loop car in Fig. 3.1.15 is unpowered and frictionless. However, its speed is not constant because of the changing gravitational potential energy. If we are careful we can still use our constant speed equations to answer the question.

3.1.6 Some nice problems

(a) Loop-the-loop

Whether the question is about toy cars, marbles or an aircraft in flight, the principle of the question is exactly the same.

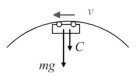

Fig. 3.1.16 Forces at the top of the loop

3.1.9 ⌄ **Self-test**

A model car track has a radius of 20 cm. Calculate the minimum speed needed at the top of the loop.

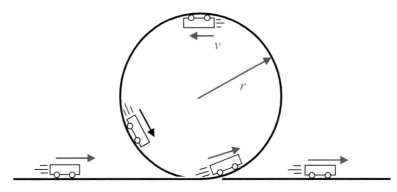

Fig. 3.1.15 Loop-the-loop cars on a track

How fast does the car (or marble, or plane) have to go to stay in the loop?

We'll examine the critical situation at the top of the curve. Supposing for the moment that the car remains on the track, what forces are acting?

- Gravity, mg, downwards

- The normal contact force, C, downwards.

As these are the only forces acting and they are both in the direction of the centre of the circle then the sum of them is the centripetal force.

$$\therefore \quad \frac{mv^2}{r} = mg + C$$

C cannot be less than zero (the track doesn't hold on to the car), so for the car to maintain contact with the track,

$$\frac{mv^2}{r} \geq mg \quad \therefore \quad v \geq \sqrt{rg}$$

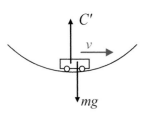

Fig. 3.1.17 Forces at the bottom of the loop

At the bottom of the loop (Fig. 3.1.17) the contact force, C', and mg now oppose each other so the resultant towards the centre is $C' - mg$.

Hence $\dfrac{mv^2}{r} = C' - mg$, $\quad \therefore \quad C' = \dfrac{mv^2}{r} + mg$

The minimum value of C' is mg as expected.

Example

Find an expression for the speed with which a car enters a loop-the-loop or radius r at the bottom, if it just remains in contact with the loop at the top.

Answer

The speed at the top is given by $v^2 = rg$. So its kinetic energy at the top is $\frac{1}{2}mrg$.

The gain in potential energy as the car climbs the loop $= 2mgr$

$\therefore$ The loss in kinetic energy as the car climbs the loop $= 2mgr$

$\therefore$ Initial kinetic energy $= \frac{1}{2}mrg + 2mgr = \frac{5}{2}mrg$

$\therefore \frac{1}{2}mv^2 = \frac{5}{2}mrg$, $\therefore$ The initial velocity $v = \sqrt{5rg}$

(b) Newton's law of gravitation

Until the 16th century it was a widely held belief that the scientific laws of the *sublunary sphere* (the geocentric universe below the sphere of the Moon, i.e. the Earth and its atmosphere[3]) were different from those of the rest of creation, which was made of *quintessence*. Newton showed that the inverse square law related the acceleration due to gravity on the surface of the Earth (i.e. 9.81 m s⁻²) with the centripetal acceleration of the Moon in orbit around the Earth.

Self-tests 3.1.10 and 3.1.11 take you through this calculation. Note that the calculation is not 100% accurate; because the Moon doesn't orbit the centre of the Earth but the centre of mass of the Earth-Moon system (see Section 4.3).

(c) The shape of a rotating planet

What is the shape of the Earth? Apart from a few wrinkles (mountains. continents, oceanic basins…) it is spherical. Well not quite. If you examine the Hubble Space Telescope image of Jupiter in Fig. 3.1.19 and compare its outline with the red circle, you'll notice that it is slightly flattened at the poles. The Earth is similarly shaped, but the deviation from a sphere is less. Why is this? We're not going to allow ourselves to say, 'Centrifugal force,' and swiftly pass on. So, what's the story?

A non-rotating planet would be spherical. This is the minimum potential energy shape. Even dwarf planets – Ceres and Pluto – have sufficient mass to make solid rocks flow into this form. So we'll start by looking at the forces on a hypothetical spherical rotating planet – it might as well be the Earth.

Study point

The same principles hold for the roller coaster (Fig. 3.1.18). This time, however, the questions is, 'What is the maximum speed to stay in contact with the track?'

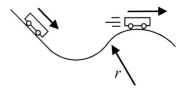

Fig. 3.1.18 Roller coaster

Self-test 3.1.10

The mean radius of the Earth is 6371 km. g at its surface in 9.81 m s⁻².

Use the inverse square law to estimate the acceleration due to the Earth's gravity at 385 000 km, the orbital radius of the Moon.

Self-test 3.1.11

The orbital period of the Moon is 27.3 days. Using the orbital radius from Self-test 3.1.10, calculate the Moon's centripetal acceleration and compare it with the answer to Self-test 3.1.10.

Fig. 3.1.19 The shape of Jupiter

We'll consider the forces on a an object of mass m on the surface of the (spherical) Earth at a latitude, λ, as in Fig. 3.1.20.

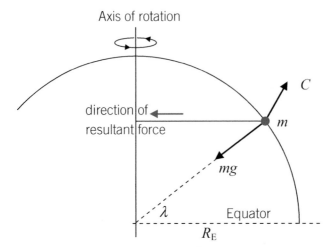

Fig. 3.1.20 Forces on an object at latitude λ.

3.1.12 Self-test

Given that the rotational period of the Earth is 24 hours and the mean radius is 6731 km. Show that the centripetal acceleration of a point on the equator is approximately 0.03 m s⁻² (i.e. about 0.3% of g).

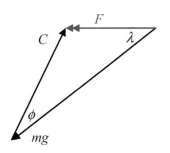

3.1.21 Resultant force

Notice that the 'normal' contact force is not directly away from the centre of the Earth. It cannot be because the resultant of C and mg is not zero but points towards the axis of rotation – it is centripetal. We'll leave the detailed analysis to a Stretch & Challenge question at the end of Exercise 3.1. However we'll anticipate a result from it. For $\lambda = \frac{\pi}{4}$, the value of C is almost identical to mg but the angle of C is 0.0017 **rad** (about 0.1°) away from the 'vertical'.

The effect of this is that the horizontal line at the position of the mass is not at right angles to mg but at right angles to C – with the effect that the Earth bulges outwards towards the equator and is flattened at the poles. Note that this undermines (by a small amount) our initial assumption of a spherical Earth! From Fig. 3.1.21 we see that the magnitude of C (i.e. the object's apparent weight) is less than mg: this effect gets smaller towards the pole. There is a secondary effect, which we haven't taken account of: because the Earth now bulges at the equator, actual value of mg is less because the surface is further away from the centre.

Exercise 3.1

1 A 3.5" hard disk drive rotates at **7200 rpm** (revolutions per minute). Calculate (a) the angular velocity of the disk, and (b) centripetal acceleration of a point on the rim.

[1" = 2.54 cm]

2 (a) Calculate the centripetal acceleration of the Solar System in its orbit around the centre of the Galaxy (see Section 3.1).

(b) Calculate the gravitational force exerted by the Galaxy on the Sun (mass = 2.0×10^{30} kg).

3 The radius of a geosynchronous orbit is **42 164 km**. Use the data for the ISS (Section 3.1.3 example) to show that these two orbits are consistent with the inverse square law of gravity.

4 A car of mass **1000 kg** travels at **20 m s^{-1}** around a circular bend of radius **80 m**.

(a) Calculate the lateral grip exerted by the tyres.

(b) The maximum grip of the tyres is **9000 N**. Calculate the maximum possible cornering speed of the car.

5 The car in Q4 drives on a circular car track which is banked as shown in the diagram.

(a) By resolving vertically, find an equation for C and G in terms of **mg**.

(b) If the car drives at **15 m s^{-1}** and the $\theta = 10°$, calculate the value of the lateral grip, G.

(c) Calculate the value of θ which would allow the car to be driven at **25 m s^{-1}** with no lateral grip (i.e. $G = 0$).

6 A load of mass **100 g** is whirled in a horizontal circle on a **50 cm** long thread, of breaking tension **20 N**. Calculate the maximum angular velocity.

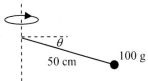

7 The load in Q6 is whirled at is now whirled at **150 rpm** and the thread makes an angle of θ to the horizontal (see diagram). Calculate θ.

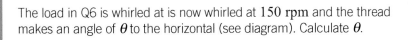

8 A stone of mass **500 g** is whirled in a sling in a vertical circle of radius **80 cm** with a frequency of **3 Hz**. With the unlikely assumption that the speed of the stone is constant, calculate the maximum and minimum values of the tension in the sling and identify the points in the circle when they occur.

9 The vertical loop in a model car set has a radius of **30 cm**. A **40 g** car enters the bottom of the loop with just enough speed to remain in contact with the loop at the top. Calculate:

(a) The speed at the top of the loop.

(b) The kinetic energy and gravitational potential energy at the top of the loop.

(c) The speed with which the car enters the loop.

10 The car in Q9 is observed to slow down to rest from **5.0 m s^{-1}** to rest over a distance of **15 m** on a horizontal track.

(a) Calculate the frictional force (assumed constant) on the car.

(b) Use your answer to part (a) to refine your answer to Q9(c), stating your assumption.

Stretch & Challenge

1. Assuming a spherical Earth of radius 6371 km, find the centripetal acceleration on an object at a latitude of $\frac{\pi}{4}$ (45°).

2. Use trig rules and the triangle of forces, Fig. 3.1.19, to show that the angle of the 'normal' contact force to the vertical is approximately 0.1°.

3. Use the cosine rule to show that at latitude λ the magnitude of the normal contact force,
$$C^2 = m^2g^2 + m^2R_E^2\omega^4\cos^2\lambda - 2m^2gR_E\omega^2\cos^2\lambda$$

4. Noting that $\frac{R_E\omega^2}{g} \ll 1$, use the binomial approximation $(1 + x)^n \approx 1 + nx$ to derive the approximate relationship between C and λ:

$$C \approx mg\left(1 - \frac{R_E\omega^2}{g}\cos^2\lambda\right)$$ and hence the measured acceleration of free fall is $g\left(1 - \frac{R_E\omega^2}{g}\cos^2\lambda\right)$

Hint: Ignore middle term in the accurate equation for C^2, because it is much smaller than either of the other two terms.

5. Research the measured values of the acceleration of free fall at different geographical latitudes.

3.2 Vibrations

What is the difference between vibration and oscillations? Not a lot really. An object *oscillates* if it moves with a regular period; we tend to use the word *vibrations* for short period oscillations but there is no hard and fast rule. In any case, the study of periodic changes – their characteristics, causes and effects – is an important focus of study in physics. They can cause disasters, often because of resonance effects such as in the collapses of Ferrybridge cooling towers (1965) or the bridge over the Tachoma Narrows; they can transmit information by acoustic, seismic, electromagnetic and gravitational waves; they form the basis of accurate timekeeping from the pendulum clock to the atomic clock and the pulsar.

A level Physics concentrates on one particular sort of oscillation called **simple harmonic motion**, usually abbreviated to shm. We shall consider this first.

3.2.1 Introduction to simple harmonic motion

The red sphere in Fig. 3.2.1 is held in stable equilibrium at point **0** between two springs under tension. We shall consider the horizontal surface to be frictionless. If we move the sphere slightly to the left, the tension in spring **A** will decrease; the tension in spring **B** will increase. If we let the sphere go there will be a resultant force on it to the right and it will therefore move back to its equilibrium position. That's what we mean by stable. What happens when it gets back to **0**? The two tensions are now equal but the sphere is moving, so it carries on to the right – building up the tension in **A** and lowering the tension in **B**. Hence decelerating until it stops and accelerates back to the left…

The tension in the springs in Fig 3.2.1 varies linearly with extension, so the restoring force, F, on the mass is directly proportional to the displacement, x, and in the opposite direction. By N2, so is the acceleration, a, i.e.

$$F = -kx \qquad \text{and} \qquad a = -\frac{k}{m}x$$

where k is the stiffness constant of the arrangement of springs (in fact in Fig. 3.2.1 it is the sum of the spring constants of **A** and **B**). Note that $k > 0$.

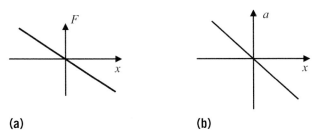

(a)   **(b)**

Fig. 3.2.3 (a) F-x and (b) a-x graphs for shm

The graphs in Fig. 3.2.3 illustrate the definition of simple harmonic motion.

— Terms & definitions —

Simple harmonic motion occurs when an object moves such that its acceleration is always directed toward a fixed point and is proportional to its distance from the fixed point.

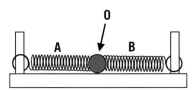

Fig. 3.2.1 Stable equilibrium

▼ **Study point**

For vibrational purposes we can consider the molecules of a (crystalline) solid to be small masses linked by springs (which represent the bonds), so our model in Fig. 3.2.1 is not too artificial. See Fig. 3.2.2.

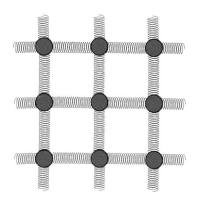

Fig. 3.2.2 Molecules as masses on springs

Self-test   **3.2.1**

In Fig. 3.2.1 the spring constants of **A** and **B** are 200 N m^{-1} and 300 N m^{-1} respectively. The mass is in equilibrium at **0**. The sphere is displaced 2.0 cm to the right and released. Calculate:

(a) the change in the tension, ΔF_A, of spring **A**,

(b) ΔF_B and

(c) the resultant force, F_{res}, on the sphere.

3.2.2　Self-test

The mass of the sphere in Self-test 3.2.1 is 200 g. Sketch graphs of (a) F against x and (b) a against x.

3.2.3　Self-test

Show that the unit of $\sqrt{\dfrac{k}{m}}$ can be written s^{-1}.

MATHS TIP

In the language of A level Mathematics, $a = -\omega^2 x$ is written $\dfrac{d^2x}{dt^2} = -\omega^2 x$. The two functions which satisfy this are:

$x = A\cos(\omega t + \varepsilon)$ and

$x = A\sin(\omega t + \phi)$ where A, ε and ϕ are constants. The two are equivalent, e.g.

$A\cos(\omega t - \frac{x}{4}) = A\sin(\omega t + \frac{x}{4})$.

MATHS TIP

When you use trigonometrical functions to analyse vibrations, always make sure that the calculator is in radian mode.

To check this, enter $\cos\pi$. If the calculator returns -1 it is in the correct mode.

3.2.4　Self-test

Use Fig. 3.2.4 to determine:

(a) the gradient of $x = A\cos 2\theta$ when $\theta = 0, \frac{\pi}{4}, \frac{\pi}{2}, \frac{3\pi}{4}$.

(b) the gradient of $x = 10\sin 2\theta$ when $\theta = \frac{\pi}{4}, \frac{\pi}{2}, \frac{3\pi}{4}$.

3.2.2 Equations of simple harmonic motion

It is convenient to write the ratio $\dfrac{k}{m}$ as ω^2. We can do this because k and m are both positive. Making this substitution gives $a = -\omega^2 x$. At this point we'll undertake a brief mathematical diversion to consider the function $x = A\cos 2\theta$. The top graph in Fig 3.2.4 is of $x = A\cos 2\theta$ (with θ in radians).

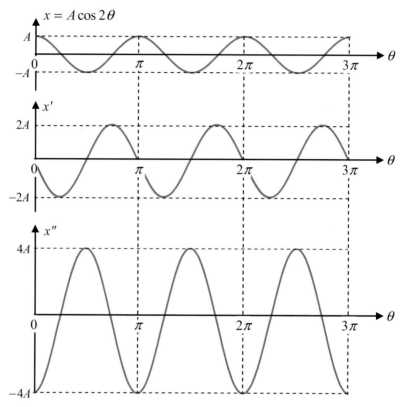

Fig. 3.2.4 $x = A\cos 2\theta$ and successive gradients

The second graph (x') is the plot of the gradient of $x = A\cos 2\theta$ against θ.

x' is the function: $\qquad\qquad\qquad\qquad x' = -2A\sin 2\theta$

The third graph (x'') takes this a stage further and is the plot of the gradient of $x' = -2A\sin 2\theta$ against θ.

x'' is the function: $\qquad\qquad\qquad\qquad x'' = -4A\cos 2\theta$

So $\qquad\qquad\qquad\qquad\qquad\qquad\qquad x'' = -4x$

Similarly, if $x = A\cos(2\theta + \varepsilon)$, then $\quad x' = -2A\sin(2\theta + \varepsilon)$ and

$\qquad\qquad\qquad\qquad\qquad\qquad\qquad x'' = -4A\cos 2\theta$

So, once again: $\qquad\qquad\qquad\qquad\quad x'' = -4x$

How does this relate to $a = -\omega^2 x$? The gradient of x is v and the gradient of v is a.

So, if $$x = A\cos(\omega t + \varepsilon), \qquad [1]$$

we can use the patterns in the above results to realise that:
$$v = -\omega A \sin(\omega t + \varepsilon) \qquad [2]$$

and $$a = -\omega^2 A \cos(\omega t + \varepsilon) \qquad [3]$$

So $$a = -\omega^2 x \qquad [4]$$

Hence, the solution to the $a = -\omega^2 x$ problem is $x = A\cos(\omega t + \varepsilon)$

We know that the extreme values of the sine and cosine are ± 1, so the maximum values x, v and a are:

$$x_{max} = A \qquad\qquad v_{max} = \omega A \qquad\qquad a_{max} = \omega^2 A$$

Example

The resultant force on a body of mass 0.10 kg is given by $F_{res} = -kx$, where x is the displacement and $k = 0.25$ N m^{-1}. The body is pulled 20 cm from its equilibrium position and released. Calculate the maximum speed and acceleration.

Answer

The maximum displacement, A, is 20 cm.

$$\omega = \sqrt{\frac{k}{m}} = \sqrt{\frac{40}{0.10}} = \sqrt{400} = 20 \text{ s}^{-1} \text{ (see Study point)}$$

$$\therefore v_{max} = \omega A = 20 \text{ s}^{-1} \times 20 \text{ cm} = 400 \text{ cm s}^{-1} \text{ [4.0 m s}^{-1}\text{]}$$

and $a_{max} = \omega^2 A = (20 \text{ s}^{-1})^2 \times 20 \text{ cm} = 8\,000 \text{ cm s}^{-2}$ [80 m s^{-2}]

3.2.3 Characteristics of simple harmonic motion

How does the equation $x = A\cos(\omega t + \varepsilon)$ relate to the amplitude, frequency, period and phase of the oscillation? As an example, consider the oscillation represented by the graph in Fig. 3.2.5. We'll see how to write the equation of the motion in the form $x = A\cos(\omega t + \varepsilon)$.

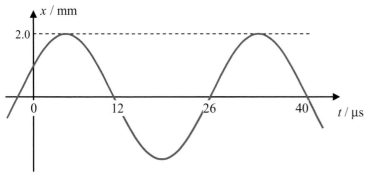

Fig. 3.2.5 Example oscillation for finding the equation

The amplitude and period are clearly 2.0 mm and 28 μs respectively. So the value of A in $x = A\cos(\omega t + \varepsilon)$ is 2.0 mm. In order to establish the other physical characteristics, we need to examine the *argument* or **phase** of the oscillation.

▼ **Study point**

We could equally write
$x = A\sin(\omega t + \phi)$

Then $v = \omega A \cos(\omega t + \phi)$
and $a = -\omega^2 A \sin(\omega t + \phi)$

So, once again, $a = -\omega^2 x$

─ *Terms & definitions* ─

The quantity ω is called the **angular frequency** or **pulsatance** of the oscillation.

UNIT: s^{-1} [or rad s^{-1}]

▼ **Study point**

ω always has a positive value so we don't need to write $\omega = \pm 20$ s^{-1}.

─ *Terms & definitions* ─

The angle $\omega t + \varepsilon$ in the equation $x = A\cos(\omega t + \varepsilon)$ is called the **phase** of the oscillation.

The angle ε is called the **phase constant**.

Self-test 3.2.5

For the oscillation in Fig. 3.2.5, write down all the times between 0 and 50 μs for which:

(a) the displacement is (i) 0, (ii) 2.0 mm and (iii) −2.0 mm

(b) the velocity is 0

(c) the acceleration is 0.

3.2.6

Self-test

Calculate the maximum values of (a) the velocity and (b) the acceleration, for the motion in Fig. 3.2.5.

(a) The period, T, and frequency, f

The oscillation repeats itself with every increase by 2π in the value of the phase. In other words, if we add T (the period) to any time t, we add 2π to the phase:

So, for any value of t $\qquad \omega(t + T) + \varepsilon = \omega t + \varepsilon + 2\pi$

$\therefore$ Simplifying $\qquad\qquad\qquad\qquad \omega T = 2\pi$

So the period, T, is given by $\qquad\qquad T = \dfrac{2\pi}{\omega}$ [5]

In this case, the period is 28 µs, so the pulsatance, ω, is given by:

$$\omega = \frac{2\pi}{T} = \frac{2\pi}{28\ \mu s} = 2.2 \times 10^5\ s^{-1} \text{ (to 2 s.f.)}$$

We can also relate the pulsatance to the frequency using $f = \dfrac{1}{T}$.

This leads to $\omega = 2\pi f$ [6]

(b) The phase constant, ε

If $\varepsilon = 0$ the graph of displacement against time is a cosine curve, i.e. it has its maximum value at time, $t = 0$. This is clearly not the case in Fig. 3.2.5. To calculate ε, we need to insert known values of the other quantities into the equation $x = A \cos(\omega t + \varepsilon)$. For example, when $t = 5$ µs, $x = 2.0$ mm. Inserting these with the already known values of A and ω gives:

$$2.0 = 2.0 \cos\left(\frac{2\pi}{28\ \mu s} \times 5\ \mu s + \varepsilon\right)$$

[Notice we have reverted to the more precise value of ω.]

Simplifying $\quad 1 = \cos(1.12\ \text{rad} + \varepsilon)$

So $\qquad 1.12\ \text{rad} + \varepsilon = \cos^{-1} 1 = \ldots -2\pi, 0, 2\pi, 4\pi \ldots$

It makes sense to choose a value of ε in the range $\pm 2\pi$. In this case this leaves us with

$$\varepsilon = 0 - 1.12\ \text{rad or } 2\pi - 1.12\ \text{rad}$$

i.e. $\qquad\qquad \varepsilon = -1.12\ \text{rad or } 5.16\ \text{rad}$

Combining the results of this and the previous section gives the equation:

$$(x / \text{mm}) = 2.0 \cos(2.2 \times 10^5 (t / s) - 1.12)$$

This is perhaps an overly fussy way of writing the equation. We could write:

$$x = 2.0 \cos(2.2 \times 10^5 t - 1.12\ \text{rad}) \qquad \text{with } x \text{ in mm and } t \text{ in s}$$

However, inserting the units in this has the advantage of clarity and allows us to express the time in other units, e.g.

$$(x / \text{mm}) = 2.0 \cos(0.22(t / \mu s) - 1.12)$$

3.2.7

Self-test

Repeat the calculation for ε by using the values, $x = 0$ when $t = -2.0$ µs. In this case you will get 4 apparent answers in the range $\pm 2\pi$ but you can rule some out, e.g. by considering the value of x at $t = 0$.

▼ **Study point**

The cosine term could also be $\cos(\omega t + 5.16)$. The difference in the two values of ε is 2π.

3.2.8

Self-test

Rewrite the equation for x using the units cm and ms.

(c) The equation for v

We have already seen that, if $x = A\cos(\omega t + \varepsilon)$, then $v(t)$ can be written as $v = -\omega A\sin(\omega t + \varepsilon)$. From the answer to Self-test 3.2.6, to 2 s.f., $\omega A = 450$ m s^{-1}, so we can write

$$v = -450\sin(2.2 \times 10^5 t - 1.12).$$

This is a perfectly acceptable way of writing the velocity equation but we can get rid of the '−' sign. If we add (or subtract) π to the phase, this shift the graph along to the left (or right) by half a cycle, which turns a positive value of v to the same negative value. This gives us

$$v = 450\sin(2.2 \times 10^5 t + 2.02)$$

(d) The relationship between velocity and displacement

Equations [1] – [3] in Section 3.2.2 give the time variations of displacement, velocity and acceleration. Equation [4], which is effectively the definition of simple harmonic motion, relates acceleration to displacement. The connection between displacement and velocity is

$$v = \pm\omega\sqrt{A^2 - x^2} \qquad [7]$$

This is derived in the mathematical skills chapter, Section M1.2(c). Equation [7] is clearly consistent with the equation for v_{max}. The speed is maximum when the displacement, x, is zero.

$$\text{maximum speed} = \omega\sqrt{A^2 - 0^2} = \omega\sqrt{A^2} = \omega A$$

3.2.4 Examples of simple harmonic motion

Any object for which the resultant force and the displacement are related by $F = -kx$, can oscillate with simple harmonic motion. We shall consider two simple systems. These derivations are not required to be learned for the examination.

(a) Oscillating mass on a spring

The spring in Fig. 3.2.6 has a constant k. The spring applies a force F_0 to the mass. The mass, m, in part (a) is in equilibrium, so we can write

$$F_0 = mg$$

Consider raising the mass and releasing it. It will oscillate vertically. Consider the moment in the oscillation, shown in part (b) of Fig. 3.2.6. The mass is at a displacement x above its equilibrium position. The extension of the spring from its natural length is reduced by x, so the force, $F(x)$, exerted by the spring on the mass is now given by

$$F(x) = F_0 - kx$$

The resultant upward force, F_{res}, on the mass is given by $F_{\text{res}} = F(x) - mg$

$$\therefore \qquad F_{\text{res}} = [F_0 - kx] - mg = [mg - kx] - mg$$

$$\therefore \qquad F_{\text{res}} = -kx$$

Self-test 3.2.9

Sketch a v-t graph for the oscillation, labelling significant points.

Hint: rather than starting from the equation, consider the amplitude, period and the phase difference between the x and v graphs (Fig. 3.2.4).

Stretch & Challenge

Use the compound angle formula for $\sin(X + Y)$ to show that $\sin(\theta + \pi) = -\sin\theta$

▼ **Study point**

In $v = \pm\omega\sqrt{A^2 - x^2}$, the maximum possible value of x is A (and the minimum is $-A$). At these points $v = 0$.

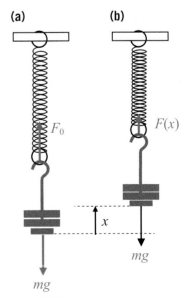

(a)　　　(b)

F_0　　$F(x)$

x

mg

mg

Fig. 3.2.6 Mass on a spring

3.2.10 Self-test

A 300 g mass oscillates on a spring with a spring constant 30 N m^{-1}. Calculate the pulsatance, frequency and period of the oscillation.

3.2.11 Self-test

A body of mass 0.20 kg is suspended from two identical springs each with $k = 25$ N m^{-2}, arranged as in Fig. 3.2.7. Assume that both springs remain under tension.

(a) Calculate the distance of the equilibrium point below the midpoint of the two anchor points.

(b) Calculate the frequency of oscillation if the mass is pulled down slightly and released.

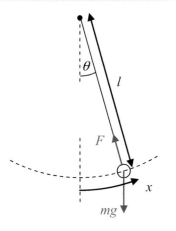

Fig. 3.2.8 Simple pendulum

So the spring oscillates with simple harmonic motion. Its pulsatance, frequency and period are given by:

$$\omega = \sqrt{\frac{k}{m}} \qquad f = \frac{1}{2\pi}\sqrt{\frac{k}{m}} \qquad T = 2\pi\sqrt{\frac{m}{k}}$$

The same principle holds for the two-spring system in Fig. 3.2.1 and the vertical arrangement in Fig. 3.2.7. The stiffness constant of the two springs is the sum of the spring constants (see Self-test 3.2.1). If friction is negligible, the sphere will execute shm with a pulsatance,

$$\omega = \sqrt{\frac{k_1 + k_2}{m}}.$$

Example

A body of mass 0.5 kg is suspended from a Hookean elastic cord with stiffness constant 50 N m^{-1}. It is raised from its equilibrium point by 5.0 cm and released at time $t = 0$. Find the variation of the body's position, x, as a function of time.

Answer

The body executes simple harmonic motion. The amplitude is 5.0 cm.

The pulsatance, $\omega = \sqrt{\frac{k}{m}} = \sqrt{\frac{50\ \text{Nm}^{-1}}{0.5\ \text{kg}}} = 10\ \text{s}^{-1}$

So the equation of motion is: $(x\,/\,\text{cm}) = 5.0\cos(10(t\,/\,\text{s}) + \varepsilon)$

The body starts off at time $t = 0$ with its maximum displacement, so the phase constant, $\varepsilon = 0$.

Hence $(x\,/\,\text{cm}) = 5.0\cos(10(t\,/\,\text{s}))$

Note: in an exam, you would not be penalised for leaving out the units in the equation, i.e. writing $x = 5.0\cos 10t$.

(b) Simple pendulum

A mass, m, on a light inextensible thread of length l (measured to the centre of mass) executes shm as long as the amplitude of the oscillations is small, e.g. a maximum angle to the vertical of 5°.

Ignoring air resistance, there are two forces on the pendulum bob: the weight, mg, and the tension, F, in the thread. The bob is constrained to travel along the arc of a circle of radius l. The displacement from the centre along this arc, is x. Considering tangential components:

Component of F along the arc = 0 [the thread is at 90° to the arc].

Component of mg along the arc = $-mg\sin\theta$ [see Study point].

∴ From N2: $ma = -mg\sin\theta$

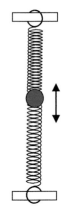

Fig. 3.2.7 Shm with two springs

For small θ (in radians): $\qquad\qquad \sin\theta \approx \theta$

Also: by definition, $\qquad\qquad\qquad \theta = \dfrac{x}{l}$

Combining these and dividing by m: $\quad a = -\dfrac{g}{l}x$

Comparing this to $a = -\omega^2 x$, this is an shm equation with $\omega = \sqrt{\dfrac{g}{l}}$

So the period, T, of a simple pendulum is given by $T = 2\pi\sqrt{\dfrac{l}{g}}$

We can therefore write $x = A\cos(\omega t + \varepsilon)$ or $\theta = \theta_{max}\cos(\omega t + \varepsilon)$, with the above value of ω. If the oscillation starts from its maximum positive position (e.g. the pendulum is pulled to the side and released) then $\varepsilon = 0$.

Question: What angle of swing is small?

Answer: $\sin(0.2\ \text{rad}) = 0.1987$ which is a difference of 0.65%. So even at this quite large swing (11.5° either side) the difference in period between an actual pendulum and the small-angle approximation is tiny. (See Stretch & Challenge).

(c) General system

As we've seen, all we need to do to work out the period of a system's oscillation is to find out its stiffness, k, and the mass, m, to be oscillated. As long as the oscillating mass can be approximated as a point mass then the period is given by $T = 2\pi\sqrt{\dfrac{m}{k}}$.

Example

When a **load** of mass 200 g is attached to the end of the cantilever in Fig. 3.2.9 the latter is depressed by 1.5 cm. The load is then pulled down by another 1.0 cm. Describe in detail the subsequent motion, stating any assumptions.

Answer

Assumptions: 1. The depression is proportional to the load. 2. The mass of the cantilever itself is insignificant. 3. There are no frictional losses.

Calculate k: $k = \dfrac{F}{x}$. $F = mg$, $\therefore k = \dfrac{mg}{x} = \dfrac{0.20 \times 9.81\ \text{N}}{0.015\ \text{m}} = 131\ \text{N m}^{-1}$.

Then $\omega = \sqrt{\dfrac{k}{m}} = \sqrt{\dfrac{131\ \text{N m}^{-1}}{0.20\ \text{kg}}} = 25.6\ \text{s}^{-1}$.

And: The frequency of the oscillation $= f = \dfrac{\omega}{2\pi} = 4.1\ \text{Hz}$.

Thus the cantilever oscillates with a frequency of 4.1 Hz and an amplitude of 1.0 cm about its equilibrium position.

The depression d from the equilibrium position at any time is given (in cm) by:
$d = 1.0\cos(25.6t)$

Self-test 3.2.12

A *seconds pendulum* is one of period 2 seconds, i.e. 1 second each way. Historically it was used in clocks.

Calculate the length of the pendulum arm required.

Self-test 3.2.13

If the pendulum thread were light but could stretch elastically, why and how would the length of the thread vary throughout the swing? [Qualitative answer]

Stretch & Challenge

The period, T, of a simple pendulum for larger angles is related to that of the small-angled pendulum, T_0, by an infinite series:

$T = T_0\left(1 + \dfrac{1}{6}\theta^2 + \dfrac{11}{3072}\theta^4 + \ldots\right)$

Use the second and fourth order approximations to compare the period of a pendulum with $\theta = 0.2$ rad and 0.5 rad with the small-angled pendulum.

200 g

Fig. 3.2.9 Loaded cantilever

Self-test 3.2.14

Calculate the maximum value of the resultant force on load on the cantilever. State when this first occurs after the load is first released.

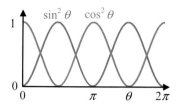

Fig. 3.2.10 Graphs of $\cos^2\theta$ and $\sin^2\theta$

MATHS TIP

The statement of Pythagoras' theorem in terms of angles, i.e. $\sin^2\theta + \cos^2\theta = 1$ and the nature of the sine² and cosine² graphs are important in AC circuit theory. See the Maths chapter and the AC option.

3.2.15 Self-test

The displacement, x, of an object of mass 5 kg object varies with time according to $x = 10 \cos 100t$. Calculate:

(a) the total energy,

(b) the PE and KE at $t = 0$

(c) the PE and KE at $t = \dfrac{\pi}{200}$ s.

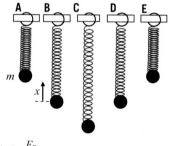

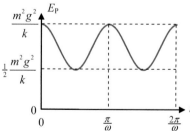

Fig. 3.2.11 Potential energy variation for an oscillating mass

3.2.16 Self-test

For the oscillating mass in Fig. 3.2.11:

(a) State the times at which the kinetic energy is zero.

(b) State the maximum value of the kinetic energy and the times at which it occurs.

(c) Copy the graph in Fig 3.2.11 and add a graph of the variation of E_k with time.

3.2.5 Energy of oscillation

As a body moves through an oscillatory cycle its kinetic energy and potential energy are constantly changing. The systems of springs and spheres in Figs. 3.2.6 and 3.2.7 have gravitational as well as elastic potential energy. So does the cantilever and mass in Fig. 3.2.9. On the other hand, the pendulum just has GPE. We'll start by just considering the general case, but we'll set the phase constant to zero.

$$\therefore \qquad x = A \cos \omega t \quad \text{and} \quad v = -\omega A \sin \omega t$$

The potential energy at x is the area under the F-x graph, which is the same as for a spring, i.e. $\frac{1}{2}kx^2$. The kinetic energy is $\frac{1}{2}mv^2$. Combining these with the above equations

$$\therefore \qquad E_p = \tfrac{1}{2}kA^2 \cos^2 \omega t \quad \text{and} \quad E_k = \tfrac{1}{2}m\omega^2 A^2 \sin^2 \omega t$$

But $k = m\omega^2$ so we can express E_p as $\qquad E_p = \tfrac{1}{2}m\omega^2 A^2 \cos^2 \omega t$.

The graphs in Fig. 3.2.10 show how $\cos^2 \theta$ and $\sin^2 \theta$ vary over one cycle of θ, from 0 to 2π. From the symmetry of the graphs it is clear that the sum of the two functions is always 1.

$$\therefore \text{ The total vibrational energy} \quad = E_p + E_K$$
$$= \tfrac{1}{2}m\omega^2 A^2 (\cos^2 \omega t + \sin^2 \omega t)$$
$$= \tfrac{1}{2}m\omega^2 A^2$$

The fact that the total energy is constant (in the absence of resistive forces) is reassuring. The actual details are a little more complicated if gravitational potential energy is also involved – but not much more complicated as the next example shows:

Example

A load of mass m is attached to a spring, with a spring constant k, which is clamped vertically. The tension in the spring is initially zero (at **A** in Fig. 3.2.11). The load is released. Sketch a graph of the variation of total potential energy (GPE + EPE) over one complete oscillation. Take the equilibrium position of the load as the zero of GPE **(B)**.

Answer

The equilibrium point is a distance A below the release point, where $A = \dfrac{mg}{k}$. The initial value of x is A so the variation of x with t is:

$$x(t) = \frac{mg}{k}\cos \omega t \qquad \text{where} \qquad \omega = \sqrt{\frac{k}{m}}$$

So the variation of GPE with time is: $E_g(t) = mgx(t) = \dfrac{m^2g^2}{k} \cos \omega t$

The elastic potential energy $E_e = \tfrac{1}{2}k \times \text{extension}^2 = \tfrac{1}{2}k\,(A-x)^2$

So the variation of EPE with time is: $\qquad E_e(t) = \tfrac{1}{2}k\left(\dfrac{mg}{k} - \dfrac{mg}{k}\cos \omega t\right)^2$
$$= \tfrac{1}{2}\frac{m^2g^2}{k}(1 - \cos \omega t)^2$$

Multiplying out the $(1 - \cos \omega t)^2$ and adding the GPE gives us

$$\therefore \text{ Total PE,} \qquad E_p(t) = \tfrac{1}{2}\frac{m^2g^2}{k}(2 \cos \omega t + 1 - 2 \cos \omega t + \cos^2 \omega t)$$

$$\therefore \text{ Simplifying} \qquad E_p(t) = \tfrac{1}{2}\frac{m^2g^2}{k}(1 + \cos^2 \omega t)$$

The maximum value of $(1 + \cos^2 \omega t)$ is 2, which occurs when $\cos \omega t = \pm 1$, i.e. twice in one cycle at positions **A**, **C** and **E**. The minimum value of $(1 + \cos^2 \omega t)$ is 1, which occurs when $\cos \omega t = 0$, i.e. at positions **B** and **D**. Hence the variation of total PE with time is as shown in Fig. 3.2.11.

3.2.6 Damped oscillations

We shall continue for the moment to focus on **free oscillations**. In all real-life situations, systems do not carry on vibrating indefinitely but gradually wind down due to the effect of dissipative forces called damping. For example, if a car drives over a small obstacle, the compression in suspension springs is increased. The springs then expand, releasing the excess elastic energy and, in the absence of damping forces the car would continue oscillating – with emetic effects. Many damping forces, e.g. air resistance and electromagnetic damping due to eddy currents (see Section 4.5.5), are proportional to the speed of motion. Engineers build in a required degree of damping to a system which can oscillate, e.g. door closures, car suspensions, the Millennium Bridge (in retrospect!). We recognise three categories, under damping, over damping and critical damping, the last of which is the aim in designing many structures.

(a) Under damping

As the name implies, the degree of damping here is light. The mass on a spring system in Fig. 3.2.12 is an example. In these cases, the equation of motion is still $A\cos(\omega t + \varepsilon)$ but the amplitude, A, decreases exponentially as $A_0 e^{-\lambda t}$, as in Fig. 3.2.13. The resulting equation is thus

$$x = A_0 e^{-\lambda t} \cos(\omega t + \varepsilon)$$

The value of ω is not precisely equal to its value in the absence of damping, i.e. $\sqrt{\dfrac{k}{m}}$, but it is very close. A car whose suspension behaved in this may would still be rather sick making.

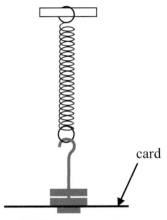

Fig. 3.2.12 Under-damped system

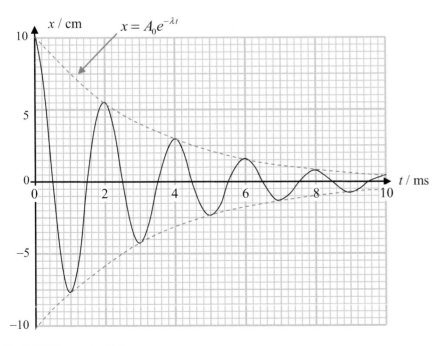

Fig. 3.2.13 Damped oscillations

Self-test 3.2.17

The value of A_0 in the graphs in Fig. 3.2.13 is pretty obvious. Use the graphs to calculate (a) ω and (b) λ.

Self-test 3.2.18

(a) Use the 0.55 figure in the 'things to notice' to state the fraction of the potential energy (PE) lost in each cycle.

(b) Without considering the PE, how could you measure the fraction of the kinetic energy lost per cycle?

 **Stretch & Challenge**

What relationship does $\sqrt{0.55}$ have to the graph in Fig. 3.2.13?

3.2.19 Self-test

Sketch a graph of the damped oscillation with the same frequency and initial amplitude but with a lighter degree of damping, with λ one half of the value in Fig. 3.2.14.

┌─ **Terms & definitions** ─┐

A system is **over damped** if it returns to its equilibrium position without oscillating.

A system is **under damped** if it oscillates with the amplitude reducing gradually to zero.

A **critically damped** system returns to equilibrium as rapidly as possible without oscillating.

▼ **Study point**

In the critically damped graph in Fig.3.2.14, the displacement decreases monotonically to zero. This is not necessarily the case for critical damping but it is so if the system is set into motion by displacing it and releasing with zero velocity.

3.2.20 Self-test

What feature of the critically damped and over damped curves in Fig. 3.2.14 shows that they are not pure exponential decays of the form $x = A_0 e^{-\lambda' t}$?

3.2.21 Self-test

Automatic door closures are normally designed to be critically damped. Why is this preferable to under damping or over damping?

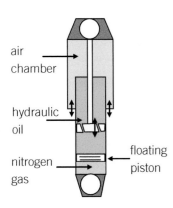

air chamber

hydraulic oil

nitrogen gas

floating piston

Fig. 3.2.15 Bicycle shock absorber

Things to notice about damped oscillation curves

1. The period of the oscillation is constant – the time for the oscillations doesn't become shorter as the amplitude decays. In Fig. 3.2.13 the period of the oscillations remains at 2 ms even when the amplitude has dropped to less than 10% of the original.

2. The amplitude goes down by the same fraction in each cycle, e.g. in Fig. 3.2.13 the amplitude after 1 cycle is 0.55 of the initial, after 2 cycles it is 0.30 (= 0.55²) of the initial.

3. The greater the value of λ, the more rapidly the amplitude of the oscillations decreases.

(b) Over damping and critical damping

As the degree of damping increases, the system initially returns to equilibrium more quickly. The pecked line in Fig.3.2.14 is a similar degree of damping to that in Fig. 3.2.13. The **green graph** is for a higher degree of damping (about 5 times the pecked graph). It still shows a tendency to oscillate (with a slightly increased period). The **red** and **blue** graphs are for still greater damping (about 10 times and 20 times) respectively.

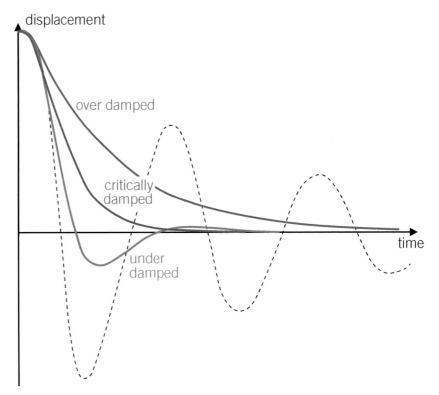

Fig. 3.2.14 Degrees of damping

For many purposes, engineers attempt to design systems which are **critically damped**. For example, consider a bike's suspension system:

• If the damping is too light, the bike will oscillate after going over a pothole in the road.

• If the damping is too heavy, jolts (e.g. from potholes) are transmitted strongly from the wheels to the rest of the bike and the rider.

Fig. 3.2.15 is a schematic diagram of a typical bicycle shock absorber which is coupled to a spring suspension. The damping effect is produced by the hydraulic oil being forced through the holes in the main piston. The role of the gas chambers is to prevent cavitation in the oil which can occur if the pressure on one side of the piston drops below zero.

3.2.7 Forced oscillations – resonance

An oscillatory system can be set into oscillation by the application of a force, be it a one-off impact such as in a car or bike suspension, a periodic driving force or an irregular external stimulus such as a moist finger rubbing around the top of a wine glass. These last two situations are not entirely dissimilar because any driving force can be considered to be composed of the sum of a number of sinusoidal oscillations of different frequencies.[1] For the moment we'll look at forced oscillations with a periodic driving force.

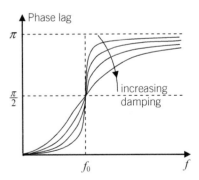

Fig. 3.2.16 Investigating forced oscillations

(a) Resonance curves

The easiest way to investigate this (you <u>are</u> advised to try this yourself) is to use the simple arrangement in Fig. 3.2.16, i.e. our old friend the mass on a spring but supported by an oscillating finger – yours! Try very low frequency oscillations (i.e. $f \ll f_0$, where f_0 is the natural frequency of oscillation of the spring mass system), very high frequency ($f \gg f_0$) and then (more difficult) sweeping the frequencies between these two extremes. You should find that:

- For $f \ll f_0$, the amplitude and phase of the oscillations are the same as that of your finger.

- For $f \gg f_0$, the amplitude of the oscillations is very small and the phase (more difficult to see) is π (i.e. 180°).

- For $f \sim f_0$ the amplitude of the oscillations is very large (in fact there is a strong possibility that the masses and spring will part company) and the phase is $\frac{\pi}{2}$ behind that of your finger.

Using more sophisticated set ups, the resonance curves in Fig. 3.2.17 and Fig, 3.2.18 can be obtained.

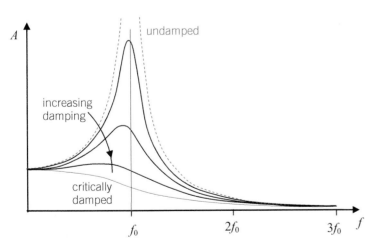

Fig. 3.2.17 Resonance curves

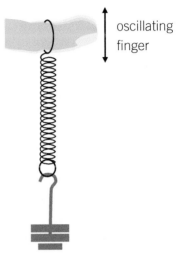

Fig. 3.2.18 Phase lags for forced oscillations

---

1 The mathematics of this is called Fourier analysis and is covered in university engineering courses.

— Terms & definitions —

A **forced oscillation** occurs when a sinusoidal driving force is applied to an oscillatory system, causing it to oscillate with the frequency of the driving force.

Fig. 3.2.19 Ferrybridge power station, November 1965

Fig. 3.2.20 Paths of the air stream across a flute mouthpiece

Fig. 3.2.21 London's Millennium Bridge

▼ Study point

One well-known anti-resonance measure is for marching troops to be ordered to break step when crossing a bridge.

3.2.22 Self-test

Two oscillations with frequencies / amplitudes of 0.8 Hz / 50 mm and 1.0 Hz / 75 mm were recorded on the Millennium Bridge. Calculate the maximum accelerations on pedestrians as a result.

(b) Resonance without periodic driving force

The Tachoma Narrows bridge failure is perhaps the most famous example of resonance causing a major structure to collapse without an apparent periodic driving force.
On 1 November 1965 the cooling towers of the newly built power station at Ferrybridge in Yorkshire also underwent catastrophic failure due to resonance. The fitful gusts of wind had a range of frequencies. Unfortunately one of the natural modes of oscillation of the air around the regular array of cooling towers, see Fig. 3.2.19, was within this range and so the air resonated.[2] This would not have mattered but for the fact that this frequency also coincided with one of the natural resonances of the cooling towers.

Woodwind and brass instruments work on a similar (but not catastrophic) principle. The lips of the trumpeter, or the reed of the clarinettist, produce 'white noise' or sound of a wide range of frequencies. Most of these are of no consequence but the ones which coincide with frequencies of stationary waves in the air column inside the instrument will resonate to produce the loud sound. Generally there will be one main note together with several harmonics which are characteristic of the instrument. These sound waves (as well as escaping to the audience) feed back to lips and reed of the instrumentalist to reinforce the same frequencies. The operation of a flute is even more akin to the power station: the air from the flautist's lips rapidly switches between the two paths shown in Fig. 3.2.20 with a wide range of frequencies. Resonance occurs in the air inside the flute which is fed back to the stream of air and reinforces an oscillation at the resonant frequency.

(c) Designing resonance out of systems

During the spin cycle of domestic washing machines, the drum rotates at up to 1200 revolutions per second (1800 in some models). A slight asymmetry in the position of the various items in the washing load, which is inevitable, will result in a periodic driving force of significant amplitude. The frequency of this driving force will vary from 0 up to 20 Hz (or 30 Hz). Hence, if any of the electrical or mechanical components of the washing machine have a natural frequency of vibration in this range, they will resonate at the appropriate point as the machine spins up and down. Eventually any such components will fail as a result. Hence, in designing washing machines, engineers try and ensure that this does not happen.

Designing resonance out is not always easy. The human eyeball has a natural rotational resonance at 18 Hz. This has been implicated in several helicopter accidents in which pilots have missed seeing power lines because of vibrations from the helicopter engine and rotors.

Engineers must also take account of human factors in designing out resonance as the story of the Millennium Bridge shows.

When we walk, as well as exerting vertical forces on the ground, we push slightly outwards with each step, which creates a roughly 1 Hz sideways forcing function on any structure we are walking on. The central span of the Millennium Bridge has a few transverse resonances in the 1 Hz range. With several hundred people on the bridge it was not anticipated that they would co-ordinate their footsteps but once slight sideways swaying occurred this is precisely what they effectively did, which increased the effect. The bridge was closed until dampers could be fitted!

---

2 The Yorkshire word for this is 'wuthering' as in *Wuthering Heights*.

(d) Useful resonance

Resonance is hardly ever a good thing in mechanical systems. There are a few notable exceptions, e.g.:

- The basilar membrane in the ear has varying dimensions and stiffness so that different parts of it respond to different input frequencies of sound, enabling us to discriminate between sounds.

- Playground swings are only satisfying once you apply the principle that the forward push be applied from the moment you start moving forwards.

Radio devices are required to respond to a restricted range of frequencies. Audio information is restricted to the range 0–20 kHz. This means that a radio station broadcasting at a frequency of, say, 92.4 MHz has a signal which ranges over 92.40 ± 0.02 MHz, so the tuned detector circuit needs to have a resonance curve which includes this range and drops to zero outside this (see Study point). Television has much more information so greater bandwidth requirements, from about 5 MHz for standard definition TV up to several GHz for some HDTV formats, so the tuned circuit needs to be designed to accommodate this.

Other examples of useful resonance are:

- Medical MRI scans (the clue is in the name – Magnetic Resonance Imaging). See Option B.

- Laser operation – the incident photon must have the same energy as the difference in the energy levels to stimulate emission. See Year 12 textbook.

(e) Microwave cooking

It is a common misconception that the effectiveness of microwave ovens is due to resonance between the microwaves and the rotation of water and other polar molecules in food. The reality is water molecules rotate at frequencies in the terahertz band (there are peaks in the absorption spectrum of water in the gaseous phase at 1.5, 2.0 and 3.0 μm), whereas microwave ovens operate at 2.45 GHz – a long way below resonance. In any case it is difficult to imagine freely rotating molecules in the liquid or solid state.

The e-m radiation passes rotational energy on to the water molecules which almost immediately lose it in collisions with other molecules.

Study point

In fact the channel separation for FM radio stations is 200 kHz to allow for a variation in frequency of the central *carrier* wave of the transmitter.

Self-test 3.2.23

Compare the frequency of the microwave oven with that of the absorption spectrum peaks of water.

Study point

If the microwave resonated with water molecules, we'd expect virtually all the heating to take place in the outer few mm of the food – which it clearly doesn't.

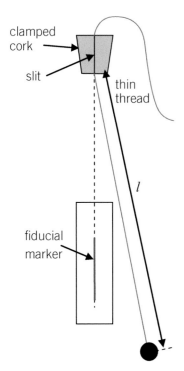

clamped cork

slit

thin thread

l

fiducial marker

Fig. 3.2.22 Pendulum set up

3.2.8   Specified practical work

(a) Measuring *g* by simple pendulum

This is, in principle, the most straightforward of the practicals that A level students are asked to do. However, it does provide a useful introduction to some basic experimental techniques and uncertainty analysis.

The arrangement is as in Fig. 3.2.22. The split cork arrangement is to allow the length of the pendulum to be adjusted easily. The cork should be orientated so that the slit is at right angles to the plane of oscillation of the pendulum. In this way the position of the top of the thread is clear. The cork should be clamped at such a height that lengths, *l*, of at least 1 m are possible. The bob should be a small heavy sphere, e.g. lead or brass.

To obtain the greatest accuracy, the oscillations should be timed between centres of oscillation, i.e. when the thread is vertical. This is because:

1. The thread is moving most quickly here and hence it is easy to judge the instant it moves past. At the extremes of the oscillation, the bob is moving very slowly and the instant of the extreme difficult to judge.

2. If the pendulum loses energy, it will still pass though the vertical line; the position of maximum displacement will gradually change. Hence a fiducial marker can be used against which to time the oscillations.

Procedure

1. Adjust the length of the pendulum to approximately (just less than) 1 m and measure length *l* using a metre rule. Note that this is to the centre of the bob.

2. Pull the bob to one side a short distance (a few cm) and release.

3. Use a stop watch to measure the time for multiple oscillations. (How many? See Study point.) The uncertainty should be quite low as you can anticipate the thread's crossing the fiducial maker at both the beginning and the end.

4. Take some repeat readings. Not many are needed – it is more a check on miscounting.

5. Repeat for a range of values of *l*.

Analysis

The period *T* is given by: $T = 2\pi \sqrt{\dfrac{l}{g}}$ $\therefore T^2 = \dfrac{4\pi^2}{g} l$

A graph of T^2 against *l* is drawn. It should be a straight line (through the origin) of gradient $\dfrac{4\pi^2}{g}$. So $g = \dfrac{4\pi^2}{\text{gradient}}$.

▼ **Study point**

The fiducial marker doesn't have to be made specially. The vertical post of the clamp stand (or even a convenient window frame) could serve!

▼ **Study point**

Rather than specifying a set number of oscillations, it makes sense to time for a given (approximate) length of time, say 20 s. In this way the precision of all the timings will be about the same.

3.2.24 Self-test

If the length of the pendulum thread above the bob (rather than to the C of G) is measured, how would the analysis be different?

(b) Investigating the damping of a spring

The arrangement in Fig. 3.2.23 can be used to investigate the way in which the oscillations decay under the influence of the air resistance on the sheet of thin card. At very low speeds, the magnitude of the drag is approximately proportional to the velocity, v, so the amplitude of the oscillations is expected to decay according to

$$A = A_0 e^{-\lambda t}.$$

Hence a graph of $\ln A$ against t should be a straight line of gradient $-\lambda$ and an intercept of $\ln A_0$ on the vertical axis.

The procedure is as follows:

1. Set up the apparatus as shown, selecting the mass to give as large an amplitude of oscillation as possible without exceeding the elastic limit of the spring.

2. Note the position of the card at equilibrium.

3. Displace the card vertically by lifting or lowering the load by as large a distance as possible so that the spring is always under tension in the subsequent oscillations. Note the initial displacement, i.e. the initial amplitude.

4. Release the load and start a timer. Record the amplitude of the oscillations at suitable intervals, e.g. every 30 s (see Study point) for a suitable period, e.g. 5 minutes.

5. The procedure should be repeated several times.

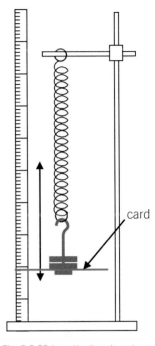

Fig. 3.2.23 Investigating damping

Analysis

If there are several sets of results with identical starting amplitudes, then the amplitude means (or totals) at each data time should be used. It is possible to combine data sets in this way even if the starting amplitudes are not the same. This is shown as follows.

If there are three data sets, $X(t)$, $Y(t)$ and $Z(t)$, with starting amplitudes X_0, Y_0 and Z_0, we expect the following to be true:

$$X(t) = X_0 e^{-\lambda t} \qquad Y(t) = Y_0 e^{-\lambda t} \qquad \text{and} \qquad Z(t) = Z_0 e^{-\lambda t}$$

Then, adding these gives: $\qquad (X(t) + Y(t) + Z(t)) = (X_0 + Y_0 + Z_0)e^{-\lambda t}$

So the variation in the sum of the amplitude against time is the same as the variation in the individual amplitude, so a graph of $\ln(X + Y + Z)$ against time should have the same form and gradient (though a different intercept).

▼ **Study point**

The oscillation is unlikely to be at maximum displacement at the notional times but the decay time is much longer than the period of oscillation so that an uncertainty of a second is unimportant.

Exercise **3.2**

1 Sketch the *a-t* curve for the oscillation in Fig. 3.2.5, labelling significant values.

2 Calculate the values of displacement, velocity and acceleration at $t = 0$ for the oscillation in Fig. 3.2.5.

3 Calculate the velocity in the oscillation in Fig. 3.2.5 when the displacement is $+1.5$ mm. Why are there two values?

4 A pendulum clock has a pendulum with a heavy bob and a light pendulum bar of length 1.200 m at $20\ °C$, at which temperature it keeps good time.

(a) Calculate the period of the pendulum at $20\ °C$.

(b) The material of the pendulum bar expands by 0.01% per degree. Calculate the fractional change in period if the clock is moved to a room at $25\ °C$.

(c) By how many seconds will the clock be 'out' at the end of a 24 hour period if it is kept at $25\ °C$?

5 Write the equation for the motion shown in Fig. 3.2.5 in the form $x = A\sin(\omega t + \phi)$.
[Hint: either start from scratch as in Section 3.2.3 or rewrite the result in this form.]

6 A pendulum bob is set into oscillation by applying a nudge to give it a speed of 10 cm s^{-1}. The subsequent period of oscillation is 1.5 s.

(a) Express the displacement, x, of the bob at time t in the form $x = A\cos(\omega t + \varepsilon)$. [Hint: $v_{max} = \omega A$.]

(b) Find the position, velocity and acceleration of the bob after 1.0 s

7 A person on the unmodified Millennium Bridge found herself oscillating laterally with amplitude of 7.5 mm and a period of 1.0 Hz.

(a) Calculate her maximum velocity and acceleration.

(b) If her position at time $t = 0$ was 5.0 mm and her velocity was negative, express her displacement in the form $x = A\cos(\omega t + \varepsilon)$.

8 In the 'pin and pendulum' set up in the diagram, a pin, **P**, is placed $l/2$ vertically below the pendulum support, so that the pendulum rotates about **P** on the left part of the swing.

(a) Write an expression for the period of the pendulum in terms of l.

(b) Sketch a graph of displacement against time for the pendulum bob, starting from a maximum displacement to the right.

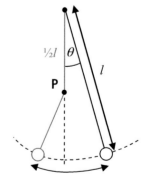

9 A large pendulum is set up in a public building. The mass of the bob is 5.0 kg, the length of the pendulum is 50 m and the horizontal amplitude of the oscillations 5.0 m.
Sketch graphs of the variation of potential energy, kinetic energy and total energy with time over a period of two cycles, starting from a point of maximum positive displacement. Use the lowest position as the point of zero PE.

10 A damped oscillating system has an initial amplitude of 20 cm and a period of 0.50 s. After 2.5 s the amplitude has become 5 cm. The initial displacement is 20 cm.

(a) Write the equation of motion for the system in the form $x = A_0 e^{-\lambda t} \cos(\omega t + \varepsilon)$.

(b) Sketch a graph of displacement against time from $0 - 2.5$ s.

(c) Calculate the displacement at $t = 0.9$ s.

11 A student timed the period of a simple pendulum to measure the acceleration due to gravity.
Results: length $= 0.850 \pm 0.002$ m   Time for 20 oscillations $= 36.95 \pm 0.05$ s
Explain whether these results are consistent with the accepted value of 9.81 m s^{-2}.

12 A group of students attempts to measure the acceleration due to gravity and also to estimate the height of a rafter in the roof of their school hall using a football pendulum. The football is glued to a length of thread the other end of which is slung over the rafter. They measure the period of the pendulum when the bottom of the ball is (a) 10.0 cm and (b) 100.0 cm above the floor. Their results are 4.21 s and 3.76 s. Given that the radius of a football is 11.1 cm, what height did they obtain for the height of the rafter?

The students estimated the period uncertainties as ± 0.01 s. Estimate the likely uncertainty in the height result.

13 A load on a spring oscillates. Because of air resistance, the amplitude of oscillations steadily decreases and the following results are obtained:

Time /s	0	30	60	90	120	150
Amplitude / cm	15.0	12.8	9.5	7.5	6.3	5.3

Use these results to show that the decay is approximately exponential and determine the value of the decay constant, λ.

Stretch & Challenge

It was shown in Section 3.2.5 that the potential energy of a vibrating system is proportional to the square of the displacement. This is a general result and is fairly obvious for a horizontal spring because the elastic potential energy stored is given by $\frac{1}{2}kx^2$. However, it is less so for a pendulum for which $E_P = mg\Delta h$.

Show that for small angles the gravitational potential energy of a pendulum bob is proportional to the square of the displacement, whether the displacement is expressed as an angle, the arc distance or the horizontal distance.

Use your results to verify that the total energy of oscillation for a simple pendulum is $\frac{1}{2}m\omega^2 A^2$.

You may find some or all of these results useful:

For small angles: $\sin\theta \approx \theta$

Series expansion for $\cos\theta$: $\cos\theta = 1 - \frac{\theta^2}{2!} + \frac{\theta^4}{4!} - \frac{\theta^6}{6!} + \ldots$

Binomial approximation: For small x, $(1 + x)^n \approx 1 + nx$

3.3 Kinetic theory

Until the early 20th century there was no direct evidence for the existence of atoms and molecules. However, physicists and chemists had developed a particle model of matter that explained a whole series of gas properties:

- The Ideal gas laws of Boyle and Charles: $\dfrac{pV}{T}$ = constant.

- Dalton's law: the pressure of mixtures of gases is equal to the sum of the pressures of the individual gas components.

- The law of multiple proportions (gases react in simple proportions, e.g. 1 volume of nitrogen reacts with 2 volumes of oxygen to give two volumes of nitrogen dioxide).

- Brownian motion: the random motion of smoke particles in a gas.

- Graham's law (the rate of diffusion of a gas is inversely proportional to the square root of the molecular mass)

In 1905, Einstein's mathematical explanation of the observed detail of Brownian motion finally provided the direct evidence.

This chapter does not attempt to follow the historical process of the development of the ideas in the 19th century. For example, Section 3.3.1 draws on 20th-century techniques.

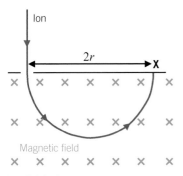

Fig. 3.3.1 Mass spectrometer

3.3.1 Atomic properties

By careful measurements of reacting masses, 19th-century chemists were able to establish the relative masses of the putative atoms of the different elements. For example, relative to hydrogen the masses of carbon, oxygen and iron are approximately 12, 16 and 56. But what are their actual masses? Also atoms are very small – but how small? And how big are the forces which they exert on one another?

(a) Atomic masses

If a beam of positive ions (H^+, He^+, Na^+) is accelerated and injected at right angles into a magnetic field, they follow a circular path. The reason for this is that, as explained in Section 4.4, they experience a force at right angles to their direction of motion, which is proportional to their charge and velocity. This is the principle of the *mass spectrometer*, which is depicted schematically in Fig. 3.3.1.

Using the ideas of Section 4.4.3(b), we can show that, if the ions have been accelerated through a voltage, V, the radius, r, of the path is given by:

$$r^2 = \frac{2mV}{B^2 q}$$

where m and q are the ionic mass and charge, respectively and B the magnetic flux density. As we know the electronic charge[1], we can calculate m.

Here is a typical set of results for singly-ionised helium (He^+):

$V = 100$ V $\quad B = 0.030$ T $\quad q = e = 1.60 \times 10^{-19}$ C $\quad r = 9.6$ cm

$\therefore$ The mass of a helium ion, $m_{He} = \dfrac{B^2 e r^2}{2V}$

$$= \frac{(0.030)^2 \times 1.60 \times 10^{-19} \times (0.096)^2}{2 \times 100}$$

$$= 6.6 \times 10^{-27} \text{ kg}$$

---

1 Measured by Robert Millikan and Harvey Fletcher in 1909.

(b) Atomic sizes

Atoms and molecules have no 'hard' surfaces, so there is no exact meaning to the *size* of a molecule. However, atoms do not penetrate each other (very far) when they collide or stick together. If they are close to each other in a crystal, we can use the distance between their centres as the sum of their radii. We can measure this by making use of the regular arrangement of particles in a crystal and treating it as a sort of diffraction grating. We then fire radiation of suitable wavelength at it and observe the interference pattern. This was pioneered by Laurence and Henry Bragg from 1913 onwards using X-rays. Beams of neutrons and electrons can also be used, using their wave properties.

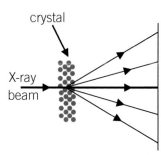

Fig. 3.3.2 X-ray diffraction

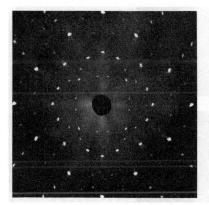 

Fig. 3.3.3 (a) X-ray and (b) neutron diffraction patterns of a sodium chloride crystal

Such images, obtained using radiation of wavelength ~ 10 pm, reveal typical values of lattice spacing in crystals of around 150–200 pm. Hence this may be taken as a typical atomic diameter.

(c) Interatomic forces

Neutral atoms and molecules exert very short-range forces on one another that are known as van der Waal's forces.[2] A typical graph of force against separation is shown in Fig. 3.3.4. The separation axis is shown in terms of the equilibrium separation, σ i.e. typically 150–200 pm. The precise form of the graph is not important for our needs but two things should be noted:

1. For $r < \sigma$ there is effectively an infinite repulsion, i.e. the particles behave like hard spheres.

2. The attractive force is very short-range: for $r > 2\sigma$ it is effectively zero.

We shall now use these ideas to develop our treatment on the behaviour of gases.

Self-test 3.3.1

Calculate the energy of:

(a) X-ray photons and

(b) neutrons,

which have a wavelength of 10 pm. Express your answers in both J and eV.

▼ **Study point**

The two-dimensional arrays of dots in the X-ray and neutron diffraction patterns arise because the crystals have planes in more than one dimension, unlike an optical diffraction grating.

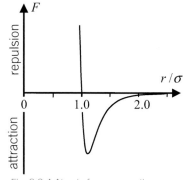

Fig. 3.3.4 Atomic force-separation curve

---

2 Strictly, the van der Waal's force is only the attractive section of the *F-r* curve.

3.3.2 Expressing the amount of a substance

(a) Moles and the Avogadro constant

There is a confusing number of ways of expressing the mass of atoms and molecules. Naturally the mass of an atom of an element such as Uranium-235 can be given in **kg**

$$m_{U\text{-}235} = 3.902\ 300 \times 10^{-25} \text{ kg (7 s.f.)}.$$

Similarly the mass of a water molecule $m_{H_2O} = 2.991\ 507 \times 10^{-26}$ kg.

These are distressingly small numbers for everyday use, so chemists and physicists have developed the idea of weighing a standard quantity of material rather than a single molecule. This standard quantity is called the **mole**. A mole of a quantity has 6.02×10^{23} (3 s.f.) basic entities, which is the number of atoms in 12 g of C-12. The number of particles in a mole is referred to as the **Avogadro constant,** N_A (see **Terms and definitions**). The quantity of material, expressed this way is referred to as the **amount** of material, with the unit mol.

Thus: 1 mol of water has 6.02×10^{23} molecules of H_2O

2 mol of iron-56 has 12.04×10^{23} atoms of Fe-56

10 mol of glucose has 6.02×10^{24} molecules of $C_6H_{12}O_6$

(b) Molar mass

The **molar mass**, M, of a pure substance (an element or compound) is the mass of one mol. We shall be almost entirely concerned with the molar masses of elements. Table 3.3.1 gives a few examples.

Substance	H	H$_2$	C-12	O$_2$	Fe-56	I-131	U-238
M / g mol^{-1}	1.008	2.016	12.000	32.00	55.93	130.91	238.05

Table 3.3.1 Molar masses of selected substances

Notice that the molar mass for an atom, in g mol^{-1} is very close to the nucleon number, A. The reason for the slight differences is explained by the **binding energy** of the nuclei – a concept which is explained in Section 3.6. The molar mass of C-12 is 12 g mol^{-1} exactly by definition.

Example

The density of Al-27 is 2.71 g cm^{-3}.

(a) Calculate the side of a cube consisting of 1 mol of Al-27.

(b) Estimate the diameter of an aluminium atom.

Answer

(a) Molar mass of Al-27, $M = 27.0$ g mol^{-1}.

$$\therefore \text{ Molar volume (i.e. volume per mol)} = \frac{M}{\rho} = \frac{27.0 \text{ g mol}^{-1}}{2.71 \text{ g cm}^3}$$

$$= 9.96 \text{ cm}^3 \text{ mol}^{-1}$$

$\therefore$ Volume of 1 mol = 9.96 cm^3. $\therefore$ Side of cube = $\sqrt[3]{9.96 \text{ cm}^3}$ = 2.15 cm.

(b) The volume of 1 mol [i.e. 6.02×10^{23} atoms] $= 9.96$ cm^3

∴ With the approximation that the Al atoms are cubes and occupy the entire volume, the width, w, of an Al atom is given by:

$$w = \sqrt[3]{\frac{9.96 \text{ cm}^3}{6.02 \times 10^{23}}} = 2.6 \times 10^{-8} \text{ cm} = 260 \text{ pm}$$

(c) Atomic / molecular mass

Frequently atomic scientists express the mass of atoms (and molecules) in the non-SI **unified atomic mass unit**[3], u. This is defined as one twelfth of the mass of an isolated C-12 atom.

From its definition it should be clear that the mass, m, of an atom expressed in u is numerically equal to the molar mass, M, expressed in g mol^{-1}.

(d) Sorting it out

Let us take the example of atomic oxygen (O). We have:

$$\text{Nucleon number (a.k.a. mass number)}, A = 16$$

$$\text{Molar mass}, M = 16.0 \text{ g mol}^{-1}$$

$$\text{Molecular mass}, m = 16.0 \text{ u}$$

$$= 16.0 \times 1.66 \times 10^{-27} \text{ kg}$$

$$= 2.66 \times 10^{-26} \text{ kg}$$

So, we can remember it this way. The nucleon number is, to a good approximation and with the addition of appropriate units, the molar mass (in **g mol^{-1}**) and the molecular mass (in **u**). To convert to **kg**, we simply multiply by **u**.

To go the other way: If we know the atomic (or molecular) mass in **kg**, we divide by **u** to get either the molar mass or the nucleon number.

Example (one for the chemists)

A hydrocarbon has a molecular mass of 4.65×10^{-26} kg. Write its molecular formula.

Answer

$$\text{Molecular mass} = \frac{m / \text{kg}}{u} = \frac{4.65 \times 10^{-26}}{1.66 \times 10^{-27}} = 28.0 \text{ u}$$

∴ Molar mass $= 28.0$ g mol^{-1}.

Carbon has a molar mass of 12 g mol^{-1} and hydrogen 1 g mol^{-1}, so there must be 2 carbon atoms and 4 hydrogen atoms in a molecule of the hydrocarbon, i.e. it must be C_2H_4 (ethene).

—Terms & definitions—

The **unified atomic mass unit**, u, is one twelfth of the mass of an isolated carbon-12 atom in its nuclear and atomic ground state.

(See Stretch & Challenge)

Stretch & Challenge

The value of u is 1.660 538 921 (73) $\times 10^{-27}$ kg. Because it can be measured to such a precision, the mass is specified for a carbon atom in its ground state and unbound to other atoms. The first excited state of C-12 is ~9 eV. How much is the mass of a C-12 atom in this state bigger than that of one in the ground state?

Self-test 3.3.4

Write A, M and m for U-235 as in Section 3.3.2(d).

Stretch & Challenge

It was proposed in 2012 to change the definition of u to:

$$u = \frac{0.001}{N_A} \text{ kg}$$

Why won't this make much difference?

Self-test 3.3.5

The volume of 1.0 mol of gas (at room temperature and atmospheric pressure) is 0.024 m^3.

(a) Calculate the volume per molecule.

(b) Use your answer to part (a) to show that the mean separation of the molecules ~3 nm.

---

3 The *Dalton* (Da) is an alternative name for u. It is often used by biochemists ('a protein of mass 120 kDa'). It is not used here.

Stretch & Challenge

Consider a sample of gas with pressure, volume and temperature values of p_0, V_0 and T_0 respectively.

(a) Allow it to change to p_1, V_0 and T_1 via the intermediate state p_1, V_1 and T_0.

Derive the relationship $p \propto T$

from $p \propto \dfrac{1}{V}$ and $V \propto T$.

(b) Now allow it to change from p_0, V_0, T_0 to p_1, V_1, T_1 via an intermediate state of your choice to derive $pV \propto T$.

▼ **Study point**

The law of multiple proportions suggests strongly that, 'Under the same conditions of temperature and pressure, equal volumes of all gases contain the same number of molecules,' which is *Avogadro's law*. The electrolysis of water to give two volumes of hydrogen to one of oxygen is a nice example of this because

$2H_2O \rightarrow 2H_2 + O_2$.

┌─ **Terms & definitions** ─┐

An **ideal gas** is one which strictly obeys the equation $pV = nRT$.

3.3.6 Self-test

Show that the unit $J\ mol^{-1}\ K^{-1}$ makes the ideal gas equation homogeneous.

▼ **Study point**

In the working of the example notice:

1. Using radius not diameter when applying $\pi r^2 h$.

2. Converting the °C to K.

3. The molar mass of oxygen is $32\ g\ mol^{-1}$ because it is O_2.

3.3.3 The ideal gas equation

Low pressure gases which are well above their boiling points behave in a very simple way, as we shall see. Such gases are referred to as **ideal gases** and we shall spend the rest of Section 3.3 considering their properties.

If we investigate the relationships between the pressure, p, volume, V, and kelvin temperature, T, of a sample of an ideal gas (such as dry air) we find that:

- at constant temperature: $\quad p \propto \dfrac{1}{V} \quad$ or $\quad pV = \text{constant}$

- at constant pressure: $\quad V \propto T \quad$ or $\quad \dfrac{V}{T} = \text{constant}$

- at constant volume: $\quad p \propto T \quad$ or $\quad \dfrac{p}{T} = \text{constant}$

In fact these are not independent relationships as any one of them can be derived by combining the other two (see Stretch & Challenge). In the same way we can combine them to give the relationship (for a fixed sample of gas):

$$pV \propto T$$

Writing this as $pV = kT$, the constant k will be proportional to the amount of gas present (double the amount at constant pressure will give double the value of pV because V is doubled). Combining this with Avogadro's law (see Study point) leads to:

$$pV = nRT$$

where $\qquad n$ is the number of moles (i.e. the amount) of gas, and

$\qquad\quad R$ is a constant called the *molar gas constant*, $8.31\ J\ mol^{-1}\ K^{-1}$

This equation is called the *equation of state for an ideal gas* or (slightly more catchily) the ideal gas equation. We are now in a position to define an **ideal gas** (see **Terms and definitions**). No gas exactly obeys the ideal gas equation but some common gases (oxygen, nitrogen, helium, methane) do so to a very good approximation under everyday conditions and as do other gases (carbon dioxide, propane, butane) at very low pressure.

This example illustrates how we can use the equation and a few pitfalls to avoid (see Study point).

Example

Calculate the mass of gas in a medical oxygen (O_2) cylinder of height 80 cm and internal diameter 15 cm. At 20 °C the gas pressure is 20 MPa.

Answer

Volume of gas, $V = \pi \left(\dfrac{d}{2}\right)^2 h = \pi \times (0.075\ \text{m})^2 \times 0.80\ \text{m} = 0.0141\ \text{m}^3$

$pV = nRT, \therefore n = \dfrac{pV}{RT} = \dfrac{20 \times 10^6\ \text{Pa} \times 0.0141\ \text{m}^3}{8.31\ \text{J mol}^{-1}\text{K}^{-1}(20 + 273.15)\text{K}} = 115.8\ \text{mol}$

$\therefore$ Mass of oxygen $= nM = 115.8\ \text{mol} \times 32\ \text{g mol}^{-1} = 3700\ \text{g} = 3.7\ \text{kg}$.

3.3.4 The behaviour of molecules in gases

Kinetic theory, seeks to explain the properties of materials (gases, liquids and solids) in terms of the motion and interactions of molecules. We'll concentrate on gases but refer to solids and liquids where appropriate.

(a) Molecular basics

We start from the assumption that gases (as well as liquids and solids) consist of large numbers of molecules (see Study point) in permanent motion.

If you've done Self-test 3.3.5 you'll have seen that molecules of a gas have plenty of space to move around in. This explains why we can compress gases easily, unlike liquids and solids. At atmospheric pressure a typical separation ~ 3 nm is about 10 times the size of a molecule. It also explains why gases diffuse into each other much more quickly than do solids and liquids.

Molecular speeds are typically a few hundred m s^{-1} (we'll calculate them later), so even with all this space, they must collide frequently with one another. They also collide frequently with the container wall.

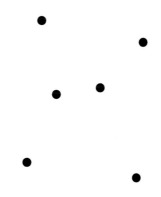
Fig. 3.3.5 Molecular spacing

(b) Molecular collisions

Making the simple assumption that the kinetic energy of the molecules of a gas depends upon the temperature of the gas and with the observation that gas in a cylinder doesn't spontaneously cool down below room temperature, we must conclude that intermolecular collisions must be elastic – at least on the average – meaning that there is no loss in kinetic energy.

Let's examine this in more detail. What does 'on the average' mean? The diagram in the Stretch & Challenge was of two monatomic molecules, e.g. helium. Could some kinetic energy be lost in the collision? It is possible for kinetic energy to be lost and transferred to one of the atoms, putting it into an excited state. On a subsequent collision this energy can be transferred back to kinetic energy (see AS textbook Fig. 2.8.3).

The scenario depicted in Fig. 3.3.6 gives a way of losing translational kinetic energy for diatomic molecules (e.g. O_2 and N_2). Rotational kinetic energy is picked up in the collision, slowing the molecules down. Again, subsequent collisions can convert this back to translational KE.[4]

So some translational kinetic energy can be lost in a collision; in other collisions it can be gained. On average the translational KE is constant. In any case, the effect of the frequent collisions is to randomise the velocities and hence the kinetic energies of the gas molecules. Fig. 3.3.7 shows the probability function, P, for molecular speeds.

---

4 Because of their very low moments of inertia (see Option C) and for quantum mechanical reasons, monatomic molecules cannot possess rotational kinetic energy.

(a) Before collision

(b) After collision

Fig. 3.3.6 Collision between diatomic molecules

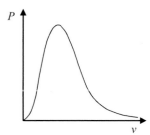
Fig. 3.3.7 Molecular speed distribution

3.3.5 Molecular motion and gas pressure

The aim of this section is to use our knowledge of molecules and Newtonian mechanics to refine the qualitative pre-A level explanation of gas pressure to make quantitative statements.

(a) Some extra assumptions

We are going to restrict ourselves to ideal gases, so to do so we'll make a couple of additional assumptions about the molecules:

1. The volume of the molecules themselves is a negligible fraction of the volume occupied by the gas.

    We can justify this by reference to Self-test 3.3.5. Also, steam isn't an ideal gas (far from it) but compare the density of liquid water at 100 °C (1000 kg m⁻³) with that of steam (0.6 kg m⁻³). Even if there were no wiggle room in liquid water, the same molecules in steam occupy less than 0.1% of the total space.

2. The molecules exert negligible forces on one another except during collisions.

    If the molecules occupy less than 0.1% of the space then their mean separation must be at least ten times their diameter. Looking back at Fig. 3.3.4, this means that most of the time they are well out of the range of intermolecular forces. And then…

3. The effect of gravity is negligible.

    At first sight this is a bit strange but the molecular speeds are so high, that in the short distances travelled between collisions, the effect of gravity is tiny. See Self-test 3.3.8.

    Taken together, assumptions 2 and 3 mean that molecules travel at a constant velocity between collisions.

(b) Pressure and molecular motion

There are several ways of deriving the equations which follow. This one is not entirely satisfactory but it does illustrate the physics ideas.[5]

Imagine a single molecule of mass m in a box, as in Fig. 3.3.8. We ask the question, 'What pressure does it exert on face **A**?'

At the instant of the diagram let the velocity c have components c_x, c_y and c_z. Eventually, the molecule will hit face **A**. Because we are assuming elastic collisions, it will bounce off with velocity $(-c_x, c_y, c_z)$

∴ Change in momentum of molecule at **A** $= -2mc_x$

After bouncing off a few more walls it will reach **A** again after a time, Δt given by

$$\Delta t = \frac{2l_x}{c_x}$$

at which point its momentum will change by $-2mc_x$ again. So its rate of change of momentum at **A** is $-2mc_x \times \dfrac{c_x}{2l_x} = -\dfrac{mc_x^2}{l_x}$.

But, from N2, rate of change of momentum = Force

∴ The mean force exerted by **A** on the molecule $= -\dfrac{mc_x^2}{l_x}$

5 For a deeper analysis see *Mathematics for A Level Physics*, by Kelly and Wood.

3.3.7 Self-test

The density of liquid water is 1.0 g cm⁻³.

The molar mass of water of 18 g mol⁻¹.

(a) Show that the 'diameter' of a water molecule ~300 pm.

(b) Estimate the separation of water molecules in steam at 100 °C and 1 atm.

3.3.8 Self-test

A molecule travels horizontally at 300 m s⁻¹. How far does it drop:

(a) as it crosses the 8 m wide lab?

(b) in the mean distance between collisions of ~0.1 μm?

Terms & definitions

The SI unit of pressure is the **pascal** (Pa) which is equivalent to N m⁻².

Engineers often express pressure in **bar**:
1 bar   = 10⁵ Pa

An everyday unit is the **atmosphere** (atm):
1 atm   = 1.01325 bar
        = 1.01325 × 10⁵ Pa

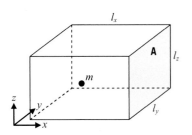

Fig. 3.3.8 Molecule in a box

▼ Study point

Can we just add up the pressure contributions from the N molecules? Won't they collide and which will change the values of c_x?

Yes, but when they collide, momentum and energy will be conserved and so the total change of momentum each second at face **A** is just the same as if they hadn't collided. See Stretch & Challenge.

∴ By N3, the mean force exerted by the molecule on **A** $= \dfrac{mc_x^2}{l_x}$

And the mean pressure, p, exerted on **A** by the molecule is given by:

$$p = \frac{\text{Force}}{\text{area}} = \frac{mc_x^2}{l_x(l_y l_z)} = \frac{mc_x^2}{V}$$

where V is the volume of the box.

We now boldly introduce a lot more molecules so there are N molecules in total and we'll call their x-velocity components $c_{1x},\ c_{2x},\ ....c_{Nx}$. Their pressures on face **A** add to give:

$$p = \frac{mc_{1x}^2}{V} + \frac{mc_{2x}^2}{V} + \dots \frac{mc_{Nx}^2}{V} = \frac{Nm\overline{c_x^2}}{V} \qquad [1]$$

where $\overline{c_x^2}$ is the mean of the c_x^2 values. See Study point on the previous page.

We'll write this pressure in terms of c rather than c_x. Applying Pythagoras' theorem for each molecule: $c^2 = c_x^2 + c_y^2 + c_z^2$

$$\therefore \quad \overline{c^2} = \overline{c_x^2} + \overline{c_y^2} + \overline{c_z^2} \qquad [2]$$

Given assumption 3 above, there is no special direction and after multiple collisions the distribution of the components of velocity in all directions would be the same, even if they started out different.

This must mean that $\overline{c_x^2} = \overline{c_y^2} = \overline{c_z^2}$

which together with equation [2] means that $\overline{c_x^2} = \frac{1}{3}\overline{c^2}$

Substituting in equation [1] → $\quad \therefore p = \frac{1}{3}\dfrac{Nm\overline{c^2}}{V}$

which is more conveniently written $pV = \frac{1}{3}Nm\overline{c^2}$ $\qquad [3]$

(c) What does $\overline{c^2}$ mean?

Imagine a group of five nitrogen molecules which are travelling along the x axis. Their velocities, c, are:

$$100\text{ m s}^{-1} \quad -200\text{ m s}^{-1} \quad -300\text{ m s}^{-1} \quad 400\text{ m s}^{-1} \quad 500\text{ m s}^{-1}$$

We'll work out various means:

1. Mean velocity, $\overline{c} = \dfrac{100 - 200 - 300 + 400 + 500}{5}$ m s^{-1} = 100 m s^{-1}

2. Mean speed, $\overline{|c|} = \dfrac{100 + 200 + 300 + 400 + 500}{5}$ m s^{-1} = 300 m s^{-1}

3. Mean square speed, $\overline{c^2} = \dfrac{100^2 + 200^2 + 300^2 + 400^2 + 500^2}{5}$ m^2 s^{-2}

$$= 110\ 000 \text{ m}^2\text{ s}^{-2}$$

Notice the difference between the mean speed and the mean velocity. All the speeds are positive – direction is not significant.

We now introduce the **rms velocity** (or speed), c_{rms} which is the square root of the mean square speed. For the above molecules:

4. Rms velocity (or speed), $c_{rms} = \sqrt{\overline{c^2}} = \sqrt{110\,000 \text{ m}^2\text{ s}^{-2}} = 331$ m s^{-1}.

This value is the most often used for molecules because the mean kinetic energy, $\overline{E}_k$, can be calculated from

$$\overline{E}_k = \frac{1}{2}mc^2_{rms}$$

Consider a 1 m long box with two 'molecules' as shown.

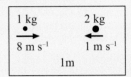

(a) Assuming no collisions, calculate the mean force exerted by the two molecules on the end faces.

(b) Let the molecules collide head on as shown. Apply the principle of conservation of momentum and energy to show that the velocities after the collision are −4 m s^{-1} and 5 m s^{-1} (right is positive).

(c) Show that the mean force exerted by the two molecules on the ends following this one-off collision is the same as in part (a).

─ *Terms & definitions* ─

The quantity $\overline{c^2}$ in equation [3] is called the **mean square velocity**. It is the mean value of the squares of the molecular speeds [squaring it makes it a scalar quantity].

Taking the square root, gives $\sqrt{\overline{c^2}}$ which is called the **root mean square velocity** (or **rms velocity**) which is written c_{rms}.

▼ **Study point**

Note the expression root mean square velocity means the (square) root of the mean square velocity. If you study AC electricity you will meet the rms current and pd too.

Self-test 3.3.9

Why is there no difference between the mean square speed and mean square velocity for a group of molecules?

3.3.10 Self-test

Notice that the rms speed for the five molecules in Section 3.3.5(c) is greater than the mean speed.

(a) Explain why this must always be the case (unless the molecules have identical speeds).

(b) Copy Fig. 3.3.7 and label suggested positions of the mean and rms speeds.

▼ **Study point**

Note that, from the definitions:

$$\overline{c^2} = c^2_{rms}$$

───── *Terms & definitions* ─────

Many pressure gauges, including car tyre pressure gauges indicate the pressure above atmospheric pressure. This is known as the **gauge pressure**.

Stretch & Challenge

The radius of an argon atom is 71 pm. The centre of a second argon atom cannot come within 2 radii of the centre of the first argon atom. 1 mole of argon atoms occupies a volume of 0.024 m³. Show that the fraction of the volume which the atoms effectively occupy is only 0.03% of the total.

Example

For the five nitrogen molecules, calculate:

(a) their mean momentum

(b) their mean kinetic energy.

Answer

(a) Mass of N_2 molecules $= 28\ u = 28 \times 1.66 \times 10^{-27}\ kg = 4.6 \times 10^{-26}\ kg$

$$\text{Mean momentum} = \text{mass} \times \text{mean velocity}$$
$$= 4.6 \times 10^{-26}\ kg \times 100\ m\ s^{-1}$$
$$= 4.6 \times 10^{-24}\ N\ s$$

(b) Mean KE $= \frac{1}{2}mc^2_{rms} = \frac{1}{2}4.6 \times 10^{-26} \times 110\,000\ J$
$$= 2.53 \times 10^{-21}\ J$$

(d) Comparison with the ideal gas equation

The equation of state for an ideal gas, $pV = nRT$, under the conditions of a constant temperature and a fixed amount of gas, reduces to

$$pV = constant$$

Is equation [3] above consistent with this? The left-hand sides are the same. The right-hand side of equation [3], $\frac{1}{3}Nm\overline{c^2}$, is proportional to the kinetic energy of the molecules, which, according to our assumption is constant at constant temperature.

How have our other assumptions figured in the derivation of [3]? We mentioned the lack of gravitational effect. The others are:

1. Negligible volume of molecules: No two molecules can be in the same place, i.e. the space occupied by a molecule is not available for any other. This means that the effective volume for the molecules would be less than $l_x l_y l_z$ if the molecules did not have negligible volume.

2. Negligible intermolecular forces: If the molecules exerted significant forces on one another, the speed of impact with the container walls would not be the same as just before impact (it could be more or it could be less because of interactions with the other gas molecules and the wall molecules).

Without being explicit during the derivation, we have assumed that both these effects are negligible.

Conclusion: The molecular model appears to reproduce the pV aspect of the ideal gas equation.

3.3.6 Molecular energy

Consider again the equation $\qquad pV = \frac{1}{3}Nm\overline{c^2}$ [1]

Equation [1] tells us that we can calculate the translational kinetic energy of the molecules if we know just the pressure and volume of a gas: we don't need to know the temperature, the number of molecules or the mass of the molecules!

The translational KE of the molecules, $E_{k\,\text{trans}} = \frac{1}{2}Nm\overline{c^2}$

But, from equation [1] $\qquad \frac{1}{2}Nm\overline{c^2} = \frac{3}{2} \times \frac{1}{3}Nm\overline{c^2}$

$$E_{k\,\text{trans}} = \frac{3}{2}pV \qquad [2]$$

Example

Calculate the kinetic energy of the molecules in a helium cylinder of volume 0.020 m³ at a pressure of 10 MPa.

Answer

$E_{k\,\text{trans}} = \frac{3}{2}pV = 1.5 \times 10\ \text{MPa} \times 0.020\ \text{m}^3 = 300\ \text{kJ}$

Note: Helium is monatomic so this energy is the total kinetic energy of the molecules. See footnote to Section 3.3.4(b).

(a) Molecular energy and temperature

We can also write $E_{k\,\text{trans}}$ in terms of temperature by substituting from equation [2] into $pV = nRT$, to give

$$E_{k\,\text{trans}} = \frac{3}{2}nRT \qquad [3]$$

So the molecules of one mole of gas have a translational kinetic energy of $\frac{3}{2}RT$. We can go further and calculate the mean translational kinetic energy per molecule:

Mean KE per molecule $= \frac{1}{2}m\overline{c^2} = \frac{3}{2}\frac{R}{N_A}T = \frac{3}{2}kT$ [4]

The constant k is called the Boltzmann constant and has a value of 1.38×10^{-23} J K⁻¹.

Example

Calculate the mean kinetic energy per molecule at 300 K. Express the answer in (a) J and (b) eV.

Answer

(a) Mean KE per molecule $= \frac{3}{2} \times 1.38 \times 10^{-23}\ \text{J K}^{-1} \times 300\ \text{K}$

$= 6.21 \times 10^{-21}\ \text{J}$

(b) Mean KE per molecule $= \dfrac{6.21 \times 10^{-21}\ \text{J}}{1.60 \times 10^{-19}\ \text{J eV}^{-1}}$

$= 0.039\ \text{eV}$

Note that this is the <u>mean</u> KE per molecule. Looking back at the molecular speed distribution in Fig. 3.3.7 it should be obvious that a small fraction of molecules would have a significantly greater energy.

▼ **Study point**

How can the molecules of argon in a 1 m³ container of argon at 1 atm and 600 K have the same kinetic energy as those in the same container at 1 atm and 300 K?

Answer: Because the mean molecular KE is double, but there are only half as many molecules!

Self-test 3.3.11

If the cylinder of helium is at 25 °C, estimate:

(a) the number of moles and

(b) the number of molecules of helium in the cylinder.

Self-test 3.3.12

Use

$R = 8.31\ \text{J mol}^{-1}\ \text{K}^{-1}$

and

$N_A = 6.02 \times 10^{23}\ \text{mol}^{-1}$

to confirm the value and unit of the Boltzmann constant given in the text.

▼ **Study point**

It is worth remembering that molecular energy at room temperature ~0.04 eV [$\frac{1}{25}$eV].

A photon with this energy is in the far infrared. This explains why room temperature objects emit radiation in the infrared.

Stretch & Challenge

Given the mean molecular energy at room temperature and considering the typical separation of the atomic energy levels (see AS textbook), explain why molecular collisions in monatomic gases are elastic.

── Terms & definitions ──

The molar heat capacity of a gas at constant volume, C_V, is the heat required to raise the temperature of 1 mol of the gas by 1 K without changing the volume.

(b) Molar heat capacity: monatomic gases

The relationship between temperature and molecular energy provides us with another test of the kinetic theory because the heat needed to raise the temperature of a gas can be measured experimentally and compared with the theoretical prediction. Equation [2] tells us that the energy of the molecules in a monatomic gas is $\frac{3}{2}nRT$, so the energy needed to raise the temperature of 1 mol of gas is $\frac{3}{2}R$, i.e. 12.5 J mol^{-1} K^{-1}. This quantity is called the **molar heat capacity at constant volume**[6], C_V. Table 3.3.2 gives a few values of C_V.

Gas	C_V / J mol^{-1} K^{-1}	C_V / R
Helium (He)	12.5	1.50
Argon (Ar)	12.5	1.50
Nitrogen (N$_2$)	20.6	2.49
Oxygen (O$_2$)	21.1	2.54
Carbon monoxide (CO)	20.7	2.49

Table 3.3.2 Molar heat capacities of some common gases

The theoretical prediction for the two monatomic gases is spot on. We'll now look at the diatomic gases.

(c) Molar heat capacity: diatomic gases

The values of the molar heat capacity for the diatomic gases, nitrogen and oxygen and carbon monoxide, are consistently greater than those for the monatomic gases – very close to $\frac{5}{2}R$ rather than $\frac{3}{2}R$. Why should this be? The answer is that the pressure of a gas is determined by the translational KE of the molecules $\frac{1}{2}Nm\overline{c^2}$, but diatomic gases can possess energy of rotation as well – see Fig. 3.3.9. As we saw earlier, Fig. 3.3.6, energy can be interchanged between the translational and rotational motion. According to *equipartition theory* each independent way of possessing energy (known as *degrees of freedom*) of the gas molecules should have the same energy. The three independent translational degrees of freedom (x, y and z motion) have $\frac{3}{2}nRT$ between them, i.e. $\frac{1}{2}nRT$ each. There are two rotational degrees of freedom too making the total molecular energy $5 \times \frac{1}{2}nRT = \frac{5}{2}nRT$. Hence the molar heat capacity at constant volume should be $\frac{5}{2}R$ – very close to the observed values.

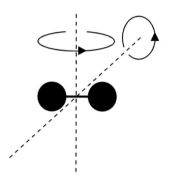

Fig. 3.3.9 Rotational degrees of freedom

Stretch & Challenge

There is no kinetic energy of rotation about the axis along the bond in Fig. 3.3.9 for the same reason that monatomic molecules have no rotational kinetic energy.

---

6 The significance of the phrase 'at constant volume' will become clear in Section 3.4 Thermal physics.

Exercise 3.3

The first seven questions are to give practice at using the mole, molar mass and the Avogadro constant.

1 The molar mass of gold is 196.97 g mol^{-1}. Calculate the number of gold atoms in a 1.000 kg gold bar.

2 A cylinder contains 47 kg of propane gas (C_3H_8). Calculate:

(a) The number of moles of propane in the cylinder.

(b) The number of molecules of propane in the cylinder.

3 The *farady* (F) is a historic unit of charge. It is the charge on a mole of electrons. Express the faraday in coulombs.

4 The density of pure salt (NaCl) is 2.163×10^3 kg m^{-3}. Calculate the number of sodium ions in a 1.000 mm^3 salt crystal.

$m(Na) = 22.990$ u $m(Cl) = 35.453$ u 1 u $= 1.660 \times 10^{-27}$ kg

5 (a) Use the data below to calculate the relative molecular mass of UF_6.

(b) State the molar mass of UF_6 (i) in g mol^{-1}, (ii) in kg mol^{-1}.

(c) 0.7% of natural uranium is U-235. Calculate the number of moles of U-235 in 1 tonne (i.e. 1000 kg) of uranium hexafluoride.

$m(U) = 238.02$ u $m(F) = 19.00$ u

6 The diagram shows the structure of a salt (NaCl) crystal. Calculate the inter-ionic distance, x, from the following data:

Density of salt $= 2.165$ g cm^{-3}
Molar mass of sodium $= 22.990$ g mol^{-1}
Molar mass of chlorine $= 35.453$ g mol^{-1}
Avogadro constant $= 6.022 \times 10^{23}$ mol^{-1}

[Hint: Calculate the volume and hence the length of the side of a 1 mole cubic NaCl crystal. From the total number of particles in the crystal, calculate x.]

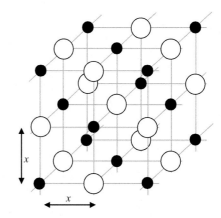

7 One way of determining the Avogadro constant is to measure the inter-ionic distance using X-ray diffraction.

Potassium chloride has the same crystal structure as sodium chloride. The inter-ionic distance is measured as 314.75 pm. The density of potassium chloride is 1984 kg m^{-3} and the molar mass of potassium is 39.101 g mol^{-1}. Use these data to calculate a value for the Avogadro constant.

8 The density of dry air at atmospheric pressure of 1.013×10^5 Pa and 15 °C is 1.225 kg m^{-3}. Use these data to calculate:

(a) The rms speed of air molecules at 15 °C.

(b) The number of moles of air molecules in 1.0 m³ of air at 15 °C.

(c) The mean molecular mass of air molecules, comparing your answer with that expected from the composition of air [78.09% N_2, 20.95% O_2, 0.93% Ar].

9 In an experiment to determine the molar gas constant, R, a student weighs an oxygen cylinder, of volume 0.0100 m³ and reads the pressure gauge. She releases some oxygen, weighs the cylinder again and reads the final pressure (after allowing the cylinder to re-equilibrate with the room). She measures room temperature. Use her results to find calculate a value for R.

Results:

Temperature of room = 18.5 °C
Initial pressure = 70.0 atm
Final pressure = 37.5 atm

Initial mass of cylinder + oxygen = 2.832 kg
Final mass of cylinder + oxygen = 2.397 kg

10 Five molecules of oxygen have speeds in m s^{-1} of 150, 250, 300, 350 and 400.

(a) Calculate (i) the mean speed, (ii) the mean square speed and (iii) the rms speed of the molecules.

(b) Calculate the temperature of oxygen with the same rms speed as that calculated in part (a).

11 An interstellar gas cloud consists mainly of atomic hydrogen, molecular hydrogen and helium, with small quantities of atomic and molecular oxygen.

(a) Compare the rms speeds of these molecules.

(b) Calculate the rms speeds of the molecules at 50 K.

12 The atmosphere exerts a pressure of 100 kPa at ground level. To a good approximation its composition may be taken as $\frac{4}{5}N_2$ and $\frac{1}{5}O_2$.

(a) Estimate the number of molecules in the atmosphere.
[Radius of Earth = 6370 km.]

(b) It has been suggested that every breath you take contains one or two molecules from the dying breath of Julius Caesar. Investigate whether this is plausible.

3.4 Thermal physics

The impulse for developing this central area of physics was the industrial revolution. The high grade chemical energy store of coal was used to pump water out of mines, transport people and goods, drive mill machinery and, eventually, generate electricity. Steam engines and internal combustion engines were developed to fulfil these needs and the questions of how to improve their power output and maximise their efficiency became central. The science of *thermodynamics* was developed in response to these requirements.

Fig. 3.4.1 LNER's 60103, The Flying Scotsman

Fig. 3.4.2 Stott Park bobbin-mill engine

Fig. 3.4.3 Steam turbine under construction

3.4.1 Thermodynamic systems

Thermodynamics starts off with the concept of the **system**. This is just the collection of particles that we are considering. A system has a *boundary*, i.e. a surface which limits its extent; the universe outside is referred to as its *surroundings*.

In the context of heat engines, such as those illustrated in Figs. 3.4.1–3, the system is a collection of gas molecules and the boundary is the walls of a cylinder and piston. The engine is designed to extract energy from the gas molecules as they collide with the piston and make it move, i.e. they do **work**.

We'll now look at how we express the energy interchanges between a system and its surroundings.

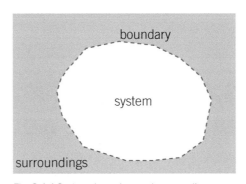

Fig. 3.4.4 System, boundary and surroundings

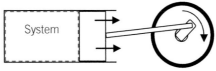

Fig. 3.4.5 Gas in a cylinder as a system

3.4.2 Internal energy, U

The first of the key concepts is that of the **internal energy** of a system. Put simply this is the sum of the energy possessed by all the particles. How can the particles possess energy? They can possess:

- Kinetic energy (translational and rotational)

- Potential energy.

3.4.1 ∨ Self-test

Calculate the internal energy of:

(a) 3 moles of a gas at a temperature of 300 K.

(b) 4 kg of helium at 0 °C.

(c) 1 kg of oxygen [α. diatomic gas – see Section 3.4.6(c)].

3.4.2 ∨ Self-test

Calculate the pressure of the gas in the model system if the mass of the piston is 1 kg and the diameter of the cylinder is 10 cm. Atmospheric pressure = 1.013×10^5 Pa.

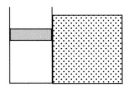

Fig. 3.4.7 Thermal equilibrium

We know from earlier work that the term *potential energy* covers a variety of different types, such as gravitational and elastic. In terms of thermodynamic systems we also have to consider the interactions between the particles, so there is also the potential energy due to:

- Intermolecular forces

- Chemical bonds

- Excited energy states in atoms

- Nuclear energy states.

Luckily, we are only going to be concerned with the *change* in internal energy, ΔU, so any forms of potential energy that stay constant need not be considered (see Study point). If the system is a sample of an ideal gas, then intermolecular forces can be ignored and we can consider U to consist entirely of the molecular kinetic energy.

i.e for a monatomic gas: $\qquad U = \frac{3}{2}nRT \quad$ [1]

which we can also write as $\qquad U = \frac{3}{2}pV \quad$ (see Section 3.4.6).

The $\frac{3}{2}pV$ form of the equation 1 gives us a way of monitoring the internal energy of a simple system. We'll use it in developing the concepts of heat and temperature using the gas in the idealised arrangement in Fig. 3.4.6. as our system, together with some thought experiments.

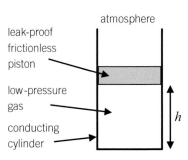

Fig. 3.4.6 Model system

3.4.3 Thermal equilibrium, temperature and heat

Section 3.4.3 underpins thermal physics but can be skipped on a first reading. You should learn the **Terms and definitions** boxes.

The only aspect of our model system that we'll measure in our thought experiments is the height of the gas column, h. This is proportional to the volume of the gas. The pressure is constant (slightly higher than atmospheric pressure – see Self-test 3.4.2) so the internal energy of the gas is proportional to h.

(a) Experiment 1

This is a very short experiment and really just helps us to define a term. We imagine our model system in thermal contact with another system, e.g. a large block of metal, as in Fig. 3.4.7. If the internal energy of the model system remains constant, we say that the two systems are in **thermal equilibrium**. How do we know that U is constant? We know because the height of the gas column stays the same.

(b) Experiment 2

This is slightly (but only slightly) more exciting. This time, our two systems are separate to start with and we bring them together.

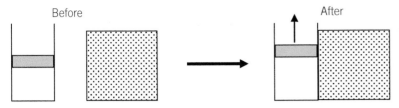

Fig. 3.4.8 Definition of heat and temperature

When we bring them together, we notice that the piston rises and then stops rising. Because $U = \frac{3}{2}pV$ we conclude that energy has moved into our model system from the metal block. This spontaneous flow of energy between two systems which are not in thermal equilibrium is called **heat**. We say that heat has flowed from the metal block into the gas (through the conducting wall of the cylinder).

We also use this experiment to define the *concept* of **temperature**. Two systems in thermal equilibrium with each other are said to be 'at the same temperature'. If heat flows from system 2 to system 1, system 2 is said to have a higher temperature, i.e. we define temperature in terms of the direction of heat flow: heat flows from a high to a low temperature.

(c) Experiment 3

This rather strange experiment is needed to underpin our inchoate concept of temperature. We start with two of our model systems, each separately in thermal equilibrium with our metal block. We then bring the model systems into thermal contact with each other. We find that they are also in thermal equilibrium.

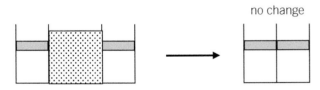

Fig. 3.4.9 Zeroth law of thermodynamics

This is the *zeroth law of thermodynamics*: two systems which are separately in thermal equilibrium with a third system will be in thermal equilibrium with each other. The Study point explains its importance.

▼ **Study point**

This definition of temperature is what is meant by its common definition as the *degree of hotness as measured on a thermometer*. The definition of temperature scales and a discussion on how temperature is measured are not covered in this book.

▼ **Study point**

We can think of the two model systems as thermometers – the height *h* is the temperature reading. They are both in equilibrium with the block, so they should be at the same temperature. If they were not in equilibrium with each other, they couldn't be. So if the zeroth law were not correct, we wouldn't be able to define, let alone measure, temperature.

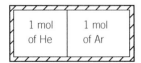
(d) Temperature and internal energy

$U \propto T$ for an ideal gas. Indeed, we can be more specific: for a monatomic ideal gas $U = \frac{3}{2}nRT$. And generally the higher the temperature the greater the value of U for any system. However, consider what happens if we heat 1 kg of ice up from −100 °C until it has all boiled away as steam. An approximate graph of temperature, θ, against heat added, Q, is shown in Fig. 3.4.10.

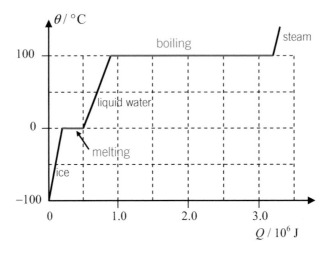

Fig. 3.4.10 The effect of heat on the temperature of water

Taking the heat added as an indication of the increase in internal energy (see Study point), we see that the temperature does not rise with internal energy during a change of state. This is because temperature is related closely to the translational kinetic energy component of the internal energy and not the total internal energy. We can say that the internal energy per mole of two monatomic ideal gases (or two diatomic gases) are equal at the same temperature (and Section 3.4.6 shows us how to compare U for a monatomic and a diatomic gas).

When we change state, e.g. liquid to gas, we need to break a lot of inter-molecular bonds in separating the molecules, so even if the temperature of the substance stays the same, the total internal energy of the gas molecules is much greater than that of the liquid.

Example

Compare the internal energy of 2 mol of argon at 0 °C with that of 3 mol of helium at 200 °C.

Answer

$$U = \frac{3}{2}nRT \therefore \frac{U_{Ar}}{U_{He}} = \frac{n_{Ar}T_{Ar}}{n_{He}T_{He}} = \frac{2 \text{ mol} \times 273.15 \text{ K}}{3 \text{ mol} \times 473.15 \text{ K}} = 0.38$$

For all systems, as the temperature falls the internal energy decreases. As the temperature drops towards absolute zero, 0 K, the internal energy falls to a minimum value. This minimum value of internal energy is, strangely, not equal to zero because of a quantum mechanical effect called zero point energy. One of the results of the motion associated with the zero point energy is that helium at atmospheric pressure does not freeze.

3.4.4 Energy transfer by work

We started this section by noting that the stimulus for developing the science of thermodynamics was to understand and improve the operation of heat engines. These engines do work on their surroundings, e.g. by driving vehicles, turning generators and pumping out mines. As we saw in the AS course:

- Work is the product of force and distance moved in the direction of the force.

- The energy transfer is equal to the work done.

Fig. 3.4.11 shows a system expanding into its surroundings. Assuming that the surroundings contain molecules (i.e. the expansion isn't into a vacuum) the system will need to exert a force on the surroundings and therefore does work as it expands. Consequently, energy is transferred to the surroundings and, unless energy is supplied to the system (e.g. by heat), its internal energy must decrease.

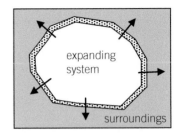

Fig. 3.4.11 A system doing work on the surroundings

(a) The work done by an expanding gas at constant pressure

The gas in the cylinder in Fig. 3.4.12 is doing work by expanding and pushing back the piston. If it exerts a force, F, and expands a distance, d, then:

$$\text{Work done, } W = F\Delta x$$

If p = the gas pressure   and A = cylinder cross-sectional area

then $F = pA$ and $W = pA\Delta x$

But $A\Delta x = \Delta V$, the increase in volume of the gas

$$\therefore\ W = p\Delta V \text{ as long as the pressure remains constant.}$$

Note that neither the orientation of the cylinder nor whether piston is frictionless are relevant. The expression $p\Delta V$ simply gives the energy transfer to the piston and whatever it is connected to (crank, dynamo, etc.); it doesn't indicate what form the energy takes, whether kinetic, gravitational potential, elastic…

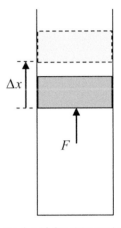

Fig. 3.4.12 Gas doing work against a piston

(b) The work done by an expanding gas with varying pressure

The equation $W = p\Delta V$ for a constant pressure expansion tells us that, if the pressure varies during the expansion, the total work done for an expansion from volume V_1 to V_2 is given by:

$$W = \int_{V_1}^{V_2} p(V)\mathrm{d}V$$

i.e. the area under the p-V graph.

▼ Study point

Surely the idea of a constant pressure expansion is not possible. In principle it is. The gas expansion in Fig. 3.4.8 was of this type. There was a source of heat which kept the pressure up as the gas expanded. In a gas turbine the pressure of the gas feed is constant.

Stretch & Challenge

(a) For an ideal gas
$$pV = nRT$$

If an ideal gas expands at constant temperature from V_1 to V_2 show that, W, the work done is given by

$$W = nRT \ln \frac{V_1}{V_2}$$

(b) Calculate the work done by the methane sample in Fig. 3.4.13 if it expanded from A to 4 m³ at constant temperature of 600 K.

3.4.5 Self-test

For the expansion of the methane sample in Fig. 3.4.13

(a) show that $T_C > T_A > T_B$, and

(b) find the ratio $\dfrac{T_B}{T_A}$, where T_A is the temperature at **A**, etc.

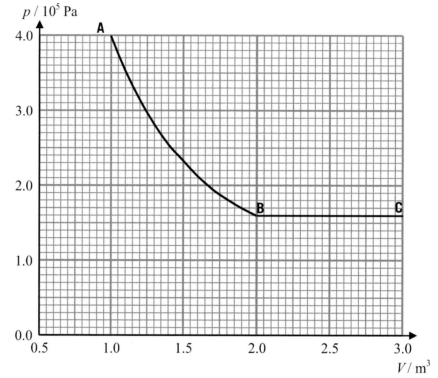

Fig. 3.4.13 The expansion of methane

3.4.6 Self-test

Calculate the mass of the methane sample in Fig. 3.4.13 given that the temperature at **A** is 600 K.

▼ **Study point**

See Section 3.4.6 for examples of the work done by expanding gases in the context of heat engines.

Example

A sample of methane (CH_4) expands as shown in Fig. 3.4.13. Calculate the work it does in doing so.

Answer

Work done	= Area under graph
	= Area under AB + Area under BC
Area under AB	= 493 small squares [see AS textbook Section 4.5.3]
	= $493 \times (0.1 \times 10^5 \text{ Pa} \times 0.05 \text{ m}^3)$
	= 246 500 J
Area under BC	= $1.00 \text{ m}^3 \times 1.6 \times 10^5 \text{ Pa} = 160\,000 \text{ Pa}$

∴ Total work done = 246 500 J + 160 000 J = 410 kJ (2 s.f.)

(c) Electrical work

The main focus of this chapter is on the work done by gases, e.g. in heat engines, but this is not the only form of work. If an energy transfer is directly caused by a temperature difference, we call it heat and give it the symbol, Q. Any other form of energy transfer is called work, W, even if it involves a rise in temperature, e.g. in the traditional way of making a fire illustrated by the survivalist in Fig. 3.4.14.

The hallmark of work is that involves an ordered transfer of energy. In order to turn the round wooden stick pressing on the wooden block, the survivalist pushes a bow backwards and forwards; all the molecules of the bow string are moving in the same direction. Heat involves a disordered transfer, whether it be by conduction (random movement of molecules / electrons colliding) or radiation (random absorption of individual photons).

Fig. 3.4.14 Working to make a fire

Now we'll consider an electrical circuit, containing a cell and a lamp, in terms of heat and work.

For the moment we'll consider the system consisting of the lamp only. There is energy being transferred in and out:

Energy in: transferred from the cell by means of an electric current. This is an ordered flow of charge, so it is work:

Electrical work input, $W_{in} = VIt$

Energy out: transferred by radiation (mainly infrared). This is disordered so it is categorised as heat, Q. Once the lamp has reached a steady temperature, $Q_{out} = W_{in} = VIt$.

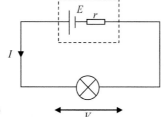
Fig. 3.4.15 Electrical circuit

We should now go on to consider what is happening to the internal energy of the lamp (and the cell) but we'll leave that until after the next section.

3.4.7

Self-test

Consider the cell (including the internal resistance) as a thermodynamic system. What are the energy inputs and outputs? Where appropriate give your answer in algebraic form.

3.4.5 The first law of thermodynamics

The law of conservation of energy applied to thermodynamic systems is referred to as the **first law of thermodynamics**.

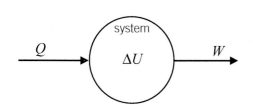

Fig. 3.4.16 $\Delta U = Q - W$

The relationship between U, Q and W is:

$$\Delta U = Q - W$$

where Q = heat input

and W = work output

Note carefully the direction of W and Q.

Question: Why has U got a Δ sign but Q and W haven't?

Answer: U, the internal energy, is a state variable of the system. At any one moment, U has a definite value. ΔU is the change in the value.

W is the work which is done by the system. The system does not 'have' a quantity of work, so its work cannot change.

Similarly, the system does not possess heat, so its heat cannot change. Heat is the flow of energy driven by a temperature gradient. It is worth looking at several examples of the application of this law.

Terms & definitions

The **First law of thermodynamics** states that when net heat, Q, flows into a system which does net work, W, on its surroundings, the internal energy, U, changes according to the relationship

$\Delta U = Q - W$.

▼ Study point

Each of Q, W and ΔU can be positive, negative or zero, e.g.

1. If you compress a gas, work is done on (not by) the gas, so $W < 0$.

2. When a cup of tea cools down, it gives out heat, so $Q < 0$.

Example 1

A 3 mol sample of monatomic gas is heated (e.g. by an electrical heater). Its temperature changes from 300 K to 400 K. Calculate the values of ΔU, W and Q if the change is at constant pressure.

Answer

$\Delta U = \frac{2}{3}nR\Delta T = 1.5 \times 3\text{ mol} \times 8.31\text{ J mol}^{-1}\text{ K}^{-1} \times 100\text{ K} = 2490\text{ J}$

$pV = nRT$, ∴ at constant pressure $p\Delta V = nR\Delta T$ (see foot note[1])

∴ $W = nR\Delta T = 3\text{ mol} \times 8.31\text{ J mol}^{-1}\text{ K}^{-1} \times 100\text{ K} = 1660\text{ J}$

∴ $Q = \Delta U + W = 2490\text{ J} + 1660\text{ J} = 4150\text{ J}$

3.4.8

Self-test

For the equation

$\Delta U = Q - W$

state the sign of:

(a) ΔU if $Q > 0$ and $W = 0$

(b) Q if $\Delta U = 0$ and $W < 0$

(c) W if $Q = 0$ and $\Delta U > 0$

(d) ΔU if $W > Q > 0$

(e) ΔU if $Q > 0$ and $W < 0$.

1 The value of W doesn't depend on the pressure. If the p were doubled, the V values and therefore ΔV would be halved, so $p\Delta V$ stays the same.

The internal energy of methane, $U = 3nRT$

(a) Use equipartition theory (Section 3.4.6) and the shape of the molecule to explain why.

(b) Calculate the values of Q, W and ΔU for sections AB and BC of Fig. 3.4.13.

3.4.9 Self-test

State (no calculation needed) the answers to Example 1 if the volume of the gas and not the pressure were constant.

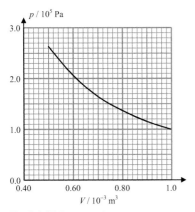

Fig. 3.4.17 Compressing a gas

Example 2

A cell of emf 9 V and internal resistance 1 Ω delivers a current of 1 A. Analyse the energy changes in a period of 1 minute in terms of the first law of thermodynamics. Assume that the cell's temperature is constant.

Answer

Terminal pd $V = E - Ir = 9\ \text{V} - 1\ \text{A} \times 1\ \Omega = 8\ \text{V}$

Work output $W_{\text{out}} = VIt = 8\ \text{V} \times 1\ \text{A} \times 60\ \text{s} = 480\ \text{J}$

Heat output $Q_{\text{out}} = I^2 rt = (1\ \text{A})^2 \times 1\ \Omega \times 60\ \text{s} = 60\ \text{J}$

∴ Heat input, $Q_{\text{in}} = -60\ \text{J}$

∴ $\Delta U = Q - W = -60\ \text{J} - 480\ \text{J} = -540\ \text{J}$

i.e. the cell has done 480 J of electrical work on the circuit and has emitted 60 J of heat, so its internal energy has decreased by 540 J.

Example 3

A well-insulated sample of an ideal gas at 300 K is compressed from a volume of $1.0 \times 10^{-3}\ \text{m}^3$ to half this volume; the pressure increases by a factor of 2.63 as shown in Fig. 3.4.17.

(a) Calculate the work done on the gas.

(b) State the heat supplied to the gas and the change in internal energy.

[There are more questions about this gas sample in Exercise 3.4.]

Answer

(a) Work done = area under graph

Using the trapezoidal rule (see AS Textbook Section 4.5.3) with $\Delta V = 0.5 \times 10^{-3}\ \text{m}^2$, the author obtained

Work done = 80 J

(b) $Q = 0$ because the gas is thermally insulated.

$W = -80\ \text{J}$ because the work is done **on** the gas

$\Delta U = Q - W = 0 - (-80) = 80\ \text{J}$

3.4.6 Heat and temperature change

When we heat a system, its temperature usually increases (see Study point). The first law of thermodynamics applies to this process, so we'll examine a typical case and see what the values of ΔU, Q and W are.

▼ **Study point**

Heating ice at 0 °C, or water at 100 °C, results in a change of state rather than temperature.

See Fig. 3.4.10.

(a) Applying the first law of thermodynamics

Consider a 1 kg sample of water at 20 °C. If we heat it up, e.g. using an electric kettle, its temperature rises. Experimentally, if we supply 200 kJ of heat its temperature increases to almost 70 °C (an increase of 50 °C).

$$\therefore Q = 200 \text{ kJ}.$$

$$\therefore \Delta U + W = 200 \text{ kJ}.$$

W is the work done against the atmosphere. $\therefore W = p_A \Delta V$, where p_A is atmospheric pressure $\sim 1.0 \times 10^5$ Pa.

Experimentally, the volume of the water increases from 1000 cm³ to about 1020 cm³, i.e. $\Delta V = 20$ cm³ [2.0×10^{-2} m³].

$$\therefore W = p\Delta V \sim 1.0 \times 10^5 \text{ Pa} \times 2 \times 10^{-2} \text{ m}^3 = 2 \text{ kJ}$$

So, because water only expands by a small fraction, W is only about 1% of Q and therefore we can write, $W \approx 0$ and $\Delta U \approx Q = 200$ kJ. Actually, water expands more than most engineering materials; for example, 1000 cm³ steel only expands by 2 cm³ over the same temperature range, so the approximations $W \approx 0$ and $\Delta U \approx Q$ are even better for other solids and liquids.

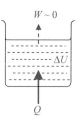

Fig. 3.4.18 Heating water

(b) Specific heat capacity

The rise in temperature, $\Delta\theta$, of an object, provided no change of state occurs is found experimentally to be closely proportional to the heat input, Q. For an element, compound or mixture of fixed composition, $\Delta\theta$ is inversely proportional to the mass of material. So, for a particular material, we can write

$$Q = mc\Delta\theta$$

where c is a constant called the **specific heat capacity** (shc) of the material. The word *specific* in this context means *per unit mass*. A verbal definition of shc would be 'the heat per unit mass needed to raise the temperature of a material per degree'. It is easier to define it using the equation.

Terms & definitions

The **specific heat capacity**, c, of a substance is defined by
$$Q = mc\Delta\theta$$

where Q is the heat input, m the mass and $\Delta\theta$ the temperature change.

Unit: J kg⁻¹ K⁻¹ (or J kg⁻¹ °C⁻¹)

Example

A 5.0 kg block of iron, initially at 20 °C, is heated for 5 minutes with a 25 W heater. Calculate the final temperature. [$c_{iron} = 450$ J kg K⁻¹]

Answer

Heat input $Q = Pt = 25 \text{ W} \times 300 \text{ s} = 7500 \text{ J}$

$$Q = mc\Delta\theta, \therefore \Delta\theta = \frac{Q}{mc} = \frac{7500 \text{ J}}{0.50 \text{ kg} \times 450 \text{ J kg}^{-1}\text{K}^{-1}} = 33.3 \text{ K} = 33.3 \text{ °C}$$

$\therefore$ Final temperature = 20 °C + 33.3 °C = 53 °C (2 s.f.)

Note: The *difference* between two temperatures is the same regardless of whether they are expressed in degrees celsius or kelvin:

$$80 \text{ °C} - 20 \text{ °C} = 60 \text{ °C} \qquad 353 \text{ K} - 293 \text{ K} = 60 \text{ K}$$

Hence the unit of c can equally be written J kg⁻¹ °C⁻¹.

▼ **Study point**

The term **specific heat capacity** is an unfortunate and misleading one. It suggests:

(a) That heat is something that can be contained, which does not agree with the definition of heat.

(b) And that a substance can only contain a certain amount of this.

The term dates back to the long-discarded 18th-century theory of heat as a fluid (called caloric) which could be added or taken away from objects.

θ_1
m_1
c_1
Q
θ_2
m_2
c_2

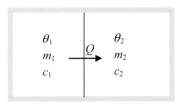

Fig. 3.4.19 Heat transfer

Study point

If the temperature of system 1 changes from θ_1 to θ then:
temperature fall = $\theta_1 - \theta$

If the temperature of system 2 changes from θ_2 to θ
temperature rise = $\theta - \theta_2$

(c) Heat transfer

We often need to consider the transfer of heat between systems. If two systems are in thermal contact, then heat will flow from the system at the higher temperature into the one at lower temperature (this was how we defined high and low temperature). This will continue until the systems are in thermal equilibrium, i.e. their temperatures are equal.

Assuming that there is no heat exchange with the surroundings we can say that the heat flow from one system is equal to the heat flow into the other, which will allow us to calculate the equilibrium temperature.

Consider the two systems in Fig. 3.4.19. θ_1 and θ_2 are the initial temperatures. Let the final temperature be θ. Then:

Heat loss by system 1 = heat gain by system 2

$$\therefore\ m_1c_1(\theta_1 - \theta) = m_2c_2(\theta - \theta_2)$$

We don't need to learn this equation but we need to be able to apply the principle of heat gain = heat loss together with $Q = mc\Delta\theta$.

Study point

In the example, it doesn't matter whether we express the masses in g or kg (or tonnes!) as long as both are in the same units. Similarly the specific heat capacities can be in kJ kg^{-1} K^{-1} or J kg^{-1} K^{-1} (or even calories per ounce per °C) as long as the units of the two sides are the same

Example

1.0 kg of water at 100 °C is poured into a 0.50 kg porcelain teapot at 20 °C. Calculate the equilibrium temperature.

$c_{\text{water}} = 4.18$ kJ kg^{-1} K^{-1}; $c_{\text{porcelain}} = 1.07$ J kg^{-1} K^{-1}

Answer

Let equilibrium temperature be θ.

Heat lost by water = heat gained by teapot

$\therefore$ Using $Q = mc\Delta\theta$:

$$1.0 \times 4.18\,(100 - \theta) = 0.50 \times 1.07\,(\theta - 20)$$

A bit of algebra leads to: $\theta = \dfrac{428.7}{4.715} = 91\ °\text{C}\ (2\ \text{s.f.})$[2]

3.4.7 Specified practical work

(a) Estimation of absolute zero using the gas laws

Hg or H_2SO_4 plug
mm scale
Dry air
HEAT

Fig. 3.4.20 Variation of volume with temperature

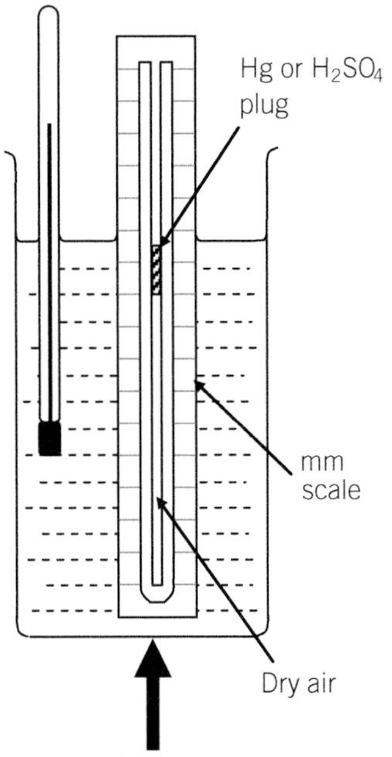

The variation of volume of a gas with temperature at constant pressure can be investigated using the apparatus in Fig. 3.4.20. The water in the beaker is heated slowly in stages using a bunsen burner. The temperature and the length of the air column in the capillary tube are measured at each stage. The volume of the air is proportional to the length of the air column.

The p-T variation is undertaken similarly using the apparatus in Fig. 3.4.21.

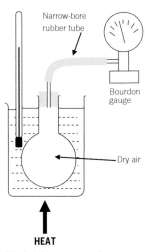

Narrow-bore rubber tube
Bourdon gauge
Dry air
HEAT

Fig. 3.4.21 Variation of pressure with temperature

2 Tea brews ('mashes' in Yorkshire) best as close to 100 °C as possible which is why you should always pre-warm the pot.

For both investigations, the dependent variable (length or pressure) is plotted against the temperature, a best-fit straight line drawn and the intercept on the temperature axis noted, as shown in Fig. 3.4.22.

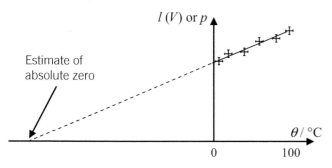

Fig. 3.4.22 Estimating absolute zero

> **Self-test** ◥ **3.4.10**
>
> The equation of the best-fit graph of p (in atm) against θ (in °C) is
>
> $p = (0.038 \pm 0.02)\theta + 0.95$
>
> Estimate the value of absolute zero in °C.

(b) Measurement of the specific heat capacity of a solid

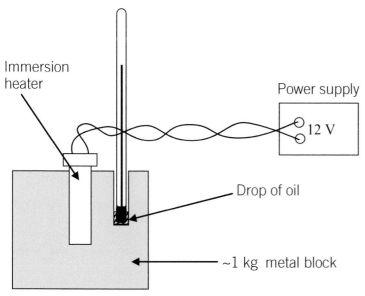

Fig. 3.4.23 Measuring the specific heat capacity of a metal

▼ **Study point**

Experiments to determine specific heat capacity of solids at A level are generally restricted to metallic solids because their high thermal conductivity allows them to reach thermal equilibrium quickly.

The drop of oil is to provide a good thermal contact between the thermometer and the metal block.

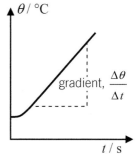

Fig. 3.4.24 Temperature-time graph for determining, c

If necessary, an initial experiment is undertaken to determine the power of the immersion heater, by measuring the current and pd and applying $P = IV$.

The immersion heater is switched on and the temperature displayed on the thermometer noted at regular intervals, e.g. every 30 s. A graph of temperature against time is plotted and the gradient of the linear portion determined (see Fig. 3.4.24).

The initial horizontal portion of the graph is a delay until the heat from the immersion heater reaches the position of the thermometer. After this has occurred

$$P\Delta t = mc\Delta\theta$$
$$\therefore c = \frac{P}{m \times \text{gradient}}$$

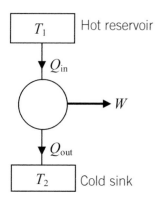

Hot reservoir

Cold sink

Fig. S&C1 Heat engine

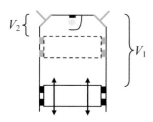

Fig. S&C2 Petrol engine cylinder

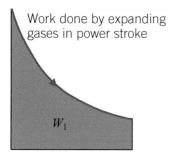

Work done by expanding gases in power stroke

W_1

Work done on mixture in compression stroke

W_2

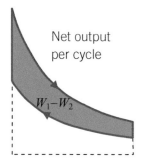

Net output per cycle

$W_1 - W_2$

Fig. S&C4 Energy inputs and outputs in an Otto cycle

Stretch & challenge – Heat engines and the second law of thermodynamics

This is a brief digression into the reason for developing the science of thermodynamics in the first place – the desire to get heat engines to work better, meaning working at greater power and with greater efficiency.

Heat engines

What is a heat engine? It is a device for using a store of internal energy to produce mechanical work by taking a working substance (steam, burnt petrol/air mixture) from a high temperature to a low temperature. Fig. S&C1 shows this schematically. The guts of the engine is the circle: essentially it grabs some of the heat from the reservoir on its way to the sink and converts it to work.

The petrol engine – the Otto cycle

The first internal combustion engine was built by the Nikolaus Otto, using coal gas instead of petrol, using an operating sequence which we call the *Otto cycle*.

The power processes in a petrol engine take place in a cylinder in which an air/petrol mixture is ignited to drive a piston. Fig. S&C2 shows a cylinder. V_1 and V_2 are the maximum and minimum volumes of the cylinder. An outline of the cycle is given in the Study point. We'll try and determine the maximum possible output of an engine based on several of these cylinders. The graph in Fig. S&C3 gives an idealised variation of pressure with volume for the gas in the cylinder.[3] Such graphs are called *indicator diagrams*.

The energy transfer in the power stroke (the work done) is the area under the line CD, W_1. This comes from (and in the ideal case this is 100% of) the internal energy of the petrol/air. But in order to achieve this we need to do compress the mixture (A→B) and this involves work W_2. So the net output is given by the area between AB and CD, ($W_1 - W_2$). This is shown in Fig. S&C4.

Power output

A petrol engine normally has several cylinders (most often four in cars) and operates at over 1000 rpm (revolutions per minute). Hence the maximum power output, P_{max}, of this engine would be given by:

$$P_{max} = nf(W_1 - W_2)$$

where n = number of cylinders

and f = engine frequency

---

3 In Fig. 3.4.16, the ratio of max to min volume is 3. This is for ease of interpretation of the indicator diagram. Most petrol engines work with a compression ratio of about 10.

Maximum efficiency

Clearly, even in this ideal engine, we cannot get out 100% of the energy we put in. The maximum possible efficiency

$$\eta_{max} = \frac{W_1 - W_2}{W_1}$$

Looking at the graphs, how can we improve the efficiency? We need to make AB as low as possible and CD as high as possible. For the same mass of mixture, this means making, T_A, the temperature of the gases at A as low as possible and T_C, the temperature at the end of the combustion as high as possible.

There is a reasonable argument that the values of work are proportional to the maximum and minimum temperatures (see question 5). Hence $\eta \leq \dfrac{T_C - T_A}{T_C}$.

In terms of Fig. S&C1, the maximum possible efficiency of a heat engine is:

$$\eta_{max} = \frac{T_1 - T_2}{T_1}$$

So for a steam engine with a steam temperature of 600 K, assuming a lower temperature of 300 K (i.e. the temperature of the surroundings):

$$\eta \leq \frac{600 - 300}{600} = 0.5$$

and this is before we start to consider energy losses such as friction.

High grade and low grade energy

The maximum efficiency equation is a universal equation. For a heat engine working between two temperatures T_1 and T_2 we cannot do better than capturing the fraction $\dfrac{T_1 - T_2}{T_1}$, or $1 - \dfrac{T_2}{T_1}$ of the heat from the higher temperature reservoir. Some energy must always be passed on as waste heat. So the higher T_1 (and the lower T_2) the better we can do: the higher the temperature of the reservoir, the more useful its energy is in terms of doing work. A bath of water at 40 °C might have more internal energy than 1 kg of steam at 500 °C but you cannot get much work out of it.

Stretch & Challenge

1. Ignoring the tiny heat input from the spark plug, explain why, Q, W and ΔU are all zero for the ignited gas mixture during BC. How can the temperature rise so much when ΔU is zero?

2. By finding the area under the relevant graphs, calculate W_1, W_2 and $W_1 - W_2$.

3. Calculate the theoretical power output of a 6-cylinder car with the above indicator diagram operating at 1800 rpm.

4. Calculate the theoretical efficiency of the engine using the results of Q2.

5. (a) What happens to the pressure at C if the temperature T_C were doubled in the Otto cycle (assume the volume is the same)?

   (b) Explain why W_1 is proportional to T_C (again, assume that the volumes are the same).

   (c) Explain why W_2 is proportional to the input temperature (for the same mass of gas).

6. A gas turbine with inlet and outlet temperatures of 1500 °C and 587 °C has an efficiency of 39.5%. How does this compare with the maximum theoretical efficiency?

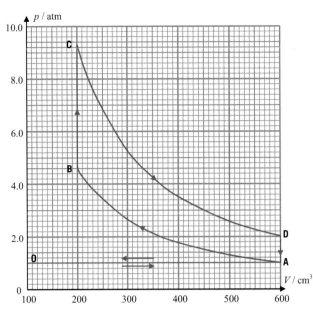

Fig. S&C3 Otto cycle indicator diagram

▼ Study point

The indicator diagram is an idealised Otto cycle.

OA: The intake stroke. A mixture of air and petrol vapour is drawn into the cylinder.

AB: The compression stroke. The piston compresses the air/petrol mixture rapidly. In the process its pressure and temperature increase.

BC: The ignition phase. The mixture is ignited (by the spark plug) and burns rapidly, increasing the temperature (and pressure) of the mixture.

CD: The power stroke. The high pressure gas mixture expands forcing back the piston (the power stroke).

DA: The valve opens, releasing the exhaust gases, which are then expelled.

AO: The exhaust stroke. The piston pushes the waste gases (the products of combustion) into the atmosphere.

▼ Study point

There are several highly technical statements of the 2nd law of thermodynamics. Perhaps the easiest one to understand is that due to Planck: It is impossible to construct an engine which produces no other effect than raising a weight and cooling a reservoir.

Exercise 3.4

1 A cylinder of volume 0.060 m³ contains helium at a pressure of 8.0 MPa.

(a) Calculate the internal energy of the helium.

(b) Explain why your answer to (a) doesn't depend upon the temperature of the helium.

(c) The mass of helium in the cylinder is 800 g. Calculate:

    (i) the temperature   (ii) the rms speed of the helium molecules.

2 An oxygen cylinder contains 3.2 kg of oxygen at 27 °C.

(a) Calculate the translational kinetic energy of the oxygen molecules.

(b) Calculate the rms speed of the oxygen molecules.

(c) The internal energy of the oxygen molecules is 620 kJ (2 s.f.). Account for the difference between this figure and your answer to part (a).

3 A rigid, insulating container has two compartments as shown, each containing 500 mol of a monatomic ideal gas. [Take 1 atm as 100 kPa]

2 m³	4 m³
12 atm	4 atm
A	**B**

(a) Calculate the internal energy in each compartment.

(b) Calculate the temperature in each compartment.

(c) The insulation between the two halves is now removed but the rigid dividing wall remains in place. Calculate:

    (i) The equilibrium temperature.

    (ii) The pressure in each compartment at equilibrium.

    (iii) The net heat flow between the two compartments.

(d) The dividing wall is now removed so the gas molecules can mix. Calculate the final temperature and pressure.

4 Consider the following situations and for the given system, state whether the quantities Q, ΔU and W from the first law of thermodynamics, written in the form $\Delta U = Q - W$, are positive, negative or zero.

(a) A sample of ideal gas is compressed rapidly, so that no heat has time to escape.

(b) The temperature of the sample of air in Fig. 3.4.20 is raised.

(c) A block of steel at 20 °C is placed into a furnace at 500 °C.

(d) A sample of gas at 2 atm in a cylinder is allowed to expand slowly against a piston, so that its temperature remains constant.

(e) A medical cylinder of oxygen is placed in a warm room after being taken out of a cold room.

(f) The gas in a cylinder is allowed to escape into a vacuum.

(g) A mixture of methane and oxygen, in a sealed, rigid thermally insulated container, spontaneously burns to produce carbon dioxide and water. [The system is all the particles initially in the methane and oxygen.]

5 It takes 2.26 MJ of heat to vaporise 1 kg water at 100 °C. The density of steam at 100 °C is 0.60 kg m⁻³.

(a) Estimate the work which the water does in expanding into steam against the atmosphere.

(b) Calculate the fraction of 2.26 MJ that your answer to part (a) represents.

(c) State the values of Q, ΔU and W in the equation $\Delta U = Q - W$.

6 Steam is bubbled through 250 g of milk to raise its temperature from 20 °C to 100 °C. The specific heat capacity of milk is 3800 J kg⁻¹ K⁻¹.

(a) Calculate the heat input into the milk.

(b) State the values of Q, ΔU and W for the milk in the equation $\Delta U = Q - W$.

(c) Use the data in Q5 to estimate the mass of steam required to raise the temperature of the milk in this way.

7 A sample of an ideal gas is taken around the sequence of pressures and volumes, ABCDA. Calculate:

(a) The ratios $\dfrac{U_B}{U_A}$, $\dfrac{U_C}{U_A}$ and $\dfrac{U_D}{U_A}$.

(b) The ratios $\dfrac{T_B}{T_A}$, $\dfrac{T_C}{T_A}$ and $\dfrac{T_D}{T_A}$.

(c) The values of ΔU, Q and W for each of the four stages A→B, B→C, C→D and D→A.

(d) The total values of ΔU, Q and W for the whole cycle.

(e) The number of moles of gas given that $T_A = 350$ K.

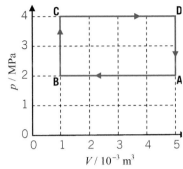

8 A hypothetical engine consists of six cylinders each of which contains a sample of the gas as in Q7. The gas in each cylinder is taken around the sequence ABCDA 1500 times per minute. Calculate the power output of the engine.

9 A student performs an experiment to determine the specific heat capacity of iron, c_i. He puts a few spatulas of iron filings into a test tube, which he heats up to 100 °C in a beaker of boiling water. He quickly tips the iron filings into a beaker of cold water and measures the temperature rise. Use the results below to calculate c_i.

Results:

Mass of beaker = 80 g

Mass of beaker + water = 250 g

Mass of beaker + water + iron filings = 400 g

Initial temperature of water and beaker = 15.0 °C

Final temperature of water and beaker + iron filings = 22.5 °C

Specific heat capacity of water = 4180 J kg⁻¹ K⁻¹

Specific heat capacity of glass = 840 J kg⁻¹ K⁻¹.

10 The graph shows the power stroke of a model steam engine. Steam under high pressure and temperature (800 kPa and 250 °C) is introduced into the cylinder and pushes the piston along by 2 cm (A→B). The piston has an area of 4 cm². At this point the steam is cut off and the steam expands, pushing the piston by another 3 cm (B→C). The lower pressure steam is vented at C. Steam is not an ideal gas but we'll assume it is for the sake of this question!

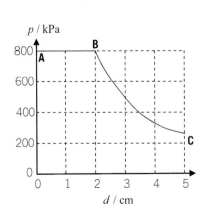

(a) Estimate the work done by the steam on the piston.

(b) Explain how you can tell from the graph that the temperature of the steam at C is lower than that at B. Calculate the temperature at C.

(c) Estimate the maximum number of moles of steam in the cylinder.

(d) The specific internal energy (i.e. internal energy per kilogram) of steam at 250 °C and 800 kPa is 2715 kJ kg⁻¹. Use your answers to (a) and (c) to estimate the efficiency of the steam engine. [Suggest why steam is no longer used as a means of locomotion.]

3.5 Nuclear decay

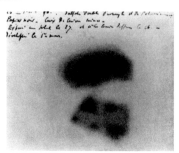

Fig. 3.5.1 Extract from Becquerel's laboratory notebook

Fig. 3.5.2 Alpha particle tracks

The story of Henri Becquerel's serendipitous discovery of radioactivity is quite well known. Briefly stated, he discovered that a sample of uranium salt had fogged a sealed photographic plate through its wrapper. This discovery was shocking because it appeared to violate the principle of conservation of energy – the uranium salt continued to affect film emulsions without any obvious energy source.[1] Even though this was a fortuitous initial observation, his scientific rigour in not just rejecting it but in investigating it thoroughly made the 1903 Nobel Prize for physics, which he shared with Marie and Pierre Curie, highly deserved. The extract from his lab notebook in Fig. 3.5.1 includes the shadow of a metal Maltese cross which had been placed between the salt and the photographic plate.

The discovery of radioactivity paved the way for research into the structure of the atom and the later development of quantum theory, so its importance cannot be overstated.

3.5.1 Nuclear decay

Following the discovery of **nuclear radiation**, by Henri Becquerel, it was quickly established that there were three kinds with different penetrating powers, which he called α, β and γ radiation. The range of each type of radiation depends upon their energy and the medium but the following are typical:

- α range: 3 cm of air; 0.02 mm of glass

- β range: 1 m of air; 0.5 mm aluminium

- γ range: 500 m of air; 20 cm aluminium; 5 cm lead

The alpha particles given out by a radio-nuclide have a well-defined energy, so they penetrate a set distance in a particular substance. The α particles, which produced the tracks in Fig. 3.5.2, have two energies, so two ranges. β particles are produced with a continuous spectrum of energies (see AS textbook, p 82) so have no clearly-defined range. We shall discuss the range of nuclear radiation in Section 3.5.2.

We now know that these types of radiation are caused by a change in the nucleus of the atom, allowing it to assume a lower energy configuration. This process is called **nuclear decay**. The atomic nucleus is composed of protons and neutrons, themselves composite particles consisting of combinations of up and down quarks. In the same way that atoms have energy levels, so do nuclei. A nucleus with excess energy can drop down to lower energy levels in various ways, which we shall now explore.

---

1 Naturally, physicists proceeded to include nuclear energy in the recognised forms of energy, thus preserving the inviolability of the principle!

(a) Alpha (α) decay

The combination of two protons and two neutrons is particularly stable. It is the composition of the nucleus of a $^4_2$He atom. The identity of the α particle with ionised helium was demonstrated in an elegant experiment[2] in 1907.

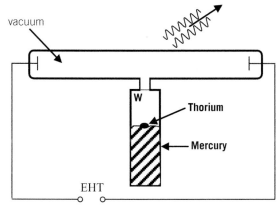

vacuum

W
Thorium
Mercury
EHT

Fig. 3.5.3 The nature of α radiation

Thorium, an α-emitter, was floated on mercury. The α particles could pass through the very thin window, **W**, into the gas discharge tube. After a few days, the characteristic lines of helium were seen in the spectrum of the radiation.

Almost 100% of naturally occurring thorium is the isotope $^{232}_{90}$Th. The decay equation is:

$$^{232}_{90}\text{Th} \rightarrow {}^{228}_{88}\text{Ra} + {}^4_2\text{He},$$

where the symbol $^4_2$He is used for the α particle. Generalising from this example, we see that in an alpha decay, from a nucleus ^A_ZX the decay products are:

- a daughter nucleus, $^{A-4}_{Z-2}$Y and

- an alpha particle $^4_2$He

α decay only occurs in heavy elements. The lightest known α-decay nuclide is telliurium-106 ($^{106}_{52}$Te) which decays to an isotope of tin.

(b) Beta (β) decay

The more common form is β^- decay, which takes place in neutron-rich nuclei. A neutron decays, via the weak interaction, into a proton with the emission of an electron and an electron anti-neutrino. Using the symbolism of particle physics:

$$\text{n} \rightarrow \text{p} + \text{e}^- + \overline{\nu}_\text{e}$$

In radioactive decay equations it is usual to use $^0_{-1}\beta$ to represent the beta particle; $^0_{-1}$e can also be used.

The equation for the decay of potassium-42 is

$$^{42}_{19}\text{K} \rightarrow {}^{42}_{20}\text{Ca} + {}^0_{-1}\beta + {}^0_0\overline{\nu}_\text{e}$$

So a β^- decay from ^A_ZX produces:

- a daughter nucleus $^A_{Z+1}$Y

- a β^- particle, i.e. an electron, e$^-$

- an electron anti-neutrino.

2 By Ernest Rutherford and Thomas Royds.

▼ **Study point**

Note that the symbols ^A_ZX in a decay equation represent nuclei rather than atoms. The sum of the values of A and Z on the two sides of the equation must be equal, i.e. the equation must balance.

Self-test 3.5.1

Uranium-235 undergoes α decay to an isotope of thorium.

Write the decay equation for this process.

Self-test 3.5.2

Write the decay equation for **Te-106**.

(Symbol for tin = **Sn**)

Self-test 3.5.3

(a) Which conservation laws are illustrated in the neutron decay equation?

(b) What indications are there that it is a weak decay?

▼ **Study point**

When writing β-decay equations it is quite common to omit the neutrino. So the K42 would be written

$^{42}_{19}\text{K} \rightarrow {}^{42}_{20}\text{Ca} + {}^0_{-1}\beta$.

The neutrino is not observed directly but its presence can be inferred from the β energy spectrum.

MATHS TIP

Physicists and chemists often use the unified mass unit, u, to express the masses of atoms and molecules.

Nuclear and particle physicists often use the energy units and the formula $E = mc^2$ to express masses in the units MeV/c^2 or GeV/c^2. See Chapter 4 for help in dealing with this proliferation of units.

Some nuclei, created in fusion processes, can decay by the emission of a positron, rather than an electron. In the context of radioactivity this is called a β^+ decay and the particle is called a β^+ particle, e.g.

$$^{13}_7\text{N} \rightarrow \, ^{13}_6\text{C} + \, ^0_1\beta \, (+ \, ^0_0 v_e)$$

Note the difference in the nuclear symbols for the β^+ and β^- particles. In order for this to happen, a proton within the N13 nucleus must transform into a neutron (with the emission of a positron and a neutrino), i.e.

$$p \rightarrow n + e^+ + v_e$$

But how is this possible without violating the conservation of mass/energy? The proton has a mass of $938.3 \text{ MeV}/c^2$; the neutron is more massive, weighing in at $939.6 \text{ MeV}/c^2$. And then there is the positron with a mass of $0.5 \text{ MeV}/c^2$, to say nothing of the kinetic energy of the neutrino. The answer lies in the nuclear energy levels (see Stretch and Challenge).

Stretch & Challenge

Just like electrons in atoms, the neutrons and protons in a nucleus each have their own sets of energy levels. This concept is illustrated in Fig. 3.5.4 for a proton-rich nucleus. The proton in the top energy level can transform and drop down into the vacant neutron energy level – but only if ΔE is enough. Suggest a minimum value of ΔE for β^+ decay to occur.

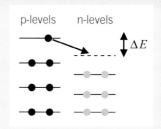

Fig. 3.5.4 Nuclear energy levels

Whatever the mechanism (!), the β^+ decay of ^A_ZX results in:

- a daughter nucleus $^A_{Z-1}\text{Y}$

- a β^+ particle, i.e. a positron, e^+

- an electron neutrino.

(c) Gamma (γ) decay

The daughter nucleus from an α or a β decay is often formed in an excited state. That means that either a proton or a neutron has a vacant lower energy state to drop down into. When it does this, the energy is carried away by a high energy photon which is indicated by γ. The proton and nucleon numbers (Z and A) of the nucleus are unchanged in the process.

(d) Some less common decay modes

The most common modes of decay are α, β and γ but there are a few others that you might come across:

Electron capture (a.k.a. K capture[3])

This occurs in some proton-rich nuclei. An electron in the 1s orbital, the lowest atomic energy level, spends a tiny fraction of its time actually inside the nucleus. The shading in Fig. 3.5.5 represents qualitatively the probability density function of the 1s orbital –

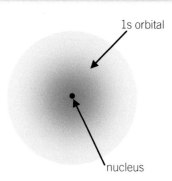

1s orbital

nucleus

Fig. 3.5.5 Nucleus and 1s orbital

---

3 Chemists historically refer to the 1s orbital as the K-shell.

this is the probability of finding the electron in any particular small region. It is actually a maximum at the centre of the atom, but because the nucleus is so tiny ($\sim 10^{-15}$ m) compared to the orbital ($\sim 10^{-10}$) the electron actually spends a tiny fraction of its time there. Nevertheless, when it does, the following reaction can occur:

$$p + e^- \rightarrow n + v_e$$

Of course, mass/energy must still be conserved (see Self-test 3.5.5).

Proton emission

This only happens in very light nuclei with a large proton excess, e.g. Li6. Neutron emission is also possible in the context of a nuclear fission event.

3.5.2 The range (penetrating power) of nuclear radiation.

The different penetrating powers of α, β and γ radiation are well known and form the basis of pre-A level experiments on radioactivity. You should remember the following:

- The penetrating power of all forms depends upon their energy: the greater the energy the greater the range.

- The denser the material the better it is at absorbing radiation of any type, i.e. the lower the range of the radiation.

- Gamma rays are the most penetrating (tens or hundreds of metres of air, several cm of dense metals, e.g. lead), then beta particles (about 1 m of air, a couple of mm of light metals), then alpha (stopped by thin paper, 3–5 cm of air, the dead epidermis of the human skin).

In order to understand these differences we need to understand the absorption mechanisms operating.

(a) The absorption of charged particles

When a high-energy charged particle, e.g. an alpha particle, penetrates a material, it is able to pass through the constituent atoms – though neither the particle nor the atoms come through unscathed.

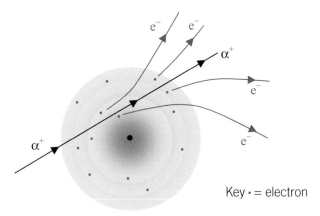

Key • = electron

Fig. 3.5.6 Effect of an alpha particle on an atom

Self-test 3.5.5

(If you've done the Stretch & Challenge this should be quite easy.) Calculate ΔE in Fig. 3.5.4 for K-capture to occur. Express your answer in (a) MeV and (b) J.

Self-test 3.5.6

State the effect on the nucleus $_Z^A X$ of (a) K capture (b) proton emission.

Self-test 3.5.7

Write the decay equation for N-11 by proton emission.

— Terms & definitions —

The **range** of a nuclear radiation is the distance it will travel before being absorbed. It depends upon the nature and energy of the radiation and the nature of the medium through which it is travelling.

▼ Study point

From Fig. 3.5.6 you will see why α particles are referred to as ionising radiation. This is also true of β and γ (and u-v and X-rays).

3.5.8 Self-test

A 5 MeV alpha particle, loses an average of 0.1 keV of kinetic energy each time is passes through at atom.

(a) Estimate the number of atoms it passes through before coming to a halt.

(b) Explain why it has a greater range in gases than in solids.

3.5.9 Self-test

For the intensity distance relationship in Fig. 3.5.7,

(a) State the fraction of the photons which penetrate 1.0 m.

(b) Use your answer to (a) to estimate the fraction which penetrate 5 m.

Stretch & Challenge

See the maths Section for the discussion on exponential decay. The intensity, I, is given by

$$I = I_0 e^{-\mu x}$$

where μ is the fraction penetrating per unit distance. For the data given, the value of μ is 0.0105 cm^{-1}. and $I_0 \sim 100$ a.u. Use these values to find x for which $I = 10$.

3.5.10 Self-test

The ionic radius of metallic aluminium ($Z = 13$) is 67.5 pm.

For lead ($Z = 82$) it is 91.5 pm. Estimate the ratio of the electron concentrations in aluminium and lead.

If an alpha particle with a typical energy of several MeV passes through an atom, it will attract electrons (because of the opposite charges). It doesn't need to collide directly with an electron – the electric attraction will cause an energy transfer and pull several electrons from the atom. The energy transfer is typically a few tens of eV. The effect on the α particle is to slow it down slightly, so it passes through the next atom a bit more slowly, increasing the ionising effect (see Study point). Eventually it runs out of energy, picks up a couple of stray electrons and becomes a helium atom.

So what about β particles? The effect is the same – it doesn't matter whether they are β^+ or β^- they will still kick out electrons. But there is one difference. Alpha and beta particles have the same energy range (~ 100 keV – 5 MeV) but electrons and positrons are much less massive, so they travel more quickly (very close to the speed of light in fact). You should be able to explain why they are more weakly ionising than alpha particles and therefore why they have a bigger range.

(b) The absorption of gamma rays

These are ionising, too, but their interaction with the electrons of atoms is different in two important ways:

1. They are uncharged so they can only affect an atomic electron if they pass very close. Therefore they pass through a large number of atoms before interacting.

2. They lose all their energy in one interaction (remember the photoelectric effect).

Suppose we have a parallel beam of gamma rays (i.e. a large number of photons) passing through a uniform material (i.e. same composition). Then suppose that a certain fraction, say 0.9 (i.e. 90%) gets through a 10 cm thickness. What about the next 10 cm? The surviving photons have not lost any energy so they have the same chance of getting through the next 10 cm. So the fraction 'surviving' after 20 cm is $0.9 \times 0.9 = 0.81$. With these figures, the intensity, distance graph looks as follows:

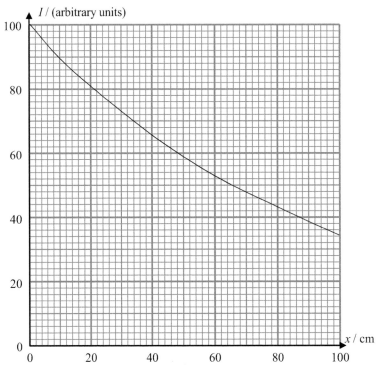

Fig. 3.5.7 Drop in intensity of a γ beam with thickness of absorber

(c) Why are denser materials more effective in absorbing radiation?

Most of the mass of a material is in the nuclei, so the denser the material, the more protons and neutrons there are. The numbers of protons and electrons are the same. Comparing a gas to a solid, there are a lot fewer atoms (in a gas) per unit volume, so there are a lot fewer electrons and it is these which absorb the radiation.

Comparing solids of different elements: atoms are all more or less the same size – with an atomic radius of something like 0.1 nm. As we increase the value of Z, the more the inner electron shells are pulled towards the nucleus (by its greater charge) but the radius is hardly changed. This means that the concentration of electrons increases with the density. Hence denser means more absorption.

Self-test 3.5.11

Explain why strongly ionising radiation has a shorter range than weakly ionising radiation.

3.5.3 Detecting and measuring nuclear radiation

Nuclear radiation is generally detected by the ionisation that it produces. Because it is only weakly ionising, γ is therefore more difficult to detect than α and β. A detector for α radiation must also be constructed with a very thin window otherwise the radiation will not be able to enter the detector.

▼ **Study point**

This very brief review of radiation detectors is for completeness, to give you some idea of how they work. **You will not be examined on Section 3.5.3.**

(a) The Geiger-Müller tube (GM tube)

The GM tube is a common detector for all three types of radiation.

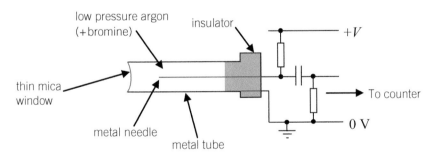

Fig. 3.5.8 The Geiger-Müller tube

The nuclear radiation ionises argon atoms. The liberated electrons are attracted to the positively charged metal needle (typical potential ~400 V) and pick up enough kinetic energy to cause further ionisation on the way. The resulting pulse of current produces a voltage pulse across the resistor which is registered on the counter.

▼ **Study point**

The GM tube is able to detect α, β and γ radiation. The α detection is directional because the α particles can only enter the tube via the mica window. Even the weakly ionising γ rays produce sufficient electrons to be registered.

The role of the bromine is to mop up excess electrons and so reduce the duration of the pulse and hence shorten the 'dead time' during which no events can be registered.

(b) Solid state detectors

Solid state detectors are diodes (see Fig. 2.7.24 in the AS book) with the n-type region connected to the positive (so-called *reverse biased*). The n region is very thin (typically 0.1 μm) so that α radiation can penetrate it into the depletion zone, d. Here it ionises the silicon, creating large numbers of electrons and holes, which are drawn to the $+V$ and earth respectively and constitute a pulse of current. This produces a pd pulse across the resistor, which is registered by a counter.

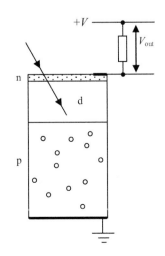

Fig. 3.5.9 Solid state detector

The advantages of solid state detectors are that they operate on low voltages (typically 3–5 V) and their dead time is very small, so they can detect high levels of radiation. Their performance does degrade with time as they accumulate radiation damage.

┌─ *Terms & definitions* ─┐

In the context of laboratory investigations into radioactive sources, **background radiation** is the nuclear radiation which is detected in the absence of the source under investigation.

3.5.12 Self-test

Calculate the mean counts per minute of the 10-minute background readings.

Stretch
& Challenge

Detailed statistical analysis predicts that, if the long-term mean count over a period is N, the distribution of a large number of individual readings will have a standard deviation of $\sqrt{N}$: ~70% of readings will be within $N \pm \sqrt{N}$ and ~95% will be within $N \pm 2\sqrt{N}$. Check the 1-min and 10-min values – and comment.

3.5.13 Self-test

Calculate the mean count rates of the one-minute background readings in batches of 10: i.e. 1st – 11th (done); 2nd – 12th, 3rd – 13th, etc. Compare the results with the 10-minute variation.

Stretch
& Challenge

A GM tube placed near a radioactive source detects 1659 counts in 10 seconds. The counting time is controlled by a manual switch. Does the random decay process or the uncertainty in handling the switch have a bigger effect on the reading?

3.5.4 The random nature of decay

(a) The variability of results

If you listen to the output speaker of a Geiger counter (i.e. a G-M tube connected to a counter), watch the rising count display or watch the needle of an analogue rate meter, you'll soon notice that the counts arrive at random times. This needs to be borne in mind when you are measuring the **background radiation** or observing a source.

The following are 20 readings, each over 1 minute, of the background radiation in Aberystwyth:
24, 20, 26, 29, 27, 24, 25, 22, 25, 21, 21, 28, 27, 25, 18, 24, 19, 27, 32, 28

The readings all fluctuate around ~25 (the mean value is 24.6). The range appears to be ± 7 counts per minute (cpm). What happens if we time for longer? These are 10 minute readings:
268, 227, 247, 258, 230, 240, 232, 226, 266, 236, 248, 242, 246, 269, 264, 230, 260, 237, 238, 267

The mean value agrees nicely with the mean of the 1-minute readings (see Self-test 3.5.12). The results have a larger spread than the 1-min readings but when we convert them to cpm we find a *smaller* spread: extreme values 22.7–26.9 i.e. ± 2. The difference is highlighted in Fig. 3.5.10.

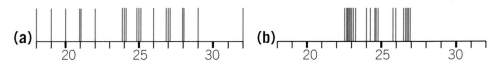

Fig. 3.5.10 Comparison of cpm variation with (a) a 1-minute and (b) a 10-minute counting period

(b) Long counting period or repeat readings?

As long as the activity of the radioactive sample is constant over the time-scale of the experiment (e.g. much shorter than the half-life) it doesn't matter whether one long reading is taken or repeat readings adding up to the same time, e.g. measure the background radiation over 10 minutes or do it 10 times over one minute (or 5 times over 2 minutes…). In all cases, the mean rate (i.e. cpm or cps) will be the total count divided by the total time. For example, the total of the first 10 of the one-minute readings is 243 counts, giving a mean rate of 24.3 cpm.

(c) Correcting for background radiation

This is done by subtracting the background count from the count with the source present (over the same period of time).

3.5.5 Distinguishing between α, β and γ radiation

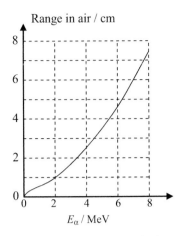

Fig. 3.5.11 Range of α particles in air as a function of energy

(a) Using penetrating power

You will recall from your pre-A level knowledge that we can use penetrating power to distinguish between α, β and γ radiation. The rule of thumb is:

- If it can't penetrate a sheet of paper or more than ~3–5 cm air it is α.

- If it can penetrate the above but is stopped by thin metal, e.g. 1–2 mm steel; 2–3 mm aluminium, it is β (either β⁻ or β⁺).

- If it gets through all of the above it is γ.

The range of any of the radiations actually depends on their energy. Fig. 3.5.11 shows how the range of α-particles varies with their kinetic energy.

Here is a slightly tough GCSE example (it would usually come with a diagram) but the numbers are easy:

Example

A GM tube is placed 2 cm from a source of nuclear radiation. It detects 525 counts per minute (cpm). With a paper absorber, the count rate falls to 400 cpm; a 2 mm thick aluminium absorber it is 225 cpm. What are the fractions of α, β and γ radiation in the emissions from the source?

The background reading was 250 counts in 10 minutes.

Answer

(We need to make certain assumptions – see Study point.)

$$\text{The background count rate} = \frac{250 \text{ counts}}{10 \text{ min}} = 25 \text{ cpm}$$

Thus the total radiation detected from the source = 525 – 25 = 500 cpm

Using the obvious symbols:

$$\alpha + \beta + \gamma + 25 = 525 \qquad [1]$$
$$\beta + \gamma + 25 = 400 \qquad [2]$$
$$\gamma + 25 = 225 \qquad [3]$$

The γ is thus 200 out of 500 = 0.40 of the total, i.e. 40%

Subtracting [3] from [2] → β = 175

So the β is 175 out of 500 = 0.35 of the total, i.e. 35%.

Then either by requiring that the fractions all add up to 1.00 (100%) or by subtracting [2] from [1], α is 0.25 of the total, i.e 25%.

Note: alternatively subtract the background at the beginning to give:

α + β + γ = 500; β + γ = 400 ; γ = 200 and then proceed as above.

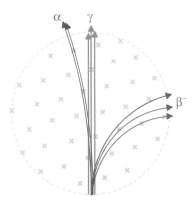

Fig. 3.5.12 Unrealistic magnetic deflection diagram

Stretch & Challenge

One 'realistic' aspect of Fig. 3.5.12 is that, whereas the tracks of the α particles are shown keeping together, those of the β particles are shown spreading out. Why should this be?

3.5.14 Self-test

The 'unrealistic' Fig. 3.5.12 shows the tracks of β⁻ particles. How would the diagram be different if β⁺ tracks were shown?

Stretch & Challenge

If the kinetic energy of a particle is expressed in MeV and its mass in MeV/c^2, suggest a consistent unit for momentum.

Express the momentum of a 1 MeV proton in this unit.

Can you use this unit to work out the radius of its track in a 0.2 T magnetic field?

—— Terms & definitions ——

The **activity**, A, of a radioactive source is the number of radioactive decays per unit time.

Unit: becquerel (Bq) $\equiv$ s^{-1}

The **decay constant**, λ, is the fraction of the atoms of a radioactive nuclide which decay per unit time.

Unit: s^{-1}

(b) Deflection of radiation in a magnetic field

Beams of charged particles constitute electric currents and so are deflected in magnetic fields. See Section 4.4. The force is always at right angles to both the magnetic field and the velocity of the particles. Thus, if the particles are travelling at right angles to the magnetic field, their path is circular, with the deflection in a direction given by Fleming's left hand rule. The radius, r, of the circular path is given by:

$$r = \frac{p}{qB}$$

where p is the momentum of the particles, q their charge and B the flux density of the magnetic field (see Section 4.4.3).

You will often come across a diagram similar to Fig. 3.5.12 to illustrate the deflections (or lack of them) for α, β and γ radiation. In one respect, the diagram gives the correct idea: α and β, having opposite charges, deflect in opposite directions and γ (being uncharged) is undeflected.

However, in another it is very unrealistic. The masses of α and β are very different (a helium nucleus is about 8000 × as massive as an electron). If the α and β have the same energy, E, their momenta are not the same. Ignoring, for the moment, the need for a relativistic treatment:

$$E = \frac{p^2}{2m} \qquad \therefore p = \sqrt{\frac{2E}{m}}$$

$\therefore$ The radius of the α particles' circle is hugely greater than that of the β particles. Allowing for the double charge on α, the ratio is $\sqrt{4000}$, i.e. about 63 (see Study point).[4]

Example

Calculate the radius of curvature of 1 MeV α particle tracks in a 0.2 T magnetic field.

Answer

It will be less confusing to work in SI units here, so we'll use the mass of the α particle in **kg**, i.e 6.64×10^{-27} kg.

$E = \dfrac{p^2}{2m}$, so the momentum is given by:

$$p = \sqrt{2mE} = \sqrt{2 \times 6.64 \times 10^{-27} \times 1.60 \times 10^{-13}} = 4.61 \times 10^{-20} \text{ N s}$$

$\therefore$ The radius of curvature, $r = \dfrac{p}{qB} = \dfrac{4.61 \times 10^{-20}}{2 \times 1.60 \times 10^{-19} \times 0.2} = 0.72$ m

3.5.6 The exponential decay of radioactive substances

In this section we shall be looking at the **activity**, A, of a radioactive material and how it changes over time. This is number of decays per second. A typical school source has an activity of ~100 kBq, i.e. ~10^5 decays per second.

---

4 For α and β particles with a kinetic energy 1 MeV, a relativistic treatment gives the radius of the α particles to be approximately 20 × that of the β particles.

(a) Activity and decay constant

The decay of an unstable atomic nucleus is a random event. It is entirely unaffected by temperature, pressure or any other physical condition. We cannot predict when an individual nucleus will decay but we can determine the probability that a nucleus will do so within any time period. For example, a $^{14}_{6}C$ nucleus has a probability of 3.94×10^{-12} of decaying each second. We say that the **decay constant**, λ, of $^{14}_{6}C$ is 3.94×10^{-12} s^{-1}. This means that, in a sample of N carbon-14 atoms, $3.94 \times 10^{-12}N$ will decay each second.

In general the relationship between N, A and λ is: $A = \lambda N$.

Example

The decay constant of U235 is 3.14×10^{-17} s^{-1}. Calculate the activity of 1 kg of U235. $[N_A = 6.02 \times 10^{23} \text{ mol}^{-1}]$

Answer

The molar mass of U235 = 235 g = 0.235 kg.

$\therefore$ The number of atoms in 1 kg of U235, $N = \dfrac{1.0 \text{ kg}}{0.235 \text{ kg mol}^{-1}} N_A = 2.56 \times 10^{24}$

$\therefore$ Activity, $A = \lambda N = 3.14 \times 10^{-17} \text{ s}^{-1} \times 2.56 \times 10^{24} = 80 \text{ MBq}$

(b) Half-life

The unpredictable, random decay leads to the half-life law. The rate of decay of a sample of single radioactive nuclide is proportional to the number of atoms; if the number remaining halves, the rate of decay halves. Hence, the time taken for the number remaining to fall to any given fraction (e.g. a half, 0.1, e^{-1}) is constant.

Suppose we have a sample of 4×10^{18} atoms of a radioactive nuclide with an initial activity of 2×10^{10} Bq. Fig. 3.5.13 shows how the number remaining, N, varies with time, where T is the (as yet unknown) half-life.

This type of graph is called an exponential decay graph (see Chapter 4). Because $A = \lambda N$, the graphs of N against t and A against t show the same characteristic half-life. Fig. 3.5.14 is the $A(t)$ graph corresponding to the $N(t)$ graph in Fig. 3.5.13.

(c) The formulae of radioactive decay curves

We normally write the initial values of N and A as N_0 and A_0. We know that after x half-lives these values have been reduced by a factor 2^x so we can write:

$$N = \frac{N_0}{2^x} \text{ and } A = \frac{A_0}{2^x},$$

or more conveniently

$$N = N_0 2^{-x} \text{ and } A = A_0 2^{-x}.$$

It is important to note that x can take any value, not just positive whole numbers. So for example, after 3.7 half-lives the activity of a sample will have fallen to 0.077 times the original (because $2^{-3.7} = 0.077$ to 2 s.f.). You should make sure you can drive your calculator to obtain this!

Self-test 3.5.15

The activity of radioactive sources used to be expressed in *curies*, a non-SI unit. The curie used to be defined as the activity of 1 g of Ra-226, which has a decay constant of 1.37×10^{-11} s^{-1}. Express the curie in becquerel.

S&C Stretch & Challenge

A short-lived radioactive nuclide has $\lambda = 1.0 \times 10^{-6}$ s. Why can we calculate the fraction of the atoms decaying in one minute by multiplying by 60 s but not the fraction decaying in one year by multiplying by 3.16×10^7 s [the number of seconds in a year]?

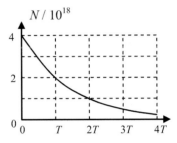

Fig. 3.5.13 Decay curve (1)

Self-test 3.5.16

Calculate the value of λ in the radioactive sample in Section 3.5.6(b).

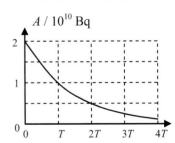

Fig. 3.5.14 Decay curve (2)

--- Terms & definitions ---

The **half-life** is the time taken for the number of nuclei of a radioactive nuclide (or the activity) to halve.

See Section 3.3 for the definitions of **molar mass**, M_r and the **Avogadro constant**, N_A

In Section M3.2 we derive the following alternative formulae in terms of time rather than the number of half-lives:

$$N = N_0 e^{-\lambda t} \text{ and } A = A_0 e^{-\lambda t}.$$

The **Stretch & Challenge** in the margin asks you to show that these formulae are equivalent to the 2^{-x} ones, but first we need to derive the relationship between λ and the half-life, $T_{\frac{1}{2}}$.

Starting from $\qquad A = A_0 e^{-\lambda t}$

By definition, after time $\quad t = T_{\frac{1}{2}}, A = \frac{1}{2}A_0$,

$\therefore$ Rearranging $\qquad e^{-\lambda T_{\frac{1}{2}}} = \dfrac{A}{A_0}$, $\qquad\qquad$ i.e. $e^{-\lambda T_{\frac{1}{2}}} = \dfrac{1}{2}$

Taking logs $\qquad\qquad \ln\!\left(e^{-\lambda T_{\frac{1}{2}}}\right) = -\ln 2$

$\therefore \qquad\qquad\qquad -\lambda T_{\frac{1}{2}} = -\ln 2,$ $\qquad$ i.e $\lambda = \dfrac{\ln 2}{T_{\frac{1}{2}}}$

Exam questions involving λ and the half-life require care with units because half-life is commonly expressed in hours, days, or years as appropriate, whereas the activity A is in Bq (i.e. s^{-1}).

Example

Calculate the activity of 1 ng of C-14, which has a half-life of 5570 years.

Answer

Decay constant, $\lambda = \dfrac{\ln 2}{T_{\frac{1}{2}}} = \dfrac{\ln 2}{5570 \times 3600 \times 24 \times 365.25} = 1.76 \times 10^{-11}$ s^{-1}

No. of atoms, $N = \dfrac{m}{M_r} N_A = \dfrac{1.0 \times 10^{-9} \text{ g}}{14 \text{ g mol}^{-1}} \times 6.02 \times 10^{23} \text{ mol}^{-1} = 4.3 \times 10^{13}$.

The activity, $A = \lambda N \therefore A = 1.76 \times 10^{-11}$ s$^{-1} \times 4.3 \times 10^{13} = 760$ Bq.

(d) Using log-linear (semi-log) graphs

We have seen that the activity of a radio-nuclide decays according to the equation $A = A_0 e^{-\lambda t}$, resulting in the familiar exponential decay graph. We can turn this relationship into a linear graph by taking logs:

$$\ln A_0 = \ln A - \lambda t \qquad \text{[See Maths tip]}$$

So a graph of $\ln A$ against t should be straight line of gradient $-\lambda$ and intercept $\ln A_0$ on the $\ln A$-axis. The advantage of a straight line graph over a non-linear one is that the eye is much better at comparing straight lines to slightly random data points than curved lines. So, in an experiment to determine the half-life of a radioactive material, a more accurate result is likely from a log-linear graph.

Sample data

A teacher monitored the decay of a sample of radon-220 over a period of 2 minutes by noting down the count in successive 10 s intervals. The results and decay graph were as follows:

Time, t / s	0	10	20	30	40	50	60	70	80	90	100	110	120
Count, C	540	495	432	361	310	290	245	220	174	165	157	125	112

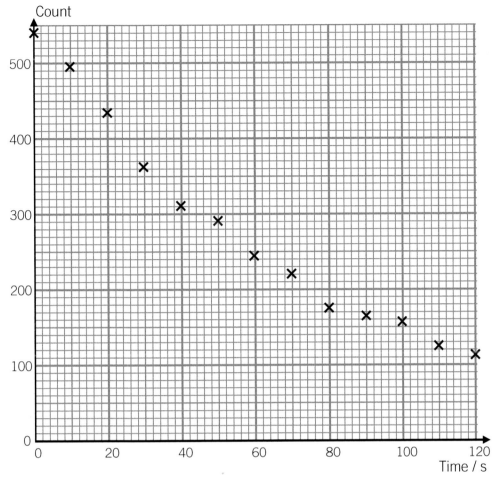

Fig. 3.5.15 Decay of radon-220

Study point

Ways of finding the half-life from Fig. 3.5.15:

First draw a best-fit decay curve then:

1. Find two times where the counts are a factor of 2^n apart, e.g. 400 and 200, 500 and 125, or

2. Choose any two widely separated points with counts C_0 and C_1. Find x from $C_1 = C_0 2^{-x}$ and calculate the half-life from the time difference, or

3. As 2 but use $C_1 = C_0 e^{-\lambda \Delta t}$ to find λ and use it to calculate the half-life.

Self-test 3.5.18

Find the half-life of Rn220 using all three methods given in the Study point.

We can use this graph to determine the half-life of **Rn220** (see Study point). A better way is to plot $\ln C$ against t, draw a best-fit straight line, determine λ from the gradient and calculate $T_{\frac{1}{2}}$. See Exercise 3.5 Q1.

Self-test 3.5.19

(a) Calculate the decay constant, λ, of I-131.

(b) Show, using $A = A_0 e^{-\lambda t}$, that the activity after 21 days is 500 kBq.

3.5.7 Calculations using the decay equations

The simplest calculations involve determining the activity, number of atoms or mass remaining after a given time, given the half-life or decay constant. This just requires inserting the data into the relevant equation:

e.g. $\qquad N = N_0 2^{-x} \quad$ or $\quad A = A_0 e^{-\lambda t}$

Study point

In answering Self-test 3.5.19 it doesn't matter whether we use seconds, minutes, hours or days as the unit of time as long as λ and t are in consistent units;

e.g. λ in s^{-1} and t in s

or λ in day^{-1} and t in days. The value of λt will always be the same so $e^{-\lambda t}$ will always have the same value. Repeat S-t 3.5.19 using a different unit of time.

Example

Iodine-131 has a half-life of 8.1 days. A sample of I131 has an activity of 3.0×10^6 Bq. Calculate the activity after 21 days.

Answer

We'll do this using half-lives. See Self-test 3.5.19 to do the same question in a different way.

$$21 \text{ days} = \frac{21}{8.1} = 2.59 \text{ half-lives.}$$

$\therefore$ Using $A = A_0 2^{-x}$, activity after 21 days $= 3.0 \times 10^6 \times 2^{-2.59}$ Bq

$$= 500 \text{ kBq}$$

Calculations which require you to calculate λ or $T_{\frac{1}{2}}$ are more complicated because we need to use the relationship between logs and exponents (see Maths tip).

MATHS TIP

Make sure you can use these relationships:

$\ln(e^x) = x;$ $\qquad e^{\ln x} = x$

$\ln(2^x) = x\ln 2;$ $\qquad e^{x\ln 2} = 2^x$

3.5.20 Self-test

Answer the I-128 example using half-life and $A = A_0 2^{-x}$.

Hint: First solve $A = A_0 2^{-x}$ for x using the given A and A_0.

Example

Iodine-128 has a half-life of 25 minutes. Calculate the time taken for a 5.0 MBq sample to decay to 10 kBq.

Answer

This time we'll work using the decay constant, λ (in s^{-1}).

$$\lambda = \frac{\ln 2}{25 \times 60} = 4.62 \times 10^{-4} \text{ s}^{-1}$$

$\therefore$ Using $A = A_0 e^{-\lambda t}$, $e^{-4.62 \times 10^{-4}t} = \frac{10 \times 10^3}{5 \times 10^6} = 0.002$

$\therefore$ Taking logs: $-4.62 \times 10^{-4}t = \ln(0.002) = -6.21$

$$\therefore t = \frac{-6.21}{-4.62 \times 10^{-4}} = 13\,500 \text{ s (3 hr 44 min)}$$

▼ **Study point**

As with the 'repeats' issue in Section 3.5.4, it doesn't matter whether all the dice are rolled at once. It can also be done with a smaller number of dice which are repeatedly rolled and counted.

3.5.8 Specified practical work

(a) Investigating random decay using a dice analogy

Many centres do not have the facilities to collect and observe short-lived radio-isotopes but the random nature of decay can be illustrated using the fact that cubical dice have a 1 in 6 chance of landing on any specific side.

A simple way to use dice to illustrate radioactive decay is to consider them to be radioactive nuclei and any that land with (say) 1 upwards have 'decayed'. So the procedure is:

3.5.21 Self-test

For the '1 is out' model with $N_0 = 1000$, calculate N_1, N_2 and N_{10}.

1. Roll a large known number (N_0) of dice.

2. Remove any that display a 1 and count the remainder (N_1).

3. Roll the remaining dice and remove the 1s.

4. Count the remainder (N_2) and repeat to obtain N_3, N_4 … .

Theory

Each die has a 1 in 6 chance of decaying, so on the average[5] the fraction remaining 'live' after any roll is $\frac{5}{6}$. After two rolls it is $\frac{5}{6} \times \frac{5}{6} = \left(\frac{5}{6}\right)^2$ and after n rolls it is $\left(\frac{5}{6}\right)^n$.

We can find the 'half-life', i.e the number of rolls needed to halve the number of undecayed dice, by solving the equation:

$$\left(\frac{5}{6}\right)^n = \frac{1}{2}$$

Taking logs and a bit of manipulation gives: $n = \dfrac{\ln 2}{\ln 1.2} = 3.8$ (2 s.f.)

This model can be varied, e.g. by letting 1 or 2 indicate a decayed die or by use of 12-sided or 20-sided dice.

Self-test 3.5.22

A student used a large number of dodecahedral (i.e 12-sided) dice as a model decay system with the '1 is out' rule. Show that the expected 'half-life' is almost exactly 8.

(b) Investigating the variation of intensity of gamma rays with distance

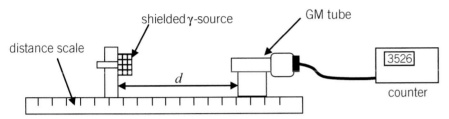

Fig. 3.5.16 Inverse square law for γ rays

Self-test 3.5.23

If the units of C and d are s^{-1} and cm respectively, state the units of k and ε.

If the separation of the source and detector is large compared with the size of the source, we expect the intensity of the gamma rays to be proportional to the inverse square of the separation. Unfortunately we know neither the exact position of the source within its shielding nor the effective position of the detector in the G-M tube. With the set-up above we therefore expect the count rate C due to the source (i.e. corrected for background) to depend upon d according to:

$$C = \frac{k}{(d - \varepsilon)^2}$$

where k and ε are constants.

Stretch & Challenge

In a S Wales school, the mean background count is 21 per minute. The count with a γ source present is 760 over a 5-minute period. Estimate:

(a) C and $\dfrac{1}{\sqrt{C}}$ in counts per second, and

(b) the uncertainties in C and $\dfrac{1}{\sqrt{C}}$.

If we take the square root of the equation and invert it we get:

$$\frac{1}{\sqrt{C}} = \frac{d}{\sqrt{k}} - \frac{\varepsilon}{\sqrt{k}}.$$

Hence, if the inverse square law holds, a graph of $\dfrac{1}{\sqrt{C}}$ against d will be a straight line.

The procedure is quite straightforward and is the subject of Exercise 3.5 Q4.

Self-test 3.5.24

Explain how a value for ε can be found from the graph of $\dfrac{1}{\sqrt{C}}$ against d.

---

5 The 'on the average' is important here. If N_1 is expected to be about 900, a variation of ±30 is also to be expected. See Stretch & challenge in Section 3.5.4.

Exercise 3.5

1 From a graph of $\ln C$ against t for the results of the Rn-220 decay in Section 3.5.6(d) find:

(a) The gradient and the intercept on the $\ln C$ axis.

(b) The decay constant, λ and half-life of the decay.

2 Consult a periodic table to complete the following decay processes:

(a) The α decay of $^{192}_{78}\text{Pt}$

(b) The β^- decay of $^{181}_{72}\text{Hf}$

(c) The β^+ decay of $^{48}_{23}\text{V}$

(d) The K-capture decay of $^{59}_{28}\text{Ni}$

(e) The decay of $^{77}_{32}\text{Ge}$ into $^{77}_{33}\text{As}$

(f) The decay of $^{65}_{30}\text{Zn}$ by the emission of a positive particle

(g) The decay of $^{65}_{30}\text{Zn}$ by the absorption of a negative particle

(h) The transformation of $^{239}_{92}\text{U}$ into a plutonium isotope by two successive β^- decays

(i) The decay of $^{17}_{7}\text{N}$ by the emission of an electron and a neutron.

3 The table gives the properties of some isotopes of neon:

Isotope	$^{18}_{10}\text{Ne}$	$^{19}_{10}\text{Ne}$	$^{20}_{10}\text{Ne}$	$^{21}_{10}\text{Ne}$	$^{22}_{10}\text{Ne}$	$^{23}_{10}\text{Ne}$	$^{24}_{10}\text{Ne}$
decay mode	β^+	β^+	stable	stable	stable	β^-	β^-

Explain the pattern in the table.

4 A group of students investigated the inverse square law for gamma radiation. They obtained the following measurements:

Background count = 130 in 5 minutes

Distance, d / cm	10	15	20	25	30	50	70
Count	755	282	155	230	275	240	365
Count time / min	1	1	1	2	3	5	10

(a) Suggest why they measured the counts over different times.

(b) Draw up another table of counts per minute, C, corrected for background radiation.

(c) Draw a graph of $\dfrac{1}{\sqrt{C}}$ against d.

(d) Use your graph to confirm the validity of the inverse square law and calculate the values of k and ε.

5 The isotopes of natural uranium have the following relative abundances and half-lives:

Isotope	$^{234}_{92}U$	$^{235}_{92}U$	$^{238}_{92}U$
Relative abundance	0.01%	0.72%	99.27%
Half-life / years	2.5×10^5	7.1×10^8	4.5×10^9

Calculate:

(a) The decay constant for each isotope.

(b) The fraction of the activity of a sample of natural uranium which is due to each of the isotopes.

(c) The activity of 1 kg of natural uranium.

6 All three uranium isotopes in Q5 decay by α emission. Explain why the total detectable emissions from 1 kg of uranium in the form of a compact lump would be much less than the answer to 5(c) and why such a lump might feel slightly warm to the touch.

7 The short half-life of U234 compared to the age of the Earth (2.56×10^9 years) means that any of this isotope that was in the Earth originally has long since decayed to homeopathic amounts (i.e. zero).

(a) Show that this statement is correct. [Hint: $2^{-10} \sim 10^{-3}$]

(b) That U234 exists currently, albeit in small amounts, is because it is constantly being produced (indirectly) by the decay of U238. The daughter products of the decay of U238 include α and β^- emitters only. Give the shortest possible decay sequence for the production of U234 from U238. [Consult a periodic table if necessary.]

8 (a) Explain why U235 cannot be produced as a result of the decay of U238.

(b) Estimate the isotopic abundances of U235 and U238 at the time that the Earth was formed.

(c) Assuming that U235 and U238 were formed in approximately equal amounts in the supernova explosion which provided the heavy elements for the nebula out of which the Solar System condensed, estimate when this supernova occurred.

9 The ratio of $^{40}_{19}K$ to $^{40}_{18}Ar$ is used by geologists to date igneous rocks (and the igneous components of sedimentary rocks) – *potassium-argon dating*. Its basis is that 10% of K40 atoms decay to the stable Ar40 with a half-life of 1.3×10^9 years. [The other 90% decay by β^- decay.] A main assumption is that igneous rocks contain no argon initially.

(a) [For chemists] Explain why the assumption is likely to be valid.

(b) Calculate the decay constant of K40.

(c) Explain qualitatively how the ratio of Ar:K depends upon the age of the rock and state a second assumption that must be made.

(d) Identify and write the equations for the two modes of decay of K40 to Ar40.

(e) Draw up a table to show the decline of the abundance of K40 in 100 million year steps up to 600 million years from an initial value of 100. Include also a column of the abundance of Ar40 over this time and another for the ratio of Ar:K.

(f) Use a graph of the potassium-argon ratio against time to estimate the age of an igneous rock with a Ar:K ratio of 0.020. If the second assumption were not strictly correct, how would this age be interpreted?

10 A GM tube is used to investigate a radioactive source. The background radiation is measured over a period of 10 minutes. The source is then placed 2 cm from the GM tube and the counts taken over 10 minutes, with and without a sheet of paper in between the source and detector.

Results:
No source present: 250 counts
No absorber: 570 counts
Paper absorber: 525 counts
Discuss whether the results show that the source is an α-emitter.

11 The photo-plates that Becquerel used were wrapped in black paper; the Maltese cross was made of thin steel. What can you conclude about the nature of the radiation that fogged the plate?

12 A GM tube is set up close to a γ-emitter. The corrected count rate with no absorber is 24.0 s⁻¹. When a 5 mm-thick lead absorber is placed between source and detector the corrected count rate drops to 18.0 cps. What count rate is expected with (a) with a 10 mm lead absorber and (b) with a 15 mm absorber?

& Challenge

For the potassium-argon dating method

(a) Show that the ratio, R of $\dfrac{Ar}{K}$ is given by $\dfrac{1-e^{-\lambda t}}{10e^{-\lambda t}}$.

(b) Show that $t = \dfrac{1}{\lambda}\ln(1 + 10R)$.

(c) Use the result to verify the answer to Q9 (f).

Sketch a graph of t against R and give its gradient when $R = 0$.

3.6 Nuclear energy

Studies in radioactivity, developments in the knowledge of atomic and nuclear structure, careful measurements of nuclear masses and Einstein's Special Theory of Relativity in the first half of the 20th century led to the ability to harness the most powerful sources of energy on the Earth: the energy in very low and very high mass nuclei. Nuclear fusion is the energy source of stars; nuclear fission is made possible by the existence of fissile nuclei which are among the debris of exploding stars. Knowledge of this section is important for the understanding of particle physics (in the AS textbook) and radioactivity. We start with a largely misunderstood topic: $E = mc^2$ and mass-energy.

Fig. 3.6.1 The Crab nebula – seeding space with new elements

3.6.1 Mass-energy

Through careful experiments in the 18th and 19th centuries, chemists developed the law of conservation of mass, according to which the mass of the products of a reaction is equal to that of the reactants. For example:

$$C \quad + \quad O_2 \quad \rightarrow \quad CO_2$$

Mass/u   12.01   +   32.00   =   44.01

At the same time physicists and chemists were developing the law of conservation of energy: if a falling ball loses 100 J of gravitational potential energy and only gains 20 J of kinetic energy, the balance (80 J) must appear in some other form, e.g an increase of the internal energy of the atmosphere and/or the ball.

▼ **Study point**

In the reaction between carbon and oxygen, the masses have been given in u, i.e. the masses of individual atoms.

Self-test   3.6.1

State the masses of the reactants and product when 1.000 mol of carbon reacts with oxygen to form carbon dioxide.

(a) $E = mc^2$

In developing his special theory of relativity Einstein concluded that mass and energy are two aspects of the same quantity, which we can call mass-energy and that this quantity is conserved. We can express mass-energy in units of mass (kg) or units of energy (J), the exchange rate between them being c^2, where c is the speed of light *in vacuo*,[1] i.e.

$$E = mc^2$$

Example

To see why chemists didn't spot this earlier, consider again the $C + O_2$ reaction. The energy released by burning 1 mol of carbon is 394 kJ (see Study point).

(a) Calculate the mass of this energy, and

(b) express this as a fraction of the reacting masses.

Answer

(a) $m = \dfrac{E}{c^2} = \dfrac{394 \times 10^3 \text{ J}}{(3.00 \times 10^8 \text{ m s}^{-1})^2} = 4.38 \times 10^{-12}$ kg

(b) Total reacting mass $= 12.01$ g $+ 32.00$ g $= 44.01$ g $= 0.004401$ kg.

$$\therefore \text{ Fraction of reacting masses} = \frac{4.38 \times 10^{-12} \text{ kg}}{44.01 \times 10^{-3} \text{ kg}} = 1.0 \times 10^{-10}$$

▼ **Study point**

In chemists' language, the *enthalpy* change, ΔH, in the reaction

$$C + O_2 \rightarrow CO_2$$

is -394 kJ mol^{-1}.

Self-test   3.6.2

Calculate the increase in mass of a 1 kg block of iron which is heated up (e.g. using a bunsen burner) from 20 °C to 520 °C. Take the specific heat capacity of iron to be 450 J kg^{-1} K^{-1}.

---

1 *In vacuo* is just the Latin for 'in a vacuum'. Scientists like to show off sometimes!

Stretch & Challenge

When the iron block in Self-test 3.6.2 was heated up either:

(a) a single layer of iron atoms was ablated (burnt off), or

(b) the outermost layer of iron atoms was oxidised.

Compare the magnitude of the mass change in each of these cases with your answer to the Self-test.

3.6.3 Self-test

The ionisation energy of hydrogen is 13.6 eV. How much greater is the combined mass of a free proton and electron than that of a hydrogen atom? Express your answer in u.

Data:

1 eV = 1.602×10^{-19} J

$c = 3.00 \times 10^{8}$ m s^{-1}

u = 1.66×10^{-27} kg

3.6.4 Self-test

Express the energy release in the 4H → He reaction:

(a) in J

(b) in eV

Compare this with the 13.6 eV energy release in the formation of H-1 from

$$p + e \rightarrow H$$

3.6.5 Self-test

Calculate the energy release in joules from the fusion of 1 kg of hydrogen.

Discussion

1. The first thing to note is that chemists would need to know the molar masses of carbon, oxygen and carbon dioxide to 10 significant figures in order to spot this mass-energy effect. Notice that the effect in Self-test 3.6.2 (a typical 'physics' process) is even smaller.

2. Where is this mass-energy? It all (i.e. 44.01 g) starts off in the reactants. It is all there in the 'hot' product (the CO_2) but the 4.38 ng is the mass of the energy that dissipates in the environment. That means that the atmosphere gets 4.38 ng heavier due to its extra internal energy; the carbon dioxide is 4.38 ng lighter because its atoms are more tightly bound together (i.e. they have a lower potential energy) than the reactants.

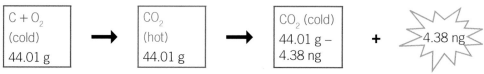

Fig. 3.6.2 Mass and energy in a combustion reaction

(b) $E = mc^2$ and nuclear energy

We've seen that $E = mc^2$ has general applicability; it does not apply only to nuclear energy. Yet $E = mc^2$ is generally seen as paving the way for the development of nuclear energy and specifically nuclear weapons. Why should this be? The reason is the energy involved in nuclear changes is a much greater fraction of the mass-energy of the particles. A very clear example is the fusion of four H1 atoms to one He4 atom. This process, the heart of the Sun's power, can be summarised as

$$4(^{1}_{1}H + ^{0}_{-1}e) \rightarrow (^{4}_{2}He + 2^{0}_{-1}e) + 2\overline{\nu}_{e}$$

where the symbols $^{1}_{1}H$ and $^{4}_{2}He$ are nuclear symbols.

A data book gives the following masses of H1 and He4 atoms as:

$$m_{H1} = 1.007\ 825\ u \qquad \text{and} \qquad m_{He4} = 4.002\ 604\ u$$

So the change in mass, Δm, is given by

$$\Delta m = m_{He4} - 4m_{H1}$$
$$= 4.002\ 604\ u - 4 \times 1.007\ 825\ u$$
$$= -0.0028\ 696\ u$$

This mass difference is about 0.7% of the original mass of the hydrogen atoms, which is much more than the fractional decrease in mass involved in chemical reactions, as the following example shows.

Example

Calculate the approximate mass loss in the fusion of 44 g of $^{1}_{1}H$ to $^{4}_{2}He$.

Answer

Molar mass of $^{1}_{1}H$ = 1.0 g. ∴ From above, fusing 4 g of $^{1}_{1}H$ results in a mass decrease of 0.028 7 g.

∴ Fusing 44 g of $^{1}_{1}H$ gives a mass decrease of $4 \times 0.028\ 7$ g ~ 0.1 g

(Compare this with the 4×10^{-9} g mass loss when burning 44 g of carbon.)

(c) The energy equivalence of u

As nuclear physicists generally express atomic and nuclear masses in **u** and energy in eV, it will help to be able to convert between the two. The conversion is:

The energy equivalent of $1\ \text{u} = 931.5\ \text{MeV}$.

You should be able to drive this relationship using the following data:

$$1\ \text{u} = 1.660538922 \times 10^{-27}\ \text{kg}$$

$$c = 299\ 792\ 458\ \text{m s}^{-1}$$

$$e = 1.60217662 \times 10^{-19}\ \text{C}$$

Self-test 3.6.6

Use the values of u, c and e to derive the equivalence

$1\ \text{u} \equiv 931.5\ \text{MeV}$ (4 s.f.)

▼ **Study point**

Be prepared for an examiner to give a slightly different value for energy equivalence, e.g. $1\text{u} \equiv 930\ \text{MeV}$.

3.6.2 Nuclear binding energy

A hydrogen-1 atom consists of a proton and an electron. If the hydrogen atom is in its ground state, it takes $13.6\ \text{eV}$ to separate these particles. We say that the **binding energy** of a hydrogen atom is $13.6\ \text{eV}$. It is another way of saying that the sum of the potential and kinetic energies of the atomic particles is $-13.6\ \text{eV}$.

It also takes energy to disassemble the nuclei of heavier atoms (i.e. ones that consist of more than just a single proton). This energy is called the **nuclear binding energy**. We can use the measured masses of the particles and $E = mc^2$ to calculate it for any nucleus, as follows:

1. We calculate the **mass deficit**, Δm of the nucleus. The nuclear mass is less than the sum of the masses of the individual protons and neutrons. The difference is the mass deficit.

2. Calculate the **binding energy** using $E = \Delta mc^2$, or the energy equivalence: $1\ \text{u} \equiv 931.5\ \text{MeV}$.

┌─ *Terms & definitions* ─┐

The **binding energy** of a system is the energy required to separate the particles of the system.

The **binding energy** of a nucleus is the energy required to disassemble the nucleus into its constituent protons and neutrons.

 **Stretch & Challenge**

The data in the $^{12}_{6}\text{C}$ binding energy example are given to 7 s.f. By looking at the calculation, discuss the number of figures actually required to calculate the binding energy to 3 s.f?

Example

Calculate the binding energy of a $^{12}_{6}\text{C}$ nucleus from the following data:

$m_{\text{p}} = 1.007\ 276\ \text{u}$, $m_{\text{n}} = 1.008\ 665\ \text{u}$, $m_{\text{e}} = 0.000\ 549\ \text{u}$

Answer

By definition, the mass of a C12 atom is 12 u exactly.

There are 6 electrons in a C12 atom

$\therefore$ Mass of a C12 nucleus, $m_{\text{C-12}} = (12 - 6 \times 0.000\ 549)\ \text{u}$

$$= 11.996\ 706\ \text{u}$$

There are 6 protons and 6 neutrons in a C12 atom.

$$6m_{\text{p}} + 6m_{\text{n}} = 6 \times 1.007\ 276\ \text{u} + 6 \times 1.008\ 665\ \text{u}$$

$$= 12.095\ 646\ \text{u}$$

$\therefore$ Mass deficit, $\qquad \Delta m = 12.095\ 646\ \text{u} - 11.996\ 706\ \text{u}$

$$= 0.098\ 940\ \text{u}$$

Leading to: Binding energy of C12 nucleus = 92.2 MeV

Self-test 3.6.7

Repeat the $^{12}_{6}\text{C}$ binding energy example by converting the mass deficit to kg and using $E = \Delta mc^2$. Give your answer in (a) J and (b) MeV.

─ *Terms & definitions* ─

A **nuclide** is a distinct variety of atom or nucleus, indicated by a specific symbol, $^A_Z X$.

3.6.3 Binding energy per nucleon

(a) Calculating the binding energy per nucleon

Imagine building up stable nuclei by adding more and more nucleons (i.e. protons and neutrons). As each extra nucleon is added, the binding energy must increase. If it didn't, that extra nucleon would not be bound to the others. Hence we'd expect the binding energy stable nuclei to increase with nucleon number, which is generally what we find. However, to determine how tightly the *individual* nucleons are bound we need to calculate the **binding energy per nucleon**:

$$BE / \text{nucleon} = \frac{\text{binding energy}}{\text{nucleon number}}$$

The binding energy per nucleon of $^{12}_6 C = \dfrac{92.2\ \text{MeV}}{12} = 7.68\ \text{MeV nuc}^{-1}$

Your attention is drawn to the table of atomic masses (Table 3.6.3) of selected nuclides at the end of Section 3.6. It contains all stable nuclides up to $Z = 20$ (calcium) and others which are used as illustrations in Self-test questions (e.g. Self-test 3.6.6) and problems in Exercise 3.6.

3.6.8 **Self-test**

Use the data in the table of atomic masses (Table 3.6.3) to calculate (a) the nuclear binding energy and (b) the binding energy per nucleon of $^{31}_{15}P$, giving your answer in MeV and MeV nuc^{-1} respectively.

─ *Terms & definitions* ─

The **nuclear binding energy curve** is a plot of the binding energy per nucleon against nucleon number for known nuclides.

▼ **Study point**

We'll see later that for nuclides above $^{62}_{28}Ni$ the stability decreases with A.

(b) The stability of low-mass nuclei

The **nuclear binding energy curve** is one of the most important diagrams in nuclear physics. Section 3.6.5 will consider it in relationship to nuclear fission and fusion. For the moment we will just examine the lower slopes of the relationship illustrated in Fig. 3.6.3, on which are plotted the binding energies per nucleon of the 20 lightest stable atoms.

We notice that there is a positive correlation of BE/nuc with nucleon number (A): as we progress up the along the Periodic table the nuclei become more and more stable in this region (see Study point).

3.6.9 **Self-test**

In Fig. 3.6.3:

(a) Use the table of atomic masses to identify the nuclides marked in red.

(b) Explain why the binding energy per nucleon is zero for $A = 1$

▼ **Study point**

Notice that there are no stable nuclides with 5 or 8 nucleons. These stability gaps (the half-life of $^8_4 Be$ is only 10^{-16} s; $^5_2 He$ is ~10^{-20} s and $^5_3 Li$ ~ 10^{-24} s) exercise an important brake on the development of stars beyond the hydrogen burning phase.

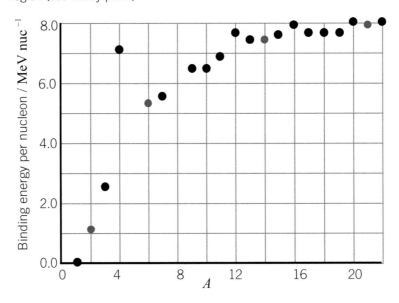

Fig. 3.6.3 Plot of stable nuclides up to $A = 22$

The next obvious feature of the plot is the spike at $A = 4$, which represents $^4_2 He$. This nuclide is particularly tightly bound. Notice also that there are smaller spikes at $A = 12$, 16 and 20: the nuclides $^{12}_6 C$, $^{16}_8 O$ and $^{20}_{10}Ne$ which are multiples of the $^4_2 He$ structure. This, together with the α decay mode of nuclei, strongly suggests that the 2p+2n combination exists within nuclei.

(c) The energetics of β⁻ decay

The nuclide $^{27}_{12}\text{Mg}$ decays by β⁻ decay to $^{27}_{13}\text{Al}$.

$$^{27}_{12}\text{Mg} \rightarrow \, ^{27}_{13}\text{Al} + \, ^{0}_{-1}\text{e} + \, ^{0}_{-1}\overline{\text{v}}_\text{e}$$

Why should this be and what is the energy release?

The clue is in the masses of the particles involved. All we have to do is work out the change in mass, Δm and convert this to energy units.

Example

Calculate the energy release in the β⁻ decay of $^{27}_{12}\text{Mg}$.

Atomic mass of $^{27}_{12}\text{Mg}$ = 26.984 35 u

Answer

Mass of Mg27 nucleus = 26.984 35 u − 12m_e

Mass of Al27 nucleus = 26.981 535 u − 13m_e

∴ Mass of products = 26.981 535 u − 13m_e + m_e = 26.981 535 − 12m_e

∴ Δm = (26.981 535 u − 12m_e) − (26.984 35 u − 12m_e) = − 0.002 815 u

∴ The energy released = 0.002 815 u × 931.5 MeV/u = 2.62 MeV

(d) The energetics of β⁺ decay

The nuclide $^{23}_{12}\text{Mg}$ (atomic mass 22.994 14 u) decays by β⁺ emission to give $^{23}_{11}\text{Na}$. How much energy is released in the decay? We can do this in the same way but it is not quite as neat!

Δm = (22.989 773 u − 11m_e + m_e) − (22.994 14 u − 12 m_e)

= −0.004 367 u + 2m_e with m_e = 0.000 549 u

= −0.003 269 u   leading to   Energy release = 3.05 MeV

However, this is not the total energy released in the decay, because the β⁺ particle is a positron, the anti-particle of the electron. This will collide with an electron and these will annihilate releasing another 2m_e of energy. So the total energy release has a mass of 0.004 367 u, i.e. 4.06 MeV.

(e) The energetics of α decay

The same principles can be applied to α decay. Because only two particles are produced (the daughter nucleus and the α particle) we can use conservation of momentum to determine the energy and hence the speed of the α particles.

Example

Using the answer of Self-test 3.6.12(b), calculate the energy of the α particles emitted in the decay of $^{180}_{74}\text{W}$.

Study point

For this part of the calculation, 3 s.f. accuracy is enough for the masses. Hence we can use the mass number as equal to the atomic mass in u.

Stretch & Challenge

Show that the energy of the α particle is given by:

$$E_a = \frac{M_{nuc}}{M_{nuc} + m_\alpha} E_{tot}$$

Use this to check the answer to the example.

Answer

Let v_α and v_n be the speeds of the α particles and nucleus.

Then total $KE = 3.86 \times 10^{-13} = \left(\frac{1}{2}4.00v_\alpha^2 + \frac{1}{2}176v_n^2\right) \times 1.66 \times 10^{-27}$

(See Study point)

Simplifying: $4.00v_\alpha^2 + 176v_n^2 = 4.65 \times 10^{14}$ [1]

Conservation of momentum: $4.00v_\alpha = 176v_n$ [2]

Eliminating v_n in [1] by substituting from [2] leads to

$$v_\alpha^2 = \frac{4.65 \times 10^{14}}{4.09} \text{ (you should check this).}$$

$$\therefore E_a = \frac{1}{2}m_a v_\alpha^2 = \frac{1}{2}4.00 \times 1.66 \times 10^{-27} \times \frac{4.65 \times 10^{14}}{4.09} = 3.77 \times 10^{-13} \text{ J}$$

Study point

To follow the electrostatics working in this section you will need to consult Section 4.2.

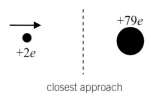

closest approach

Fig. 3.6.4 Head-on α collision with a gold nucleus

Nuc	r/fm	Nuc	r/fm
B11	2.41	Sn120	4.65
O16	2.70	Ce140	4.88
Mg24	3.06	W184	5.37
Ca40	3.48	Au197	5.43
Fe56	3.74	Pb208	5.50
Se80	4.14	U238	5.86

Table 3.6.1 Selected nuclear radii

Study point

The data for atomic nuclei obtained from high energy electron beam experiments are easier to interpret than α particle data because they (the electrons) only feel the electromagnetic force; α particles are also influenced by the strong force when they come within a few fm of the nucleons.

3.6.13 Self-test

Taking $k = 1.1$ fm, show that the density of a nucleus is approximately 3×10^{17} kg m^{-3}.

3.6.4 Physical properties of nuclei

(a) Nuclear radius

The discovery of the atomic nucleus by Rutherford, Geiger and Marsden was by the scattering of α particles by gold atoms. The number of alpha particles, N, of kinetic energy E_k, scattered from the gold nuclei was found to depend upon the angle of scattering, θ, according to:

$$N \propto \frac{1}{E_k^2 \sin^4(\theta/2)}$$

which is in agreement with the mathematical predictions for a point nucleus (i.e. one having zero size). From this, we can calculate an upper limit for the size of a gold nucleus by considering the head-on collision of an alpha particle ($Z = 2$) with a gold nucleus ($Z = 79$) (see Study point).

Assuming a 5 MeV alpha particle, the electrostatic potential energy at the closest approach is 5 MeV, i.e. 8.0×10^{-13} J.

Applying the potential energy formula[2]

$$8.0 \times 10^{-13} = \frac{2 \times 79 \times (1.6 \times 10^{-19})^2}{4\pi\varepsilon_0 d}$$

gives us 4.6×10^{-14} m. So we conclude that the radius of the gold nucleus must be less than this. Later experiments with accelerated alpha particles showed a departure from the Rutherford formula above about 25 MeV, which leads us to an estimate of 9×10^{-15} m. The most accurate measurements are made using the diffraction of high-energy electrons from nuclei and a few values are given in Table 3.6.1. Notice that the figure for gold is about half that mentioned above (see Study point).

(b) Nuclear density

Analysis of the data in Table 3.6.1 (see Exercise 3.6 Question 11) gives us the relationship $r / \text{fm} = kA^3$ for the heavier nuclides, with a value of $k \sim 1.1$ fm. This suggests that the density of nuclei is approximately constant, so we can picture the protons and neutrons in nuclei as rather like marbles in spherical bags. If the radius of the bag is multiplied by 1.5, the number of marbles is multiplied by $1.5^3 \sim 3.5$ (e.g. Fe56 and Au197).

2 Note that we are ignoring the recoil of the gold nucleus here. It makes very little difference to the outcome. You might like to check this – it'll take a page of algebra!

From this we can obtain a figure for the density of the nuclear material of about $3 \times 10^{17} \text{ kg m}^{-3}$. This is left as an exercise (Self-test 3.6.13). A pea-sized object of nuclear material would have a mass of over 30 million tonnes.

(c) Nuclear composition and stability

We cannot give a full picture of this huge topic but we can make some introductory remarks. The N-Z stability plot, Fig. 3.6.5, contains a wealth of detail which can only be revealed at a larger scale. It is worth using a search engine to find a large scale version. Here are a few of the features:

- There are no stable nuclides with more than 83 protons.

- The ratio of neutrons to protons in stable light nuclei ($Z < 20$) is very close to 1, e.g. $^{4}_{2}\text{He}$, $^{16}_{8}\text{O}$, $^{24}_{12}\text{Mg}$. For heavier stable nuclei, the ratio increases with Z. $^{209}_{83}\text{Bi}$ has 126 neutrons and 83 protons: a ratio of 1.5.

- Nuclides above the line of stability (i.e. an excess of neutrons) are β^- unstable; below the line is the β^+ instability region.

Apart from $^{1}_{1}\text{H}$, all stable nuclei contain neutrons; combinations of more than one proton are never stable. Why not? Protons are positively charged so repel one another according to Coulomb's law. All hadrons attract at short distances (less than 4–5 fm) with a force called the **nuclear force**. The strong force can only stabilise groups of nucleons if there are uncharged neutrons as well as protons.

As we saw in the Stretch & Challenge in Section 3.5.1(b), β^+ decay can occur when there is an unoccupied neutron energy level at least 1.8 MeV below the highest occupied proton energy level (electron capture decay is also possible). Similarly β^- decay arises if there is an unoccupied proton energy level of low enough energy.

Because of the Coulomb repulsion of the protons, their energy levels are further apart than those of the neutrons. For stability against β decay, the highest occupied neutron and proton energy levels must be approximately the same, so a stable nucleus needs to have more neutrons than protons.

As an example, consider the isotopes of niobium (Table 3.6.2). We conclude that the energy levels of the 41st proton and 52nd neutron are approximately the same, but the 53rd neutron is at a higher energy level and is able to decay into a proton occupying the 42nd position. Similarly, if there are only 51 neutrons, there must be a vacant neutron energy level at least 1.8 MeV below the top proton energy level allowing β^+ decay.

For nuclei with more than 83 protons, the separation between the protons on opposite sides of the nucleus is such that the nuclear force cannot overcome the coulomb repulsion, whatever the number of neutrons. Hence there are no stable nuclei with $Z > 83$.

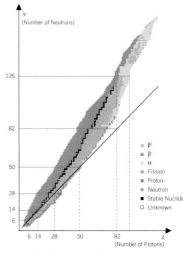

Fig. 3.6.5 N-Z stability plot

Terms & definitions

Quarks are held together in hadrons by the strong force. Outside a hadron there is a residual short-range interaction called the **nuclear force** which is equivalent to the van der Waal's force between neutral atoms.

Nuclide	decay	½ life
$^{97}_{41}\text{Nb}$	β^-	72 min
$^{96}_{41}\text{Nb}$	β^-	23 h
$^{95}_{41}\text{Nb}$	β^-	35 d
$^{94}_{41}\text{Nb}$	β^-	2.0×10^4 y
$^{93}_{41}\text{Nb}$	stable	–
$^{92}_{41}\text{Nb}$	β^+	3.5×10^6 y
$^{91}_{41}\text{Nb}$	β^+	680 y
$^{90}_{41}\text{Nb}$	β^+	14.6 h
$^{89}_{41}\text{Nb}$	β^+	2.0 h

Table 3.6.2 Decay of isotopes of niobium

▼ **Study point**

Note that binding energy per nucleon is not a continuous function of A. The nucleon number can only take integer values and so the 'curve' is actually a set of closely spaced points. There are also several different nuclides for each value of A, not all of them unstable.

3.6.5 Nuclear fission and fusion

(a) Nuclear binding energy curve

The extraction energy by splitting or merging nuclei (i.e. fission or fusion) is made possible by the form of the binding energy curve: the plot of binding energy per nucleon against nucleon number, Fig. 3.6.6. The red arrows indicates the direction of the nuclear changes in the two processes.

▼ **Study point**

Features of the binding energy curve:

- When $A = 1$, BE nuc^{-1} = 0.
- Large spike of 7 MeV at $^4_2$He.
- Spikes at multiples of $^4_2$He.
- Maximum of 8.8 MeV nuc^{-1} for A between 56 and 62.
- Uranium isotopes have a BE of 7.6 MeV nuc^{-1}.
- Fission products of uranium have a BE of ~8.5 MeV nuc^{-1}.

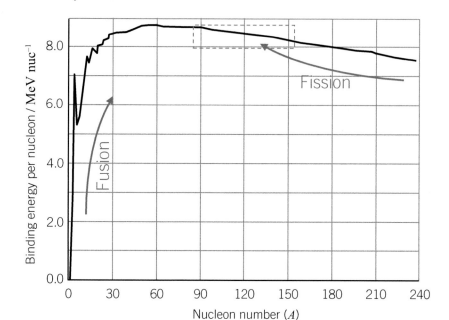

Fig. 3.6.6 Nuclear binding energy curve

We examined the low mass end of the curve in Section 3.6.3(b) and you should compare the two plots for consistency. There are many important features of the curve with which you should become familiar: the Study point gives the main ones.

3.6.14

Self-test

Show that the binding energy per nucleon of $^{235}_{92}$U is 7.6 MeV nuc^{-1} (to 2 s.f.)

Atomic mass = 235.043 930 u

Example

$^{62}_{28}$Ni has an atomic mass of 61.928 345 u. Calculate its binding energy per nucleon.

Answer

Nuclear mass of $^{62}_{28}$Ni = atomic mass − 28 × electronic mass

$$= 61.928\ 345\ u - 28 \times 0.000\ 549\ u$$

$$= 61.912\ 973\ u$$

∴ Mass deficit $= 28m_p + 34m_n - 61.912\ 973 = 0.585\ 365\ u$

∴ BE per nucleon $= \dfrac{0.585\ 365\ u \times 931.5\ \text{MeV u}^{-1}}{62\ \text{nuc}} = 8.79\ \text{MeV nuc}^{-1}$

(b) Nuclear fission

Fig. 3.6.5 shows a sprinkling of very high mass nuclides which can undergo **spontaneous fission**. The only **primordial nuclide**, which can decay in this way[3] is $^{235}_{92}U$. This nuclide decays almost 100% by α emission but in 7 out of 10^{11} of the decays of U235, the nucleus splits into two large fission fragments (the *decay products* or *daughter nuclei*) and several neutrons, e.g.

$$^{235}_{92}U \rightarrow {}^{125}_{50}Sn + {}^{107}_{42}Mo + 3{}^{1}_{0}n$$

The two daughter nuclei, Sn125 and Mo107, are both β^- radioactive. This is because they have an excess of neutrons: U235 has a neutron to proton ratio of 1.55 against 1.3–1.4 which is the stable range for stable nuclides with middle values of nucleon number.

If a U235 nucleus absorbs a neutron, e.g. from a fission event, the nucleus is likely to split in a process called **induced fission**. This is much more likely to happen with **thermal neutrons**. Fast neutrons are much less likely to induce fission. A chain reaction is possible, in which the neutrons from fission events go on to induce further fissions. This can be:

- A *controlled chain reaction* in a nuclear reactor, in which excess neutrons are absorbed by control rods so that only an average of one emitted neutron goes on to induce further fission and the neutrons are slowed down (*thermalised*) by a moderator and the energy extracted as heat.

- An *uncontrolled chain reaction* in a nuclear weapon.

The energy released in nuclear fission can be calculated by the difference in the binding energy of the U235 (7.6 MeV nuc^{-1}) and the fission products (8.5 MeV nuc^{-1}).

Example

Estimate the energy release in the fission of a nucleus of $^{235}_{92}U$.

Answer

The difference in the binding energy per nucleon = $8.5 - 7.6$ MeV nuc^{-1}

$$= 0.9 \text{ MeV nuc}^{-1} \text{ (1 s.f.)}$$

There are 235 nucleons in $^{235}_{92}U$

$\therefore$ The energy release = ~ 0.9 MeV nuc$^{-1} \times 235 = 200$ MeV (1 s.f.)

The <u>immediate</u> energy release is divided up as:

- ~83% as kinetic energy of the daughter nuclei (see Exercise 3.6 Q8).

- ~2.5% as kinetic energy of fast neutrons

- ~3.5% as gamma rays.

This energy (~89% of the total) becomes internal energy of the surrounding material (see Study point). The remaining 11% is emitted over time in the beta decay of the neutron rich daughter nuclides, of which about half is in the beta particles and half in the electron anti-neutrinos. There is a huge range of half-lives in the fission products from the milliseconds to several 10^5 years. The spent fuel rods need keeping in cooling ponds until the short lived fission products have decayed.

---

3 This is not quite true. $^{232}_{90}Th$ minerals have some evidence of spontaneous fission products.

Terms & definitions

A **primordial nuclide** is one which has existed in its current form since the Earth formed, i.e. all the stable nuclides plus very-long lived unstable nuclides, e.g. $^{235}_{92}U$, $^{238}_{92}U$ and $^{40}_{19}K$.

Spontaneous fission is the splitting of a nucleus without stimulation by an incident neutron.

Induced fission is the splitting of a nucleus caused by the absorption of a neutron.

Thermal neutrons are relatively slow-moving. They have a kinetic energy of the order of $\frac{1}{40}$eV, equivalent to kT at room temperature.

Self-test 3.6.15

Estimate the energy released from 1 kg of nuclear fuel in a fission reactor. Assume the nuclear binding energy of the fuel and fission products are 7.6 MeV nuc^{-1} and 8.5 MeV nuc^{-1} respectively. Show your working.

Hint:

(a) Estimate the number of nucleons in 1 kg of material.

(b) Calculate the difference in the binding energy of 1 kg or fuel and 1 kg of fission products.

▼ Study point

The figure of 200 MeV for the energy release per fission event is a useful figure to remember. It is also worth estimating the energy release per mol and per kg.

▼ Study point

From a thermodynamics point of view, if we include the nuclear energy in the internal energy of the fuel, then all the fission has done is to change one form of internal energy into another.

3.6.16 **Self-test**

Calculate the energy yield of the tritium production reaction in the Study point.

Atomic mass of $_1^3\text{H}$ = 3.016 049 u

3.6.17 **Self-test**

The figure of 3 fm was obtained from the sum of the radii of the deuterium and tritium nuclei. Show that this figure agrees with the kA^3 formula.

> **▼ Study point**
>
> The Sun gets by with a core temperature of about 13 million K. However, there are rather a large number of hydrogen nuclei there, so even with a weak reaction the power output is 6×10^{24} W!

(c) Nuclear fusion

The binding energy curve – Fig. 3.6.6 – shows us that, if we combine low mass nuclei the resulting nucleus is more tightly bound, so we obtain an energy output. Actually, the local peak of $_2^4\text{He}$ allows for energy to be extracted from above this point too (see the **Li6** reaction in the Study point). The simplest possible reaction is $_1^1\text{H} + _1^1\text{H} \rightarrow _1^2\text{H} + _1^0\text{e}^+ + \nu_e$.

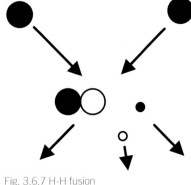

Fig. 3.6.7 H-H fusion

This reaction has several drawbacks: it is relatively low yield (2 MeV including the subsequent e^+e^- annihilation), energy is lost in the neutrino and it is a weak interaction (and therefore unlikely to happen). This last point is the most serious. This reaction is the base of the proton-proton chain in the Sun, where an individual proton in the core has a lifetime of about 10^9 years before making a successful collision!

A more promising reaction is:

$$_1^2\text{H} + _1^3\text{H} \rightarrow _2^4\text{He} + _0^1\text{n},$$

where $_1^3\text{H}$ is tritium, consisting of a proton and two neutrons. Approximately 1.5 atoms in 10^4 of hydrogen is deuterium (a primordial nuclide, dating from the big bang), so it is present in large amounts in sea-water. Tritium needs to be made artificially, which is easily done in a fission reactor (see Study point)

The reaction is controlled by the strong interaction and so, as long as we can bring the reactants close enough together (~3 fm), they should fuse. The problem is that they are both positively charged so we have to overcome the Coulomb repulsion. This means raising the reactants to a high temperature. We can estimate the temperature needed as follows:

Assume the nuclei need to be brought to 3 fm. The potential energy is:

$$E_p = \frac{1}{4\pi\varepsilon_0}\frac{Q_1Q_2}{r} = 9 \times 10^9 \frac{(1.6 \times 10^{-19})^2}{3 \times 10^{-15}} = 8 \times 10^{-14} \text{ J } [0.5 \text{ MeV}].$$

In order to achieve this, the thermal energy of the reactants each needs to be half this. So, using $E_k = \frac{3}{2}kT$, for the mean kinetic energy to be this high needs:

$$T = \frac{2}{3}\frac{E_k}{k} = \frac{2}{3}\frac{4 \times 10^{-14} \text{ J}}{1.38 \times 10^{-23} \text{ J K}^{-1}} \sim 2 \times 10^9 \text{ K}$$

Actually we don't need 2000 million K because we don't need the <u>mean</u> kinetic to be this high: we just need a small but significant fraction to have this. Engineers have achieved temperatures of 10^8 K and observed fusion taking place; however, they have not yet maintained the required density for long enough for the fusion to be self sustaining.

Nuclide	mass / u	Nuclide	mass / u	Nuclide	mass / u
1H2	2.014102	9F19	18.9984	16S34	33.96786
2He3	3.01603	10Ne20	19.99244	16S36	35.96709
2He4	4.002604	10Ne21	20.993849	17Cl35	34.96885
3Li6	6.015126	10Ne22	21.991384	17Cl37	36.9659
3Li7	7.016005	11Na23	22.989773	18Ar36	35.96755
4Be9	9.012186	12Mg24	23.985045	18Ar38	37.96572
5B10	10.012939	12Mg25	24.98584	18Ar40	39.962384
5B11	11.009305	12Mg26	25.982591	19K39	38.96371
6C12	12.000000	13Al27	26.981535	19K41	40.96183
6C13	13.003354	14Si28	27.97693	20Ca40	39.96259
7N14	14.003074	14Si29	28.97649	20Ca42	41.95863
7N15	15.000108	14Si30	29.97376	20Ca43	42.95878
8O16	15.994915	15P31	30.973763	20Ca44	43.95549
8O17	16.999133	16S32	31.972074	20Ca46	45.95369
8O18	17.99916	16S33	32.97146	20Ca48	47.95236

Table 3.6.3 The atomic masses of the stable nuclides for $Z = 1$–20.

▼ **Study point**

As well as calculating the binding energy you can use this table to search for interesting patterns, e.g. the number of stable nuclei with odd and even values of Z, the number of nuclei with odd and even values of Z and N. Notice that the least nucleon number with two different stable nuclides is 36 – $_{16}^{36}$S and $_{18}^{36}$Ar.

Note that 4Be9 means $_{4}^{9}$Be.

Exercise 3.6

1   A coal-fired power station, running continuously during the winter months, generates electricity with an efficiency of 35%. Its power output is $2.0\ \text{GW}$. Calculate:

(a) The thermal energy by burning the fuel in a two-month (60-day) period.

(b) The mass difference between the reactants (coal + oxygen) and the products of the burning (carbon dioxide and water vapour).

2   The Sun loses 4 million tonnes (1 tonne = $10^3\ \text{kg}$) per second in its radiation. Calculate:

(a) The power output of the Sun.

(b) The power incident on the Earth.

     [Radius of Earth's orbit = $150\ \text{million km}$; radius of Earth = $6400\ \text{km}$.]

3   Some nuclei decay by electron capture in which an inner electron (1s orbital) is absorbed. $^{7}_{4}\text{Be}$ (atomic mass $7.016\ 930\ \text{u}$) decays by electron capture into a nuclide of mass $7.016\ 004\ \text{u}$.

(a) Write the equation for the decay.

(b) Explain which of the nuclear interactions (strong, e-m, weak) is responsible for the decay.

(c) Calculate the energy released in the decay and explain which particle carries this energy.

4   The nuclide $^{241}_{95}\text{Am}$ (atomic mass $241.056\ 69\ \text{u}$) decays into a nuclide of mass $237.048\ 173\ \text{u}$, with a half-life of 460 years.

(a) Write the decay equation.

(b) Calculate the energy released in the decay.

(c) Use the principle of conservation of momentum to calculate the recoil speed of the nucleus.

(d) A smoke detector contains $1\ \text{mg}$ of $^{241}_{95}\text{Am}$. Calculate its activity.

5   The primordial nuclide $^{40}_{19}\text{K}$ (atomic mass $39.963\ 998\ \text{u}$) has three modes of decay: β^- (89.28%), electron capture (10.72%) and β^+ decay (0.001%), with a half-life of 1.3×10^{-9} years.

(a) Identify the daughter nuclides in the three modes of the decay and write the decay equations.

(b) Calculate the energy released in each decay mode.

6   The nuclear fuel in uranium fission reactors contains a large proportion of the non-fissile U238. A U238 absorbs a neutron and the resulting nucleus decays in two stages into the fissile isotope of plutonium, $^{239}_{94}\text{Pu}$.

$$^{238}_{92}\text{U} + {}^{1}_{0}\text{n} \rightarrow {}^{239}_{92}\text{U} \xrightarrow{\beta^-} {}^{239}_{93}\text{Np} \xrightarrow{\beta^-} {}^{239}_{94}\text{Pu}$$

(a) Write decay equations for the decay of U239 and Np239.

(b) Use the atomic masses of the nuclides to calculate the energy released in the two β^- reactions.

(c) Explain briefly why only a fraction of this energy release is absorbed within the reactor.

     Masses: $^{239}_{92}\text{U} = 239.054\ 293\ \text{u}$; $^{239}_{93}\text{Np} = 239.052\ 939\ \text{u}$; $^{239}_{94}\text{Pu} = 239.052\ 163\ \text{u}$

7 After about 1 billion years on the red giant branch, the helium-rich core in a solar-mass star achieves a high enough temperature and pressure to start 'burning' helium in the triple-alpha reaction:

$$_2^4He + _2^4He + _2^4He \rightarrow _6^{12}C$$ Calculate the energy released in this reaction.

8 One of the possible fission reactions in a U-235 reactor is:

$$_{92}^{235}U + _0^1n \rightarrow _{55}^{140}Cs + _{37}^{92}Rb + 4_0^1n$$

Rb-92 decays by successive β^- emissions: Rb → Sr → Y → Zr (stable)

Cs-140 decays by successive β^- emission: Cs → Ba → La → Ce (stable)

Nuclide	atomic mass / u
$_{37}^{92}Rb$	91.919 730
^{92}Zr	91.905 041
$_{55}^{140}Cs$	139.917 282
^{140}Ce	139.905 439
^{235}U	235.043 9

Use the table of atomic masses to calculate:

(a) The energy given off in the initial fission.

(b) The energy given off in the subsequent β^- decays.

(c) The energy given off by the fission of 1 kg of U235.

9 The proton-proton fusion chain in the Sun starts with the following two steps:

$$_1^1H + _1^1H \rightarrow _1^2H + _1^0e^+ + \nu_e \quad [1]$$

$$_1^1H + _1^2H \rightarrow _2^3He + \gamma \quad\quad [2]$$

69% of the subsequent energy release is from the following reaction:

$$_2^3He + _2^3He \rightarrow _2^4He + 2_1^1H \quad [3]$$

Calculate:

(a) The energy in MeV release in each step.

(b) The total energy released in the fusion of 1 kg of hydrogen [include the e⁺e⁻ annihilation].

10 Estimate the energy released in the fission fragments of a nuclear fission by considering the change in their electric potential energy as they separate.

Consider the tin and molybdenum nuclei as they are about to separate in the fission event in Section 3.6.5 (b):

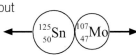

Calculate:

(a) The charge on each nucleus.

(b) The radius of each nucleus (Hint: see Section 3.6.4).

(c) The potential energy of the two charges and hence their combined kinetic energies as they separate. [They are spherically symmetric so you can treat them as point charges, separated by the sum of their radii.]
You should obtain an answer which is (at least to a fair approximation) in agreement with the calculation in 3.6.5(b).

(d) [Stretch & Challenge] The final speed that each of the fragments achieves.

11 Use Table 3.6.2 to test the relationship $r = kA^{\frac{1}{3}}$ and determine a value for k. Is it advantageous to draw a $\ln r - \ln A$ graph or one of r against $A^{\frac{1}{3}}$?

Overview:
Unit 4 Fields and options

4.1 Capacitance `p98`

- The definition of capacitance, $C = \dfrac{Q}{V}$.
- The structure and capacitance of parallel plate capacitors, $C = \dfrac{\varepsilon_0 A}{d}$; the effect of dielectrics.
- How capacitors store energy, $U = \dfrac{1}{2}QV$.
- Electric field (E field) in a parallel plate capacitor, $E = \dfrac{V}{d}$.
- The capacitance of capacitors in series and parallel.
- The processes and equations for charging and discharging capacitors; the time constant, RC.

PRACTICAL WORK
- Investigating the charging and discharging of a capacitor to determine the time constant.
- Investigation of the energy stored in a capacitor.

4.2 Electrostatic and gravitational fields of force `p115`

- Definitions and equations for field strength and potentials in electrostatic and gravitational fields.
- The gravitational field due to a spherically symmetric body.
- Electric and gravitational field lines.
- Electric and gravitational equipotential surfaces.
- Combining fields due to a number of point masses or charges.
- The conditions under which the equation $\Delta U_p = mg\Delta h$ is valid.

4.3 Orbits and the wider universe `p131`

- Newton's law of gravitation and its application to circular orbits.
- Kepler's laws of planetary motion; the derivation of K3 for circular orbits.
- Determination of the mass of a central body from the paths of orbiting objects; evidence for dark matter and the hypothesised role of the Higgs boson.
- The centre of mass and mutual orbit of binary objects.
- The application of the Doppler relationship to determine a star's radial velocity; its use in determining the masses of orbiting objects.
- The Hubble constant, its relationship to the age of the universe; the critical density of a flat universe.

4.4 The magnetic field `p149`

- The forces on current-carrying conductors and moving charges in a magnetic field.
- How a Hall voltage is produced and its use in measuring B fields.
- B fields due to long straight wires and long solenoids; the effect of an iron core in a solenoid.
- The forces between current-carrying conductors.
- The motion of charged particles in uniform electric and magnetic fields; linear accelerators, cyclotrons and synchrotrons.

PRACTICAL WORK
- Investigation of the force on a current in a magnetic field.
- Investigation of magnetic flux density using a Hall probe.

4.5 Electromagnetic induction `p168`

- The definitions of magnetic flux and flux linkage.
- Faraday's and Lenz's laws of electromagnetic induction and their application.
- The generation of an emf in a linear conductor moving across a magnetic field.
- The generation of an emf by a rotating coil in a magnetic field (qualitative).

Unit 4   Fields and options

This unit has a single main theme of fields and a set of four options, of which you should study one.

The study of fields of force is one of the basic pillars of physics. It is used to describe influences which extend through space, so that the effect takes place remotely from the cause. Three types of field are studied:

- **Gravitational field**. All material particles exert forces on one another which are proportional to their masses. These interactions are described in terms of a gravitational field which surrounds each particle. Historically this was the first field to be described in Newton's theory of gravity.

- **Electrostatic field**. All charged particles also exert forces on one another, in fields which are very similar to gravitational fields – they both decrease as the inverse square of the distance. The electric field is the origin of the adhesive forces between atoms and molecules.

- **Magnetic field**. This is a more complicated field than the other two, being caused by charged particles in motion. Its properties are useful in electrical motors and generators.

In addition to these basic areas of study, there are sections which focus on applications of the fields:

- **Capacitors** are devices for storing electrical energy and operate on the basis of the electrostatic fields between metal plates.

- **Orbits and the wider universe** describes how the motion of astronomical objects in combination with the knowledge of gravitational fields allows us to gain insights into the structure of the universe

Content

4.1 **Capacitance**

4.2 **Electrostatic and gravitational fields of force**

4.3 **Orbits and the wider universe**

4.4 **The magnetic field**

4.5 **Electromagnetic induction**

Practical work

Practical work is integral to any physics course. Unit 4 provides a wealth of opportunities for students to hone their practical skills as well as to develop their understanding of the contents.

4.1 Capacitance

Terms & definitions

A capacitor consists of two conductors (so-called 'plates') separated by an insulator known as a **dielectric**.

4.1.1 Self-test

Why are *two* insulating strips needed in the 'rolled up' capacitor of Fig. 4.1.2?

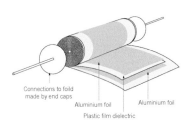

Connections to foild made by end caps
Aluminium foil   Aluminium foil
Plastic film dielectric

Fig. 4.1.2 One type of capacitor

▼ Study point

All the cylindrical capacitors in Fig. 4.1.1 are electrolytic except the silvery one.

Supercapacitors (usually of capacitance more than 1 F) have dielectrics that are only about two atoms thick. There's more than one type and the technology involved is as much chemical as physical.

Capacitors are simple devices used widely in electronics. We learn about their behaviour. Then we introduce the fundamental idea of an *electric field*, and explain how a uniform electric field can be produced in a capacitor. We finish by studying the motion of a charged particle in a uniform electric field.

4.1.1 Capacitors and capacitance

What is a capacitor? This is easily stated – see **Terms and definitions**. What does a capacitor do? It stores equal and opposite charges on its plates when there is a pd between the plates.

In practice almost all capacitors made are of the parallel plate type, with an insulating material of constant thickness (often less than 0.1 mm) between the plates. The circuit symbol for a capacitor is, appropriately, as shown in Fig. 4.1.3 (without '+' and '-' signs).

Capacitors come in all shapes and sizes (Fig. 4.1.1). A cylindrical case is likely to contain two strips of metal foil (the plates) and two strips of thin plastic (the dielectric) rolled up like a Swiss roll (Fig. 4.1.2). See also the **Study point**. Capacitors are usually marked with the maximum pd that can be applied safely.

The capacitors with exposed plates, to the right in Fig. 4.1.1, are 'variable'. One set of plates can be moved to change the area of overlap between them and another set. The dielectric is air.

Capacitors with a capacitance of more than about $1\mu F$ (see Section 4.1.2) and less than about 0.1 F are usually *electrolytic. T*he dielectric is a thin film of oxide on one plate. The capacitor must be connected the right way round in a circuit, so that the oxide-coated plate is at a positive potential relative to the other plate, or the film is destroyed.

Fig. 4.1.1 Capacitor smörgåsbord

Capacitance

If a battery or dc power supply is connected across a capacitor, as in Fig. 4.1.3(a), some free electrons are removed from one plate and transferred via the battery to the other. The plates gain equal and opposite charges (on facing surfaces). This 'charging' stops (Fig. 4.1.3(b)) when the pd between the plates is equal to the emf of the battery. It usually takes a small fraction of a second, unless a resistor is included in series or the battery has a high internal resistance.

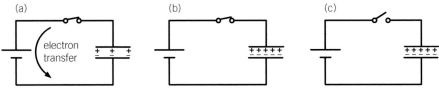

Fig. 4.1.3 Charging and isolating a capacitor

If we then disconnect the battery, as in Fig. 4.1.3(c), the charges are isolated on the plates, maintaining the pd between the plates.

We can investigate how the charge ($\pm Q$) on the capacitor plates is related to the pd, V, between them using the apparatus in Fig. 4.1.4(a). We charge the capacitor by switching the two-way switch to the left briefly, and then to the right, so the capacitor discharges into the coulomb meter. We repeat for different pds, selected using the variable dc power supply. Fig. 4.1.4(b) is a typical graph of the results.

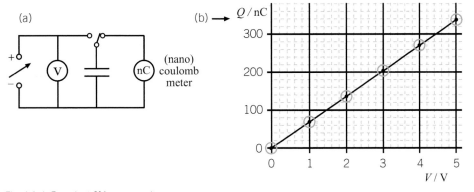

Fig. 4.1.4 Q against V for a capacitor

The charge stored on either plate is proportional to the pd, V. This is the case for any normal capacitor. So we can write

$$Q = CV$$

where C is a constant for the capacitor, called its **capacitance**, and is the gradient of the graph of Q against V.

Capacitors are made with a huge range of values, from a few picofarads (pF), used mainly in circuits involving high frequency alternating currents, up to hundreds of farads, used for backing up computer memories' power supplies or helping batteries to cope with spikes of demand for current. Time to brush up your prefixes for powers of ten!

▼ **Study point**

Don't confuse C (italic type) standing for the quantity, capacitance, with C (upright type) standing for the unit of charge, coulomb.

Self-test 4.1.2

Calculate the capacitance of the capacitor to which Fig. 4.1.4 applies. Give your answer in both nF and μF.

Self-test 4.1.3

Determine the pd needed to store a charge of 4.8 μC on a 160 μF capacitor. What should be checked before applying the pd?

▼ **Study point**

It's not surprising to find that the capacitance is proportional to the plate area, A, since it's the plates that have the charge. Something will be said about the inverse relationship between C and d in Section 4.1.6 (**Study point**).

▼ **Study point**

The formula for a parallel plate capacitor with dielectric is
$$C = \frac{\varepsilon_0 \varepsilon_r A}{d}$$
in which ε_r is a factor (with no units) taking account of the dielectric.

For a vacuum, $\varepsilon_r = 1$ (exactly), by definition. For air $\varepsilon_r = 1.00059$ (that is 1.00 to 3 s.f.!). For polypropene, $\varepsilon_r = 2.3$. For certain ceramics, ε_r is over 1000.

▼ **Study point**

A dielectric increases capacitance because of an effect called 'dielectric polarisation': a temporary shift in the positions of nuclei and electrons within the dielectric's molecules, due to forces from the *electric field* (Section 4.1.5) set up by charges on the plates.

4.1.2 The capacitance of a parallel plate capacitor

How do we design a capacitor to have a particular capacitance? There is a neat formula, testable by experiment, for the capacitance of a parallel plate capacitor (Fig. 4.1.5) with a vacuum or air between its plates (as long as their separation, d, is much less than their side length, $\sqrt{A}$). It is

$$C = \frac{\varepsilon_0 A}{d},$$

where ε_0 is a constant with a rather scary 19th-century name: *the permittivity of free space*. Most of us call it 'epsilon nought'. By experiment,

$$\varepsilon_0 = 8.85 \times 10^{-12} \text{ F m}^{-1}$$

Note that A is the area of a single surface of either plate.

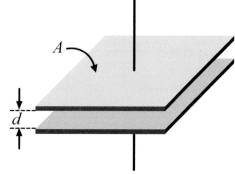
Fig. 4.1.5 Parallel plate capacitor

The dielectric

Most capacitors used in electronics have (solid) dielectrics, such as films of tantalum(V) oxide, thin sheets of polypropene or thin slices of ceramic material. A dielectric does more than maintain the insulating gap between the plates: it makes the capacitance more than it would be if the gap had nothing (or just air) inside it. In other words it makes the capacitance larger than $\frac{\varepsilon_0 A}{d}$; often by many times. See **Study points**.

4.1.3 Capacitor combinations

Combinations of capacitors behave electrically like single capacitors. This can sometimes be useful when a certain value of capacitance is needed, but no single capacitor of this value is available.

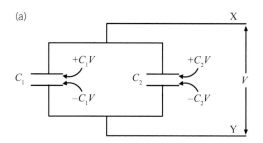 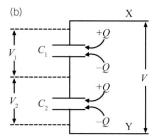

Fig. 4.1.6 (a) Capacitors in parallel (b) Capacitors in series

We'll connect capacitors together in parallel and series, as in Fig. 4.1.6(a) and (b) and find a formula for the equivalent capacitance in each case, just as we did for resistors in series and parallel in the AS/Year 12 textbook.

(a) Capacitors in parallel

From Fig. 4.1.6(a), each capacitor has the same pd, V, across it. The charges on the plates of C_1 and C_2 are therefore $\pm\, C_1 V$ and $\pm\, C_2 V$. So the charge that flowed through wires X and Y is

$$Q = C_1 V + C_2 V \qquad\qquad \text{i.e. } Q = (C_1 + C_2)V$$

So if we wanted to replace the parallel combination with a single equivalent capacitance, C, we'd choose:

$$C = \frac{Q}{V} = \frac{(C_1 + C_2)V}{V}, \qquad\qquad \text{i.e. } C = C_1 + C_2$$

Even if there are more than two capacitors in parallel we simply add individual capacitances. See **Equations** in the margin.

Notice that the formula for **capacitors in parallel** is of the same form as the formula for **resistors in series**.

A useful consistency check is to consider two parallel-plate air-spaced capacitors, with the same value of d, connected in parallel.

Then $\qquad C = C_1 + C_2$

So $\qquad C = \dfrac{\varepsilon_0 A_1}{d} + \dfrac{\varepsilon_0 A_2}{d} = \dfrac{\varepsilon_0(A_1 + A_2)}{d} = \varepsilon_0(A_1 + A_2)d$

So the **capacitors** behave like a single capacitor with plate area $(A_1 + A_2)$, as one would expect.

(b) Capacitors in series

When a pd, V, is applied as shown in Fig. 4.1.6(b), each capacitor's plates gain equal and opposite charges. These charges are the same for both capacitors, because the 'island' plates (the plates connected together) are insulated from everything else and must have a *total* charge of zero – assuming that neither capacitor had any charge before connection.

Because pds in series add,

$$V = V_1 + V_2 = \frac{Q}{C_1} + \frac{Q}{C_2} \qquad \text{so} \qquad \frac{V}{Q} = \frac{1}{C_1} + \frac{1}{C_2}$$

Since the charge that flowed through the wires X and Y is Q, the equivalent capacitance of the combination is $C = \dfrac{Q}{V}$.

Therefore $\qquad \dfrac{1}{C} = \dfrac{1}{C_1} + \dfrac{1}{C_2}$

If there are more than two capacitors we simply go on adding reciprocals.

Again, we notice that the formula for **capacitors in series** is of the same form as the formula for **resistors in parallel**.

Self-test 4.1.5

Express the farad in terms of the base SI units, m, kg, s and A.

EQUATIONS

For capacitors **in parallel**

$$C = C_1 + C_2 + \ldots$$

For capacitors **in series**

$$\frac{1}{C} = \frac{1}{C_1} + \frac{1}{C_2} + \ldots$$

Stretch & Challenge

Perform a consistency check, modelled on the one above, by considering two parallel plate air-spaced capacitors, with the same plate area, A, in series.

Self-test 4.1.6

Calculate the equivalent capacitance of a 22 µF and a 33 µF capacitor connected (a) in parallel and (b) in series.

▼ **Study point**

For two capacitors in parallel a little algebra is a worthwhile investment…

$$\frac{1}{C} = \frac{1}{C_1} + \frac{1}{C_2} = \frac{C_2}{C_1 C_2} + \frac{C_1}{C_2 C_1}$$

that is $\dfrac{1}{C} = \dfrac{C_1 + C_2}{C_1 C_2}$

We now have a single fraction on each side of the equation. The reciprocals of these fractions are also equal, so

$$C = \frac{C_1 C_2}{C_1 + C_2} = \frac{\text{product}}{\text{sum}}$$

WARNING! This doesn't work for more than two capacitors at a time.

4.1.7 ▼ Self-test

Suppose that, in Fig. 4.1.7(c), 10.0 V is applied between X and Y.

Calculate the charge and pd for (a) the 22 nF and (b) the 47 nF capacitor.

▼ **Stretch & Challenge**

Draw a circuit containing two 100 µF capacitors and one 47 µF capacitor whose combined capacitance is 60 µF (to 2 s.f.).

Show that your combination *does* have this capacitance.

▼ **Study point**

There's no simple, wholly satisfactory way of treating significant figures in a case like this, so we retained 3 and rounded to 2 at the end, as data was only to 2 s.f. Spot the apparent inconsistencies in the final answers!

(c) Capacitor combination calculations

As examples we find the equivalent capacitance for (a) and (b) in Fig. 4.1.7,

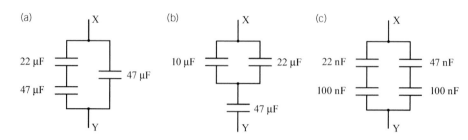

Fig. 4.1.7 Three combinations of capacitors

The equivalent capacitance in Fig. 4.1.7(a)

We first deal with the left-hand side: 22 µF and 47 µF in series. As for two resistors in parallel, so also for two capacitors in series, there's a simpler and less accident-prone version of the 'reciprocal of sum of reciprocals' formula. It's derived in the **Study point**. We'll use it now…

$$C = \frac{\text{product}}{\text{sum}} = \frac{22 \text{ µF} \times 47 \text{ µF}}{69 \text{ µF}} = 15 \text{ µF}$$

[Make sure you understand how the units work out. Write 22 µF as 22×10^{-6} F and so on if this helps.]

The full combination is therefore 15 µF in parallel with 47 µF.

So, finally, $C = 15 \text{ µF} + 47 \text{ µF} = 62 \text{ µF}$

The equivalent capacitance in Fig. 4.1.7(b)

For the capacitors in parallel, the combined capacitance is 10 µF + 22 µF = 32 µF. But this is in series with 47 µF.

So overall capacitance , $C = \dfrac{\text{product}}{\text{sum}} = \dfrac{32 \text{ µF} \times 47 \text{ µF}}{79 \text{ µF}} = 19 \text{ µF}$

Charges and pds for each capacitor in Fig. 4.1.7(b)

Suppose a pd of 12.0 V is applied between X and Y. If required, we can work out charges and pds for all the capacitors, as we now show…

For the combination, $Q = CV = 19 \text{ µF} \times 12.0 \text{ V} = 228 \text{ µC}$.

Since the combination behaves as a 32 µF capacitor in series with a 47 µF capacitor, the charges on these are both ±228 µC.

The pd across the 47 µF is therefore:

$$V = \frac{Q}{C} = \frac{228 \text{ µC}}{47 \text{ µF}} = 4.85 \text{ V}.$$

This leaves 12.0 V – 4.85 V = 7.15 V across the 32 µF combination.

So charge on 10 µF capacitor = $CV = 10 \text{ µF} \times 7.15 \text{ V} = 71.5 \text{ µC}$, and charge on 22 µF capacitor = $CV = 22 \text{ µF} \times 7.15 \text{ V} = 157 \text{ µC}$.

Conclusion (see Study point):
10 µF: 7.2V, 72 µC; 22 µF: 7.2V, 160 µC; 47 µF: 4.9 V, 230 µC.

4.1.4 Energy stored in a charged capacitor

One way of demonstrating that a charged capacitor stores energy is to use the energy to produce light (and infrared!), as in Fig. 4.1.8. After charging the capacitor (switch to the left), we can discharge it through the lamp (switch to the right).

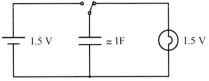

Fig. 4.1.8 Demonstrating energy stored in a charged capacitor

It's easy to work out *how much* energy is stored. We consider a capacitor discharging through any conducting device (Fig. 4.1.9(a)). We've shown a resistor, but what follows doesn't require Ohm's law to apply.

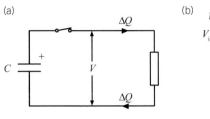
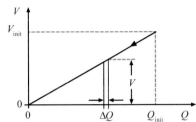

Fig. 4.1.9 Capacitor discharging through a resistor

Electrons flow off the negative plate through the resistor to the positive plate. As the charge, Q, on the plates decreases, the pd, V, between them falls, as in Fig. 4.1.9(b). This is simply a consequence of $V = Q/C$.

The energy, ΔU, transferred from the capacitor and dissipated in the resistor when a small portion, ΔQ, of charge, flows is

$$\Delta U = V\Delta Q$$

in which V is the (mean) pd between the capacitor plates *while that small portion of charge is flowing*. [Remember that pd, V, means energy converted from electrical PE per unit charge passing.]

$V\Delta Q$ is the 'area' of the narrow vertical strip drawn under the graph, and the total energy released by the capacitor is the sum of the areas of these strips, that is the area under the whole graph line, which is $\frac{1}{2} Q_{init} V_{init}$ (that is $\frac{1}{2}$ base × height for the triangle, but with units of $C \times V = J$).

Now that we've taken account of the changing pd and charge, we can drop the 'init' subscripts, and write the energy stored simply as

$$U = \tfrac{1}{2}QV.$$

Using the defining equation for capacitance ($C = \dfrac{Q}{V}$) we can express U in terms of a single variable, either Q or V. See **Equations** in the margin.

Self-test 4.1.8

The graph shows how the energy stored in a capacitor depends on the pd.

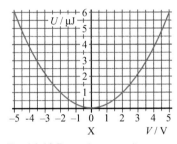

Fig. 4.1.10 Energy in a capacitor

(a) Describe the shape of the curve.

(b) Determine the capacitance.

(c) Explain briefly whether or not the capacitor is likely to be an electrolytic capacitor.

EQUATIONS

The energy stored in a capacitor of capacitance C when there is a pd V between its plates and a charge $\pm Q$ on them is

$$U = \tfrac{1}{2}QV = \tfrac{1}{2}CV^2 = \tfrac{1}{2}\frac{Q^2}{C}$$

These are all equivalent. Use whichever suits the case in hand.

4.1.9

Self-test

In a prototype torch, a 'supercapacitor', of 60 F, charged to a pd of 2.4 V, is discharged through an LED (with a current-limiting resistor in series) until the pd has fallen to 1.8 V, when the light is deemed to be too dim. If the mean power supplied is 0.30 W, calculate the length of time the LED was on.

EQUATION FOR CAPACITOR DISCHARGING

At time t the charge, $\pm Q$, on the plates of a capacitor is given by

$$Q = Q_0 e^{-\frac{t}{RC}}.$$

Q_0 is the charge at time $t = 0$, C is the capacitance and R is the resistance through which it discharges.

┌─ **Terms & definitions** ─┐

The **time constant**, τ, of an RC system undergoing exponential decay is the time taken for the charge to fall to $\frac{1}{e}$ $(= 0.37)$ of its original charge. $\tau = RC$

Stretch & Challenge

If you know how to differentiate exponentials, and how to apply the *chain rule*, show that...

if $Q = Q_0 e^{-\frac{t}{RC}}$, then $\frac{dQ}{dt} = -\frac{Q}{RC}$.

▼ Study points

- In the equation $Q = Q_0 e^{-\frac{t}{CR}}$, the index, $-\frac{t}{RC}$ is a negative number without units, because the unit of RC, like the unit of t, is s.

- e is the natural number 2.718... .

- The numerical value of $e^{-\frac{t}{CR}}$ is found using the e^x (inverse $\ln x$) facility on a scientific calculator.

  For example, if $R = 680$ kΩ, $C = 4.7$ μF, $t = 9.6$ s, then $-\frac{t}{RC}$
  $$= -\frac{9.6 \text{ s}}{4.7 \times 10^{-6} \text{F} \times 6.8 \times 10^5 \Omega}$$
  $$= -3.00$$

  giving $e^{-\frac{t}{RC}} = 0.0496$

- Q_0 is the initial value of Q. We can confirm this by remembering that anything raised to the power of zero is 1, so when $t = 0$,

  $$Q = Q_0 e^{-\frac{t}{RC}} = Q_0 \times 1 = Q_0.$$

4.1.5 Charging and discharging a capacitor through a resistor

We shall discover a characteristic behaviour – one that is shared by several systems, both mechanical and electrical. We'll study discharging first. This might seem back-to-front, but the discharging case is simpler.

(a) Discharging

We'll suppose that the capacitor has been charged in advance and that, at time $t = 0$, the resistor is connected across it (by closing the switch in Fig. 4.1.11). At any time after that, we know that

pd across resistor = pd across capacitor

Therefore $IR = \dfrac{Q}{C}$.

We shall rely on the resistor obeying Ohm's law, so that R (like C) is a constant. Now for the key step. The current I is the rate of flow of charge, in other words (see Section 4.1.1) the rate at which the negative plate loses negative charge and the positive plate loses positive charge.

So $I = -\dfrac{\Delta Q}{\Delta t}$.

Putting this into the previous equation and dividing both sides by $-R$:

$$\frac{\Delta Q}{\Delta t} = -\frac{Q}{RC}.$$

(a)

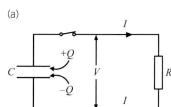

Fig. 4.1.11 CR discharge

Two remarks about this neat equation:

- The top lines on each side have the same unit (**C**). Therefore the bottom lines must have the same unit as each other, implying that the unit of RC is **s**. Can this possibly be right? Checking directly...

  Units of $RC = \Omega\,\text{F} = \text{V A}^{-1} \times \text{C V}^{-1} = \text{A}^{-1}\,\text{C} = \text{A}^{-1} \times \text{A s} = \text{s}.$

  RC is called the **time constant**, τ (Greek letter tau), of the circuit.

- The rate of decrease of charge is proportional to the charge itself. When the charge has halved, the rate of decrease will have halved.

Given that $Q = Q_0$ when $t = 0$, the special relationship between $\dfrac{\Delta Q}{\Delta t}$ and Q is possible only if Q varies with t in one particular way. It is this...

$$Q = Q_0 e^{-\frac{t}{RC}}.$$

This is called an **exponential decay** equation. The **Study points** explain how to interpret it.

The significance of the time constant, τ

As RC has the units of time, it makes sense to ask: what value does Q have when $t = RC$, that is when the capacitor has been discharging for a time RC? We find

$$Q = Q_0 e^{-\frac{t}{RC}} = Q_0 e^{-\frac{RC}{RC}} = Q_0 e^{-1} = 0.368\, Q_0.$$

In other words the capacitor has 37% of its initial charge.

The time constant, then, is a measure of how quickly (or slowly) the charge decays. [For radioactive decay, *half-life* serves a similar purpose.]

Determining t knowing Q, Q_0, R and C

The exponential decay equation can be presented as

$Q = Q_0 e^{-\frac{t}{RC}}$ or, in inverse form, as $-\dfrac{t}{RC} = \ln\left(\dfrac{Q}{Q_0}\right)$

$\ln\left(\dfrac{Q}{Q_0}\right)$ is short for 'the logarithm to the base e of $\left(\dfrac{Q}{Q_0}\right)$', which simply means the power to which e has to be raised to give $\ln\dfrac{Q}{Q_0}$.

As an example we determine the time it takes for Q to halve…

$-\dfrac{t}{RC} = \ln\left(\dfrac{Q}{Q_0}\right) = \ln\left(\dfrac{1}{2}\right) = -\ln 2 = -0.693$ (see **Study point**).

So $t = (\ln 2)\,\tau$ that is $t = 0.693\,RC$

The exponential decay curve

Fig. 4.1.12 is a typical graph of Q against t for a capacitor discharging through a resistor. (Ignore the tangent line for the time being.)

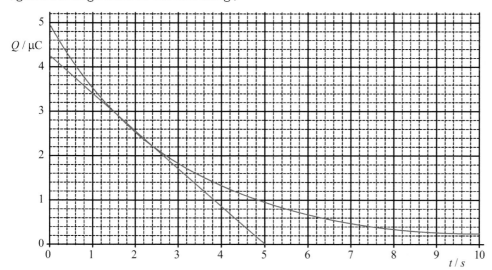

Fig. 4.1.12 Exponential decay curve for a CR circuit

The feature of the curve that confirms the decay as **exponential** is that, in equal time intervals, Q always decays to the same fraction of its value at the start of the interval. An important example is a time interval of τ, for which, as we've seen, the fraction is 0.368…

At $t = 0$, $Q = 5.00\ \mu C$, so, at time $t = \tau$, $Q = 0.368 \times 5.00\ \mu C = 1.84\ \mu C$

The graph shows this to be at a time of 3.0 s. So $\tau = 3.0$ s.

Over an interval of **3.0 s**, *starting at any time*, we find that the charge at the end of the interval is 37% of the charge at the start. Do check this.

Variation of pd and current with time

As the charge, Q, decays exponentially with time, so also do the pd, V, and the current, I, (see Fig. 4.1.13). In detail…

Putting $Q = CV$ and $Q_0 = CV_0$ into $Q = Q_0 e^{-\frac{t}{RC}}$ and dividing by C,
$V = V_0 e^{-\frac{t}{RC}}$ in which V_0 is the pd at time $t = 0$.
Putting $V = IR$ and $V_0 = I_0 R$ into $V = V_0 e^{-\frac{t}{RC}}$ and dividing by R,
$I = I_0 e^{-\frac{t}{RC}}$ in which I_0 is the current at time $t = 0$.

▼ **Study point**

$\ln\left(\dfrac{1}{2}\right) = -\ln 2$ follows from the definition of a logarithm as a power. You should convince yourself that for any positive value of x, $\ln\left(\dfrac{1}{x}\right)$ $= -\ln x$.

[This is a special case ($n = -1$) of the rule: $\ln(x^n) = n \ln x$.]

Self-test 4.1.10

A 1.5 µF capacitor is given a charge of 18 µC. At time $t = 0$, a 2.2 MΩ resistor is connected across it. Calculate:

(a) The time constant.

(b) The charge remaining after 10.0 s.

(c) The time taken for the charge to fall to 4.5 µC.

Self-test 4.1.11

The graph of Fig. 4.1.12 was plotted using a 0.50 µF capacitor. Calculate:

(a) The initial pd.

(b) The resistance (knowing τ).

Self-test 4.1.12

(a) In Fig. 4.1.12 a tangent has been drawn to the graph at $t = 2.0$ s. Calculate its gradient.

(b) State what electrical quantity this gradient represents.

(c) Determine this quantity by another method… Read the value of Q at $t = 2.0$ s and use the value of τ already found (see main text.)

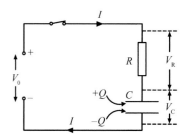

Fig. 4.1.13 Capacitor charging through a resistor

4.1.13 Self-test

A resistor of 150 kΩ is connected across a charged capacitor of 6.8 μF.

(a) Calculate the time for its charge to fall to half of its original value.

(b) Calculate the time for the energy to fall to half of its original value.

(c) What is the relationship between these times?

EQUATION FOR CAPACITOR CHARGING

At time t after a pd, V_0, is connected across an (uncharged) capacitor in series with a resistor, the charge, $\pm Q$, on the plates of a capacitor is given by

$Q = Q_0(1 - e^{-\frac{t}{RC}})$ in which $Q_0 = \frac{V_0}{R}$

Stretch & Challenge

If you know how to differentiate exponentials, and how to apply the *chain rule*, show that…

if $Q = Q_0(1 - e^{-\frac{t}{RC}})$

then $\frac{Q}{C} + R\frac{dQ}{dt} = V_0$ as required.

4.1.14 Self-test

When the capacitor in Fig. 4.1.13 is fully charged it stores energy $\frac{1}{2}Q_0V_0$. Charge Q_0 has passed through the source of V_0, which must therefore have supplied energy Q_0V_0. What has happened to the other $\frac{1}{2}Q_0V_0$?

4.1.15 Self-test

A steady p.d of 9.0 V is applied across a 2.2 μF capacitor in series with a 1.5 MΩ resistor. Calculate:

(a) The time taken for the pd across the capacitor to reach 5.7 V.

(b) The pd an equal time later.

(b) Charging

We suppose that the capacitor has no charge on its plates initially, but that, at time $t = 0$, the switch is closed, so that a fixed pd, V_0, is applied across the RC combination (see Fig. 4.1.13). At any time after $t = 0$, the pds across R and C add up to V_0.

In symbols, $V_C + V_R = V_0$ so $\frac{Q}{C} + IR = V_0$ that is $\frac{Q}{C} + R\frac{\Delta Q}{\Delta t} = V_0$.

Note that, this time, $I = +\frac{\Delta Q}{\Delta t}$, as the capacitor is charging.

We can work out the general shape of the graph of Q against t by some very simple reasoning…

• When $t = 0$, $Q = 0$. So $V_C = 0$ (because $Q = CV$). So $V_R = V_0 - 0 = V_0$.

  In other words the resistor has the full supply pd across it. So $I = +\frac{V_0}{R}$. This is the initial rate at which Q is rising.

• As Q rises, V_C rises, so V_R falls (since $V_R = V_0 - V_C$). Therefore I falls, so Q's *rate* of rise decreases, approaching zero as V_C approaches V_0. The maximum possible charge, Q_0, is simply $Q_0 = CV_0$.

The equation for Q in terms of t is given in the margin. Note that it contains the same exponential factor, $e^{-\frac{t}{RC}}$ as for the discharging case. Remember that $e^{-\frac{t}{RC}} = 1$ when $t = 0$, but approaches 0 for $t \gg RC$.

Significance of the time constant, τ

In the case of charging, the time constant represents the time taken for Q to reach $Q_0(1 - e^{-1})$, that is for it to reach 63% of its eventual value.

Graph of Q against t for a capacitor charging through a resistor

Fig. 4.1.14 is plotted for the same RC combination and for the same value of maximum charge, Q_0, as Fig. 4.1.12.

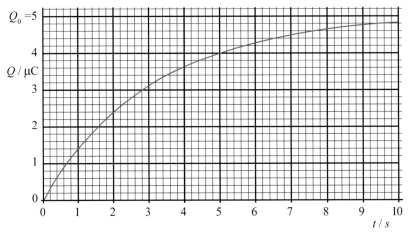

Fig. 4.1.14 Q against t for a capacitor charging through a resistor

Variation of pds and current and with time

It's easy to show (in the order given) that, if $V_0 = \frac{Q_0}{C}$, $I_0 = \frac{V_0}{R}$, then

$V_C = V_0(1 - e^{-\frac{t}{CR}})$, $V_R = V_0 e^{-\frac{t}{CR}}$, $I = I_0 e^{-\frac{t}{CR}}$.

4.1.6 Electric fields

(a) The idea of an electric field

There is an electric field between the plates of a charged capacitor, around a plastic comb which has been rubbed, around the top sphere of a Van de Graaff generator, in the air between clouds and the ground (and between the clouds themselves) in a thunder-storm, in the water around certain fish when they send out electrical pulses …

We say that *there is an electric field at any point, P, where a charged body (at rest) experiences a force proportional to its charge.*

We shall call a charged body used to test for the presence of an electric field a *test charge*. Ideally it would have a tiny charge (so as not to displace the very charges that are giving rise to the field!), and would occupy very little space, so as to respond to the field 'at a point'.

In Fig. 4.1.16 the test charge (hardly ideal) is a small, light sphere, coated with graphite and given a positive charge by touching the positive plate (to which it loses some electrons). It then experiences a force to the right when dangled in the gap – until it touches the negative plate. There is clearly an electric field between the plates!

(b) Electric field strength

Having defined what we *mean* by an electric field in terms of a test charge experiencing a force, we extend the idea in a very natural way to define the *Electric field strength* vector – see **Terms and definitions**.

Note that if we double the test charge, the force will double, so we get the same value for E. If we change from a positive test charge to a negative one the force direction will reverse, but dividing F by a negative quantity (q) reverses the direction again! In other words, E, defined as F/q , tells us about the field in which we place q, independently of the magnitude or sign of q itself. That's just what we want.

(c) Setting up a uniform electric field

Not only do electric charges *respond* to electric fields, they also *create* electric fields around them. We study the field due to a 'point' charge in Section 4.2. Now, though, we consider the field due to a special configuration of charges – those on the metal plates of a parallel plate capacitor, when a battery has been connected to them (Fig. 4.1.17).

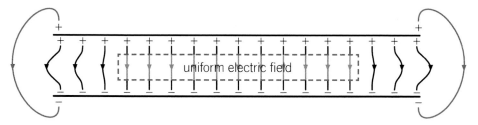

Fig. 4.1.17 Electric field between charged parallel plates

Terms & definitions

The electric field strength, E, at a point, P, is defined as

$$E = \frac{\text{force on test charge, } q, \text{ at P}}{q} = \frac{F}{q}$$

It is a vector whose direction is that of the force on a positive test charge.

UNIT: N C^{-1} = V m^{-1}
(see **Study point**).

▼ **Study point**

We'll soon see how V m^{-1} arises naturally as a unit for E. In the meantime note that…

$V\,m^{-1} = (J\,C^{-1})\,m^{-1} = (N\,m\,C^{-1})\,m^{-1}$
$= N\,C^{-1}$.

Fig. 4.1.15 An 'electric eel'

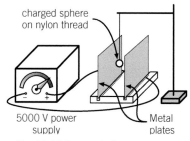

charged sphere on nylon thread

5000 V power supply — Metal plates

Fig. 4.1.16 Demonstrating an electric field

Self-test 4.1.16

A small sphere with a charge of −1.2 nC is placed at a point P above a Van de Graaff generator. The sphere experiences a force of 1.8×10^{-3} N directly upwards. Determine the electric field strength at P.

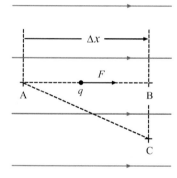

Fig. 4.1.18 Work done on q

If the gap between the plates is small compared with the length of side of the plates (if they are square) or their diameter (if they are discs), the field in the gap is *uniform*, except near the edges. See **Terms and definitions**. This uniformity is represented in Fig. 4.1.17 by parallel and evenly spaced *electric field lines*. More about these in Section 4.2.

(d) Potential difference

The charged sphere in Fig. 4.1.16 will pick up speed as it goes across the gap. As it gains kinetic energy it must lose some other sort of energy. This is *electrical potential energy*: see **Terms and definitions**.

Suppose that a test charge, q, is displaced by Δx from A to B in the direction of a uniform electric field (Fig. 4.1.18). It loses electrical potential energy (EPE) equal to the work done on it by the field.

So Change in EPE of q = $-$(force on $q \times \Delta x$)

Dividing both sides by q,

$$\Delta V = -E\,\Delta x$$

in which $E = \dfrac{\text{force on } q}{q}$ = electric field strength in x direction

and $\Delta V = \dfrac{\text{change in EPE of } q}{q}$ = *potential difference* between A and B

There are several points to note:

- ΔV, like E, is defined so as not to depend on the magnitude or sign of the test charge, q, (as long as q isn't too large).

- We would have reached the same value for ΔV by taking q on any route between A and B. For example, the work done on q, divided by q, in going from A to C would also be $E\Delta x$, because only the displacement component, Δx, in the direction of E counts. No further work is done in going from C to B. We say that the pd between two points in an electric field due to static charges is *path-independent*.

- The equation $\Delta V = -E\,\Delta x$ can be re-arranged as

$$E = -\frac{\Delta V}{\Delta x} = \textit{potential gradient}$$

In particular if we apply a pd V between parallel plates distance d apart, the magnitude, $|E|$ of the field strength in the gap is

$$|E| = \frac{V}{d}$$

For example, for the plates in Fig. 4.1.17 if $V = 1500$ V, $d = 7.5$ mm:

$$|E| = \frac{V}{d} = \frac{1500\,\text{V}}{7.5 \times 10^{-3}\text{m}} = 2.0 \times 10^{5}\,\text{V m}^{-1}. \text{ [The field is downwards.]}$$

Note how the units, V m^{-1}, emerge naturally, as promised earlier.

(e) Electric fields in conductors

In the last section we went back to basics to develop the idea of *potential difference* (pd) in the context of an electric field. But this is the very same pd that features in the study of electric circuits, implying that there's an electric field inside any conductor with a pd across it. Hence the forces experienced by the mobile charge carriers in it, making them drift through the conductor (despite collisions with ions).

If – as in the plates of Fig. 4.1.17 – there is *no* movement of charge in a conductor, there can't be any macroscopic electric field in it (that is a field extending over regions which are large compared with an atomic diameter) – or the free charges *would* be moving! Since $\Delta V = -E\Delta x$, and E is zero throughout the conductor, there is no pd between any points in the conductor, even if it carries a static charge. 'Static' is the key word.

4.1.7 Motion of a charged particle in a uniform electric field

We'll assume that forces on the particle, other than from the electric field, are negligible. This generally requires the particle to be in a vacuum, so there are no collision forces (resistive forces) on it. There's usually no need to worry about gravitational forces: see **Study point**.

The force, qE, on a charged particle is the same in magnitude and direction wherever the particle is in a uniform E field. This implies that:

- Its component of velocity at right angles to the field doesn't change.

- Its component of velocity in the direction of the field changes at a constant rate, given by:

$$\text{acceleration} = \frac{\text{Resultant force}}{\text{mass}} = \frac{qE}{m}.$$

So the particle moves like a resistance-free projectile, except that the acceleration is unlikely to be 9.8 m s^{-2}, and may well not be downwards. The particle's path is either straight or parabolic.

(a) Straight line motion

As an example, suppose a particle of mass m and (positive) charge q is released from rest near the positive plate in the set-up of Fig. 4.1.19. How long will it take to reach the other plate, and at what speed will it arrive?

To find the time taken we use $x = ut + \frac{1}{2}at^2$

in which $x = d$, $u = 0$, $a = \dfrac{qE}{m} = \dfrac{qV}{md}$

Making these substitutions, and re-arranging, we find $t = \sqrt{2d \times \dfrac{md}{qV}} = \sqrt{\dfrac{2m}{qV}} \times d.$

For the final speed we'll first use $v^2 = u^2 + 2ax$

Substituting as before, we get $v = \sqrt{2\dfrac{qV}{md}d}$, that is $v = \sqrt{\dfrac{2qV}{m}}$

For a given particle, we note that the final speed depends only on V, not on d. In fact it doesn't even depend on the field being uniform, as we can show by deriving the equation again by a *very useful argument*…

(b) Speed change related to potential difference

If the particle experiences no forces other than from the electric field, then, however it moves in the field (in a straight line or curve),

KE gained by particle = EPE lost by particle

So, if particle starts from rest, $\frac{1}{2}mv^2 = qV$, so $v = \sqrt{\dfrac{2qV}{m}}$

in which v is the speed acquired due to moving through a potential difference (fall) of V. We can, of course, apply the same principle to a particle that gains EPE and loses KE. See **Study point** for a more formal general approach.

▼ **Study point**

Consider a proton in a (weak) electric field of 10 V m^{-1} near the Earth's surface…

Force from electric field $= qE$
$= 1.60 \times 10^{-19} \text{ C} \times 10 \text{ N C}^{-1}$
$= 1.6 \times 10^{-18} \text{ N}$

Force from gravitational field $= mg$
$= 1.67 \times 10^{-27} \text{ kg} \times 9.81 \text{ N kg}^{-1}$
$= 1.6 \times 10^{-26} \text{ N} \ (qE \times 10^{-8})$

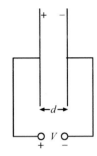

Fig. 4.1.19 Gap to be crossed

Self-test 4.1.18

Explain why multiplying the final speed, v, by the journey time, t, for the example of straight line motion in an E field, gives $2d$ rather than d.

Self-test 4.1.19

An electron enters a uniform electric field of 750 V m^{-1}, travelling at a velocity of $2.2 \times 10^7 \text{ m s}^{-1}$ in the direction of the field. Calculate how long it takes to return to its point of entry, and the total distance it travels.

▼ **Study point**

In general, for a particle going from A to B in an electric field

$\frac{1}{2}mv_B^2 - \frac{1}{2}mv_A^2 = -q\Delta V$

in which ΔV is the potential difference (rise) in going from A to B. The value of q (as well as of ΔV) needs to be entered with the correct sign.

▼ **Study point**

The electric field between the plates in Fig 4.1.20 is only very approximately uniform and the field extends significantly beyond the edges of the plates. This is because the distance between the plates isn't really small compared with their length and breadth.

4.1.20

Self-test

This question is about the electron beam shown in Fig. 4.1.20. Assume a uniform E field between the plates.

The electrons leave the gun at a speed of 3.20×10^7 m s^{-1}.

Calculate:

(a) V_{acc}.

(b) The time taken for an electron to traverse the screen (left to right).

(c) The acceleration in the y-direction responsible for the observed y-displacement (20mm).

(d) The deflecting pd, V_y, given that the separation, b, of the deflecting plates is 45 mm.

(e) The y velocity component of an electron when it reaches the right hand edge of the screen.

(f) The magnitude and direction of the velocity of an electron when it reaches the right hand edge of the screen.

[$e = 1.60 \times 10^{-19}$ C

$m_e = 9.11 \times 10^{-31}$ kg]

(c) Motion in two dimensions

One way of demonstrating this uses the apparatus of Fig. 4.1.20, and we'll take it as our example.

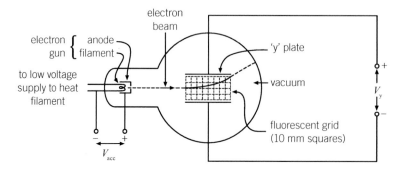

Fig. 4.1.20 Deflection of electrons in a transverse electric field

The electron gun

This produces a fine beam of fast electrons. The electrons are emitted from a metal filament heated to a high temperature by passing a current through it. They are accelerated towards a cylindrical metal 'anode' by means of a pd, V_{acc}, applied between filament and anode as shown. Many electrons are captured by the anode and returned straight to the filament via the V_{acc} power supply, but some escape through a small hole in the anode's end-cap, to form the beam. The speed, v, of these escapees is easily found, using $\frac{1}{2}mv^2 = qV_{acc}$. See (b) above. [We neglect the relatively small KE with which the electrons are emitted from the filament.]

Parabolic path

The electron beam enters a transverse (crosswise) electric field between parallel plates with a pd, V_y, applied between them. We shall assume that the field is uniform and doesn't extend beyond the edges of the plates – but see **Study point**.

In this region the electrons' x-wise velocity, v_x (= v), component, remains unchanged, and equal to the speed v at which they entered the field. So their x-wise displacement component time t after entering will be

$$x = v_x t \qquad \text{so} \qquad x = vt$$

The (y-direction) acceleration is $\dfrac{(-e)}{m}E_y = \dfrac{(-e)}{m}\left(-\dfrac{V_y}{b}\right) = \dfrac{eV_y}{mb}$

in which b is the distance apart of the plates.

Applying $s = ut + \frac{1}{2}at^2$ to the y component of motion, $y = \dfrac{eV_y}{2mb}t^2$

The key thing to understand here is that, for the y motion, the initial velocity is zero.

The electron beam (Fig. 4.1.20) brushes against a fluorescent screen between the deflecting plates, revealing the path of the electrons. We know that, according to our assumptions, this is part of a parabola because we can eliminate t between the x and y equations to give

$$y = \dfrac{eV_y}{2mb}\left(\dfrac{x}{v}\right)^2 \qquad \text{that is} \qquad y \propto x^2.$$

4.1.8 Specified practical work

We outline two 'classic' investigations of capacitor behaviour.

(a) Investigating the discharging of a capacitor through a resistor

It's possible to confirm simply and very convincingly that the decay of charge is exponential. Having charged the capacitor (switch to the left in Fig. 4.1.21), we let it discharge through the resistor (switch to the right), recording the pd, V, (see **Study point**) at roughly equally spaced times, t.

Planning

Here are some points to consider:

- How does one choose the values of C and R? The time constant, RC, needs to be long enough for ten or so pairs of (V, t) readings to be taken comfortably while the pd drops to (say) a tenth of its original value. If using a data-logger its capabilities need to be known.

- The voltmeter display, especially if digital, may not be readable if it's changing rapidly. It's worth considering doing the charge and discharge procedure afresh for each measurement, stopping the discharge at a different time (switch central) before reading the pd. The voltmeter's position would have to be different from that shown in Fig. 4.1.21!

Using the results

If we plot V (or CV) against t, we expect a graph like Fig. 4.1.12. The time constant, τ, can be determined as explained in Section 4.1.5, starting from various points on the graph, to check that the decay is exponential.

A less fiddly and more accurate way of checking the relationship and determining τ is to plot $\ln (V/\text{volt})$ against t. (We have to divide V by its unit before taking the logarithm, because only numbers have logarithms!) This gives a straight line graph, because, as shown earlier,

$$V = V_0 e^{-\frac{t}{RC}}.$$

In inverse form $\ln \left(\dfrac{V}{V_0}\right) = -\dfrac{t}{RC}$ that is $\ln \left(\dfrac{V/\text{volt}}{V_0/\text{volt}}\right) = -\dfrac{t}{RC}$.

Because $\ln \dfrac{a}{b} = \ln a - \ln b$ we have $\ln (V/\text{volt}) = -\dfrac{t}{RC} + \ln (V_0/\text{volt})$

Comparing this with $y = mx + c$, we'd expect that plotting $\ln (V/\text{volt})$ as y and t as x would give a straight line of gradient $m = -\dfrac{1}{RC}$.

Determining C from the graph gradient

R is the resistance of the parallel combination of resistor and voltmeter. The resistor's resistance can be measured with a digital multimeter. A typical resistance for a digital multimeter on it 'volts' range is $10 \text{ M}\Omega$. Knowing the value of R, we can determine C, as

$$C = -\frac{1}{R \times \text{gradient}}$$

Charging a capacitor through a resistor

A possible circuit is given Fig. 4.1.22. In this case $V_R = V_0 e^{-\frac{t}{RC}}$. Plotting $\ln (V_R/\text{volt})$ against t should therefore give a straight line graph.

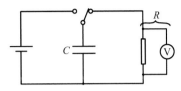
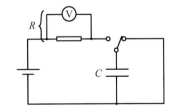

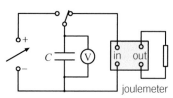

Fig. 4.1.23 Measuring energy stored in a capacitor

(b) Investigating the energy stored in a capacitor

Method 1: Using a joulemeter

The capacitor is charged (switch to the left in Fig. 4.1.23) to a measured pd and discharged through a resistor (switch to the right). The energy, ΔU, supplied to the resistor is measured by the joulemeter, interposed between capacitor and resistor as shown.

The capacitor value needs to be chosen, together with the range of pds, so that the energy stored is within the joulemeter's capabilities. For convenience the resistor value needs to be such that the time constant, RC, is no more than a few seconds.

The capacitor will not discharge completely in a finite time. One way to deal with this is deliberately to stop the discharge when the pd has fallen to the same arbitrary small value, V_{min}, each time. Then

$$\Delta U = \tfrac{1}{2}CV^2 - \tfrac{1}{2}CV_{min}^{\;2}$$

A graph of ΔU against V^2 should therefore be a straight line of gradient $\tfrac{1}{2}C$ and intercept $-\tfrac{1}{2}CV_{min}^{\;2}$.

Method 2: By measuring a temperature rise

A joulemeter will deliver reliable readings with little effort, but how it does so may be, to some extent, a mystery. A less 'black box' approach is to discharge the capacitor through a resistor in the form of a coil of wire and to measure the temperature rise.

Example

The coil shown in Fig. 4.1.24 was wound from copper wire of total mass 0.077 kg and resistance 2.35 Ω. It was wrapped in thermal insulation. A digital thermometer was used to measure the temperature at its centre.

Fig. 4.1.24 Coil (thermal insulation removed) and thermometer

An initial temperature reading, θ_1, was taken. Then the coil was connected across a supercapacitor of nominal value 5.0 F (and maximum allowed pd 5.4 V), pre-charged to a pd of 3.00 V, until the pd fell to 1.00 V. This was done 5 times in quick succession and the maximum temperature, θ_2, reached at the end was recorded. The procedure was repeated for starting pds of 4.00 V and 5.00 V. Here are the results.

V/volt	$(V/\text{volt})^2-1.00^2$	$\theta_1/°C$	$\theta_2/°C$	$(\theta_2-\theta_1)/°C$	$\dfrac{(\theta_2-\theta_1)}{V^2-(1.00\text{ volt})^2}$ /°C V^{-2}
3.00	8.0	15.3	17.6	2.3	0.28
4.00	15.0	16.1	20.4	4.3	0.29
5.00	24.0	14.4	21.9	7.5	0.31

If the energy stored in the capacitor is proportional to the pd squared, the ratio in the last column should be constant. See **Study point**.

4.1.21

Self-test

For method 2

(a) Show that, if heat loss is neglected, then $mc(\theta_2 - \theta_1)$ $= 5 \times \tfrac{1}{2}C[V^2 - (1.00\text{ V})^2]$ in which m is the mass of the copper and c is the specific heat capacity of copper.

(b) Use the results in the table to determine a value for C. You will also need these: $m = 0.077$ kg $c = 385$ J kg^{-1} °C^{-1}.

▼ **Study point**

Informal analysis suggests an uncertainty in the ratio in the last column, for $V = 3.00$ V, of about 10% due to random errors. Although the figure may be better for $V = 4.00$ V and 5.00 V, systematic errors due to heat loss are probably greater for higher pds.

Exercise 4.1

1 A parallel plate capacitor is made from two rectangular metal plates of size $1.0 \text{ m} \times 2.0 \text{ m}$, held slightly apart by small pieces of plastic of thickness 0.1 mm. The two plates are connected to a 500 V DC supply. Ignoring the dielectric effect of the plastic pieces, calculate:

(a) the capacitance

(b) the charge on each sheet of metal

(c) the energy stored

(d) the intensity of the electric field between the plates.

2 Two pieces of aluminium foil of the same area as the plates in Q1 are rolled together to make a capacitor, separated by two sheets of plastic wrap (as in Fig. 4.1.2) of thickness 12.5 μm. Rolling the plates effectively doubles the surface area because charge is stored on both faces of each. Estimate the capacitance of the arrangement ignoring the dielectric effect of the plastic (which increases the capacitance by a further factor).

3 Calculate the effective capacitance of two capacitors of value C and $2C$ connected (a) in series and (b) in parallel.

4 Two initially uncharged capacitors of value C and $2C$ and a resistor R are connected in the circuit shown. The switch is then closed. The power supply has negligible internal resistance.

(a) State the pd across each component, and the current, immediately after the switch is closed.

(b) After the charging current has dropped half its initial value:

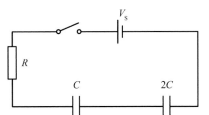

    (i) the current

    (ii) the pd across the resistor

    (iii) the charge in each of the four capacitor plates

    (iv) the pd across each capacitor

    (v) the energy stored in each capacitor

    (vi) the power dissipated in the resistor.

(c) State the quantities in part (b) after charge has stopped flowing.

5 What values of capacitance can you obtain using 3 capacitors of value 10 μF, 22 μF and 33 μF?

6 A capacitor is allowed to discharge through a 1.0 kΩ resistor. It takes 10 s for the pd across the capacitor to fall from 15 V to 6 V. Calculate the capacitance.

7 A capacitor is charged to 25 V and allowed to discharge through a resistor. The energy transfer is measured by an energy meter, which indicates an energy of 35 mJ when the capacitor is fully discharged. Calculate the capacitance.

8 A capacitor of value 220 μF is charged to 10 V and the power supply disconnected. The capacitor is then connected across a 100 μF capacitor. Calculate:

(a) The initial charge on the 220 μF capacitor and the energy stored.

(b) The final charge on each capacitor.
[Hint: the final pds across the capacitors are the same.]

(c) The final energy stored.

9 A capacitor is made from two metal plates held on insulating handles, a distance d apart as shown. The capacitor is charged by momentarily connecting it to a battery. The capacitance, pd, charge and energy of the capacitor are C, V, Q and U respectively. The plates are pulled apart to a separation of $2d$. State and account for the new values of capacitance, pd, charge and energy.

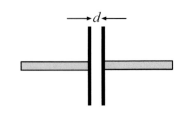

Assume that the capacitor equation $C = \dfrac{\varepsilon_0 A}{d}$ is valid.

10 The procedure in Q9 is repeated but this time the capacitor remains connected to the battery. State and account for the values of capacitance, pd, charge and energy after the plates are pulled apart.

11 A 4700 µF capacitor is charged to 12 V. It is allowed to discharge through a 100 Ω resistor.

(a) Calculate the initial charge stored on the capacitor plates.

(b) Write an equation for the discharge current, I as a function of time.

(c) Sketch a graph of I against t and show how this graph relates to the answer to part (a).

(d) Calculate the initial energy stored in the capacitor.

(e) Write an equation for the energy stored as a function of time, $U(t)$.
    [Hint: First write $V(t)$.]

(f) Write an equation for the power dissipated in the resistor as a function of time, $P(t)$.

(g) Draw a graph of P against t and show how this graph relates to the answer to part (d).

4.2 Electrostatic and gravitational fields of force

The electrostatic forces between charged bodies and the gravitational forces between all bodies with mass are, we now believe, very different in origin. One pointer is that we don't have positive and negative mass; gravitational forces are always attractive. Conveniently, though, for bodies that can be treated as point-like, the strength of the force over a wide range of distances is govered by an equation which is essentially the same in both cases – an *inverse square law*. That's why we've included both fields in this section. We start with forces between stationary charged bodies, because we can make good use of the idea of *electric field strength*, introduced in the last section.

4.2.1 Forces between stationary point charges: Coulomb's law

In the early 1780s, Charles-Augustin de Coulomb (Fig. 4.2.1) measured the forces between small, charged, conducting spheres. He used a 'torsion balance' (Fig. 4.2.2) in which the moment of the repulsive force on one of the spheres was balanced by the moment due to the twisting of a fibre. He found that:

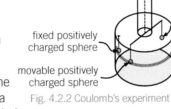

knob to adjust twist in fibre
fibre
counter-balance
fixed positively charged sphere
movable positively charged sphere

Fig. 4.2.2 Coulomb's experiment

- When the separation between the centres of the spheres was doubled the force went down by a factor close to 4. When the separation was tripled the force went down by a factor close to 9.

- When the charge on either of the spheres was halved, the force halved, and so on. [Coulomb halved the charge on a sphere by touching it on to an identical, initially uncharged, sphere.]

Generalising from his results, Coulomb suggested that an exact 'inverse square' law applied for *point charges*, that is charged bodies that are very small in size compared with their separation. The law is stated in **Terms and definitions** using SI units (not around in Coulomb's day!). The law has been amply confirmed by more accurate, but less direct, means.

Using the Coulomb's law equation

- If Q_1 and Q_2 are both positive or both negative, F clearly comes out as positive, so we can interpret a positive F as a repulsive force. If the charges are of opposite signs, F will be negative: attractive!

- The equation holds almost exactly if the charges are separated by air at ordinary densities. Only when the separating medium is a solid or liquid insulator is the force substantially reduced (because the charges will be surrounded by 'haloes' of opposite charge due to displacement of charge in the molecules of the medium). This is beyond A-level.

- We can write Coulomb's law as $F = \dfrac{1}{4\pi\varepsilon_0} \times \dfrac{Q_1 Q_2}{r^2}$. Clearly $\dfrac{1}{4\pi\varepsilon_0}$ is itself a constant.

  Using values of ε_0 and π to four significant figures, and rounding to three, we find that

  $$\frac{1}{4\pi\varepsilon_0} = 8.99 \times 10^9 \text{ F}^{-1} \text{ m}$$

Terms & definitions

Coulomb's law

The force between point charges, Q_1 and Q_2, in a vacuum and separated by distance r is given by

$$F = \frac{Q_1 Q_2}{4\pi\varepsilon_0 r^2}$$

in which ε_0 is a constant called the permittivity of free space.

By experiment,
$\varepsilon_0 = 8.85 \times 10^{-12} \text{ F m}^{-1}$

[NB $\dfrac{1}{4\pi\varepsilon_0} = 9.0 \times 10^9 \text{ F}^{-1} \text{ m}$ (2 s.f.)]

Self-test 4.2.1

In Section 4.1.2, ε_0 was shown to have the units F m^{-1}. Yet the Coulomb's law equation demands that its units be $\text{C}^2 \text{ N}^{-1} \text{ m}^{-2}$. Show, step by step, that $\text{F m}^{-1} = \text{C}^2 \text{ N}^{-1} \text{ m}^{-2}$.

Fig. 4.2.1 Charles-Augustin de Coulomb

Self-test 4.2.2

(a) In a simple theoretical 'model' of the hydrogen atom, the electron orbits the nucleus at a distance of 5.29×10^{-11} m. Calculate the force on the electron.

(b) Using a similar model for the He^+ helium ion (proton number = 2) the force on the one electron orbiting the nucleus is 8.00 times the force of part (a). Calculate the distance of the electron from the nucleus.

— Terms & definitions —

The magnitude of the electric field strength at distance r from a point charge, Q, or from the centre of a spherically symmetric charge distribution of total charge Q, and outside the charge distribution, is

$$E = \frac{Q}{4\pi\varepsilon_0 r^2}.$$

The direction of the field is radially outwards from Q if Q is positive, and radially inwards if Q is negative.

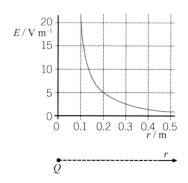

Fig. 4.2.3 E against r for a point charge

4.2.3 ⌄ Self-test

Calculate the charge, Q, whose field strength is plotted in Fig. 4.2.3.

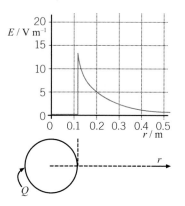

Fig. 4.2.4 E against r for a charged spherical surface

▼ **Study point**

A test charge for measuring electric field strength should be a very small charge. Why? [Think of it being used to measure the field due to a charged conducting sphere.]

For most purposes it's fine to use the very memorable 9×10^9 F^{-1} m, which is actually correct to 2 s.f. So why on earth do we state Coulomb's law as we do, rather than writing $\frac{1}{4\pi\varepsilon_0}$ as a single constant? Recall that we first met ε_0 in the parallel plate capacitor equation – which has no 4π in it (Section 4.1.2). By including 4π in the Coulomb's law equation we avoid having to have it in the parallel plate equation. The 4π is more appropriate in the Coulomb's law equation as this deals with spherically symmetric force fields around point charges. A deeper question is why the two equations are connected at all. We'll hint at the answer in Section 4.2.6.

4.2.2 The electric field strength due to a point charge or equivalent

Electric field strength, E was defined (as force per unit charge on a small testing charge) in Section 4.1.6.

In the Coulomb's law equation we can regard one of the charges as the source of an electric field and the other as a testing charge. Calling these charges Q and q respectively, the equation becomes

$$F = \frac{Qq}{4\pi\varepsilon_0 r^2} \qquad \text{that is} \qquad \frac{F}{q} = \frac{Q}{4\pi\varepsilon_0 r^2}$$

But $\frac{F}{q}$ is, by definition, the magnitude of the electric field strength, E, at distance r from Q. So we have the equation in **Terms & definitions**.

Fig. 4.2.3 shows how E varies with distance r, according to this equation. The value of Q was chosen to make the graph easy to plot – see Self-test 4.2.3. Two points are worth mentioning:

- A doubling of r results in a quartering of E and so on. Even a sketch-graph of an inverse square law relationship should use this as a basis.

- E goes to infinity as r goes to zero. Physicists call a point where something of interest can't be calculated a 'singularity'. (It is unlikely that there exist any true point charges for which the inverse square law holds right down to $r = 0$.)

Outside a spherically symmetric charge distribution, such as a conducting sphere with charge spread evenly on its surface, the field strength is the same as if all the charge were concentrated in the centre. (In the space *enclosed by* a charged spherical surface, the field strength due to the surface charge is zero!) Compare Figs 4.2.3, 4.2.4.

In fact, if a spherical conductor is given charge, free charge carriers will move in such a way that the charge *will* spread evenly over the surface by mutual repulsion. However, any nearby external charges will disturb the symmetry. For example, in Coulomb's experiment, it's not, in the first instance, the *spheres* that repel, but the charges on them, and being effectively mobile these will move so that the far sides of the spheres have more charge than the facing sides. So the spheres don't quite behave as point charges at their centres! See Study point.

4.2.3 Electrical potential, V

(a) Definition

How close can an α particle with 7.7 MeV of kinetic energy approach a compact body of charge 1.26×10^{-17}C? This was a key calculation in Rutherford's estimate of the size of an atomic nucleus. Like other similar calculations it is made much easier by using the concept of potential, which we now develop…

We start by defining the *potential energy* (PE) of a testing charge, q, at a point P in an electric field. See **Terms and definitions.**

How do we set about calculating this when the field is due to a point charge, Q? First recall that *Work* means force × distance in direction of force. The force on q is qE, radially outwards from Q. But E keeps getting less as we go from P to infinity (Fig. 4.2.3), because $E = \dfrac{Q}{4\pi\varepsilon_0 r^2}$.

We must therefore calculate the work done on q in small steps at a time (distance Δr), *so* small that E hardly changes over any one step. Then …

Work done on q as q moves distance Δr further from $Q = qE\,\Delta r$.

As shown in Fig. 4.2.5, $E\Delta r$ is represented on a graph of E against r by the area of a narrow strip (shaded) 'under' the curve at the appropriate r. The *total* work done on q is q multiplied by the sum of the areas of all the narrow strips from P to infinity. So

$$\begin{array}{l}\text{PE of } q \text{ at P} \\ \text{(due to } Q) \end{array} = q \times \left(\begin{array}{c}\text{area under } E \text{ vs } r \text{ graph} \\ \text{from P to infinity} \end{array}\right)$$

There is a mathematical technique (see second **Study point**) for calculating the area. It comes out to be $\dfrac{Q}{4\pi\varepsilon_0 r_P}$. The subscript 'P' has now served its purpose, so we can drop it and say…

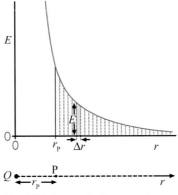

Fig. 4.2.5 Area under E vs r graph

For q at a distance r from Q: $\text{PE} = q \times \dfrac{Q}{4\pi\varepsilon_0 r}$

or, introducing an important idea in electric field theory…

For q at a distance r from Q: $\text{PE} = qV$

in which $V = \dfrac{Q}{4\pi\varepsilon_0 r}$ and is called the electrical **potential** at distance r from Q.

For any field due to any distribution of charges, the potential, V, at a point is defined by the equation $\text{PE} = qV$,

i.e. $V = \dfrac{\text{PE of test charge}}{q}$

Terms & definitions

The **potential energy** of a test charge, q, at a point P in an electric field is the work done by the field on q, as q goes from P to infinity*.

The **potential**, V, at a point P in an electric field is the work done by the field *per unit charge* on a test charge as it goes from P to infinity.

UNIT: JC^{-1} = volt (V)

So, for test charge q, $\text{PE} = q \times V$

The potential due to a point charge, Q, at distance r from Q is

$V = \dfrac{Q}{4\pi\varepsilon_0 r}$

* 'Infinity' means so far from the charge(s) giving rise to the field that the field strength is negligible.

▼ Study point

Some people define potential energy and potential at P in terms of work done by some external force (equal and opposite to the force of the electric field on q) as q goes *from infinity to* P. These definitions are entirely equivalent to ours.

▼ Study point

If you've met integral calculus this is for you…

Shaded area in Fig. 4.2.5 is

$$\int_{r_P}^{\infty} E\, dr = \frac{Q}{4\pi\varepsilon_0} \int_{r_P}^{\infty} \frac{1}{r^2} dr$$

$$= \frac{Q}{4\pi\varepsilon_0}\left[-\frac{1}{r}\right]_{r_P}^{\infty} = \frac{Q}{4\pi\varepsilon_0}\left[0 - \left(-\frac{1}{r_P}\right)\right]$$

which tidies up to give the result quoted in the main text.

4.2.4 Self-test

In Fig. 4.2.6 a tangent has been drawn to the *V* against *r* graph at *r* = 0.20 m.

(a) Determine the gradient of this tangent.

(b) Read off the value of *E* at *r* = 0.20 m from Fig. 4.2.3, and comment on whether it is in agreement with $E = -\frac{\Delta V}{\Delta x}$.

Stretch & Challenge

Show that the *difference* in potential between points B and A at distances $(r + \Delta r)$ and r from a charge Q is

$$\Delta V = V_B - V_A$$
$$= \frac{Q}{4\pi\varepsilon_0}\left\{\frac{-\Delta r}{r(r + \Delta r)}\right\}$$

Show further that, if Δr is extremely small compared to r, this equation collapses to $\Delta V = -E\Delta r$, in which E is the field strength at distance r from Q.

4.2.5 Self-test

The top sphere (radius 0.12 m) of a Van de Graaff generator has a potential of −80 kV. Calculate:

(a) the charge on the sphere, stating your assumption(s).

(b) the electric field strength just outside the sphere.

Fig. 4.2.7 Van de Graaff generator and man living dangerously

(b) Graph of *V* against *r* for a point charge

Fig. 4.2.6 is just such a graph. Note the inverse proportionality between *V* and *r*: when *r* is doubled, *V* halves and so on, in accordance with the equation $V = \frac{Q}{4\pi\varepsilon_0 r}$. Fig. 4.2.6 should be compared with Fig. 4.2.3, in which *E* is plotted against *r* for the same value of *Q* and an *inverse square* law applies.

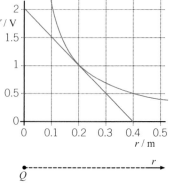

Fig. 4.2.6 *V* against *r* for a point charge

(c) Potential gradient

In Section 4.1.6 we derived the equation

$$E = -\frac{\Delta V}{\Delta x} = -\text{ potential gradient}$$

In the form $\Delta V = -E\Delta x$ it is based on the now familiar reasoning that

$$\frac{\text{work done on } q}{q} = \frac{\text{force on } q \times (\text{small}) \text{ distance gone by } q \text{ in direction of force}}{q}.$$

In Section 4.1.6, we were concerned with uniform fields, where Δx can have any value. However, we can apply the potential gradient equation to non-uniform fields, provided that $\frac{\Delta V}{\Delta x}$ is taken as the value that this fraction approaches as Δx becomes smaller and smaller. This makes it equal to the gradient of a tangent drawn to a graph of *V* against distance in direction of *E* (that is the *r* direction for our point charge, *Q*).

Self-test 4.2.4 recommends a consistency check you should make on Figs 4.2.6 and 4.2.3 using $E = -\frac{\Delta V}{\Delta x}$. A more complete check, on the underlying equations for *V* and *E*, is suggested in **Stretch & Challenge.**

(d) The zero of potential

According to $V = \frac{Q}{4\pi\varepsilon_0 r}$, *V* approaches zero as *r* becomes very large. This is a direct consequence of our defining potential in terms of work done on a testing charge going from a point *to infinity*, rather than to some other point at a finite distance (r_0, let's say) from *Q*. Had we chosen the r_0 option, the equation for potential would be

$$V = \frac{Q}{4\pi\varepsilon_0 r} - \frac{Q}{4\pi\varepsilon_0 r_0}$$

in which the second term is a constant. In fact only *differences* in potential (pds) have any physical significance, and either formula for potential would give the same value for the pd between any two particular points. We choose the option that gives us the simpler equation.

A reason sometimes given for taking the potential to be zero at infinity is that at infinity the (zero strength) field can do no more work on a positive testing charge. The argument fails, though, if *Q* is negative so its field is directed inwards *towards Q*!

Note that if *Q* is negative, then, according to $V = \frac{Q}{4\pi\varepsilon_0 r}$, *V* is negative everywhere and approaches zero for very large *r*.

(e) Using the idea of potential: an example

How close can an α particle with 7.7 MeV of kinetic energy approach a compact body (nucleus) of charge 1.26×10^{-17} C? This was the question posed at the beginning of Section 4.2.3. We can now answer it.

Fig. 4.2.8 α particle approaching a nucleus (not to scale!)

The closest approach will be when the α is heading straight at the nucleus and momentarily stops (at point P, say), having been continuously retarded by the repulsive force from the nucleus. We shall assume that the α is infinitely far from the nucleus initially, and that the nucleus is stationary throughout. Since energy is conserved,

Initial electrical PE + initial KE = electrical PE at P + KE at P

So

$$0 + 7.7 \times 10^6 \text{ eV} = 2e \times \frac{1.26 \times 10^{-17} \text{ C}}{4\pi\varepsilon_0 r_P} + 0$$

Here we have remembered that the charge on the α is $+2e$. A neat trick is to divide through by e, as 7.7×10^6 eV $= 7.7 \times 10^6$ V $\times e$. Dividing and re-arranging,

$$r_P = \frac{2}{7.7 \times 10^6 \text{ V}} \times \frac{1}{4\pi\varepsilon_0} \times 1.26 \times 10^{-17} \text{ C}$$

leading to $r_P = 2.9 \times 10^{-14}$ m

The significance of this result is explained briefly in the first **Study point**.

4.2.4 E and V due to more than one point charge

In principle we can calculate the electric field strength and potential at a point P due to any spatial distribution of charges. Here's the procedure:

• View the charge distribution as a collection of point charges.

• Calculate E and V at P due to each point charge.

• Add the field strengths at P as *vectors* and the potentials at P as *scalars*.

Actually *carrying out* the procedure and establishing the results for even something as simple as a spherical charge (see Study point) is a little beyond A-Level. We shall be content with finding field strengths and potentials due to just two or three point charges…

Example

Calculate the field strength and potential at point B in Fig. 4.2.9 (a) due to the charges shown at A and C.

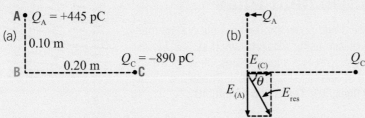

Fig. 4.2.9 Adding E and V due to two point charges

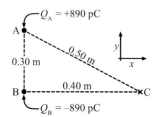

$Q_A = +890$ pC

A

0.30 m

0.50 m

y

x

0.40 m

B

$Q_B = -890$ pC

×C

Fig. 4.2.10 Two point charges

Answer

The field strength at B due to Q_A is

$$E_{(A)} = \frac{455 \times 10^{-12} \text{ C}}{4\pi\varepsilon_0 (0.10 \text{ m})^2} = 400 \text{ V m}^{-1}\downarrow$$

Because $Q_C = -2\,Q_A$, but Q_C is twice as far from B as Q_A, the field strength at B due to Q_C is: $E_{(C)} = 200$ V m$^{-1}\rightarrow$

The resultant field at B is obtained by vector addition (Fig. 4.2.9 (b)). Applying Pythagoras' theorem gives $E_{res} = 447$ V m^{-1}; $\theta = 63°$.

$$E_{res} = \sqrt{400^2 + 200^2} \text{ V m}^{-1} = 447 \text{ V m}^{-1} \text{ and } \theta = \tan^{-1}\frac{400}{200} = 63°.$$

The potential at B if Q_A were the only charge would be

$$V_{(A)} = \frac{455 \times 10^{-12} \text{ C}}{4\pi\varepsilon_0 \times 0.10 \text{ m}} = 40 \text{ V}$$

Because $Q_C = -2\,Q_A$, but Q_C is twice as far from B as Q_A, $V_{(C)} = -40$ V.

The potential at B is is the scalar sum of the potentials due to Q_A and Q_B.

Thus $V = V_{(A)} + V_{(C)} = 40 \text{ V} + (-40 \text{ V}) = 0$

Speculating from the example

A positively charged particle (charge q, mass m) released from rest at B, would initially gain velocity, at a rate $\frac{qE}{m}$, in the direction of the resultant field. This implies that it gains KE and loses electrical PE – at first.

As the particle changes its position, though, the field strength changes, and it is no trivial business to work out the path it will follow. Will it, for example, avoid capture by Q_C? To find out, we could use a computer-executed step-by-step numerical method, calculating acceleration, velocity, displacement over and over again for very short time intervals.

If the particle's trajectory took it far from the charges at A and C, its net change in electrical PE since it left B would approach zero, since $V = 0$ at B and, by definition, at infinity. Therefore the particle's kinetic energy would approach its value at B – zero! What would happen then?

4.2.5 Electric field lines and equipotential surfaces

These are huge aids to visualising how electric field strength and potential vary from point to point in an electric field.

(a) Electric field lines

In principle we could show the field direction at as many points as we please using small arrows (Fig. 4.2.11). It is usually clearer to select a few 'chains' of such arrows, running head of one to tail of the next. These are **electric field lines**. See **Terms and definitions**.

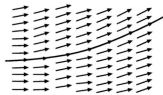

Fig. 4.2.11 Electric field line

Figs 4.2.12 (a) and (b) show some field lines for 'isolated' positive and negative point charges. The lines are radially outward for the positive charge and radially inward for the negative. Ignore the broken lines for the time being.

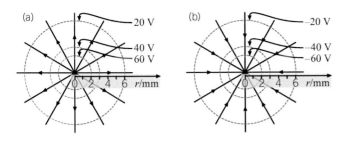

Fig. 4.2.12 Electric field lines and equipotential surfaces due to
(a) an isolated positive charge   (b) an isolated negative charge

Although, in the first instance, electric field lines tell us the direction of the field, they have other properties....

Where the field lines come closer together, the field is stronger. This is clearly the case for the isolated point charges, but holds true for the field lines from any configuration of charges, such as the equal and opposite charges in Fig. 4.2.13.

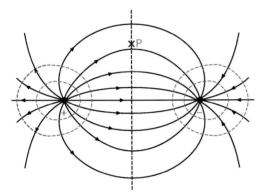

Fig. 4.2.13 Electric field lines due to equal and opposite charges (an electric *dipole*)

When the only charges present are stationary, electric field lines start on positive charges and end on negative charges as in Fig. 4.2.13. This is even true for 'isolated' charges. In Fig. 4.2.12 (a), field lines starting on the positive charge will end on negative charges 'induced', by electron displacement, on surrounding objects, specifically on the surfaces that face the charge. If these induced charges are too distant to distort the radial field near the 'original' charge, we call it 'isolated'. The negative charge in Fig. 4.2.12 (b) is also isolated in this sense, but the field lines ending on it do originate on induced positive charges elsewhere!

(b) Equipotential surfaces

The name says it all: see **Terms & definitions**. For an isolated point charge the equipotential surfaces are spherical surfaces centred on the charge. The surfaces (shown in section as broken lines in Figs 4.2.12 and 4.2.13) are by convention drawn at equal intervals of potential. Note these additional properties :

- The surfaces are closer together where the field is stronger. This follows from the equation $\Delta V = -E \Delta x$. Re-arranging, we have $\Delta x = \dfrac{\Delta V}{E}$, (*accurate* only when ΔV and Δx are very small).

- The electric field lines are at right angles to the equipotential surfaces as they pass through them. See Figs 4.2.12, 4.2.13. This is a little like the lines of greatest slope on a hillside being at right angles to the contours (lines of constant height).

Self-test 4.2.10

Calculate the electric field strength at P in Fig. 4.2.13. P is 50 mm from each charge. The charges are of ±0.15 nC, and are 80 mm apart.

▼ **Study point**

An electric field surrounds an area through which there is a changing *magnetic* field, such as the core of a working transformer. In this case the electric field lines are closed loops – not starting and finishing. This is one type of electromagnetic induction; see Section 4.4. Note that *moving* charges are involved (in setting up the changing magnetic field).

┌─ *Terms & definitions* ─┐

An **equipotential surface** in an electric field is an imaginary surface on which all points are at the same potential.

Self-test 4.2.11

Calculate the potential difference between the equipotential surfaces at 1.0 mm and 10.0 mm from the charge in Fig. 4.2.12 (a).

Self-test 4.2.12

Why are the oval equipotential surfaces around either charge in Fig. 4.2.13 closer on the 'facing' sides of the charges than on the far sides?

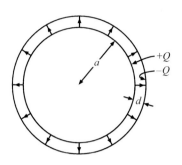

▼ **Study point**

Our result for the capacitance of a concentric sphere capacitor is the same as that quoted for a parallel plate capacitor in 4.1.2. Perhaps this is not surprising, as our condition $d << a$ can be read as $a >> d$ (relatively large radius), but a proper derivation for a parallel plate capacitor is a little beyond A-Level.

┌─ **Terms & definitions** ─┐

Newton's law of gravitation

Every body with mass attracts every other body with mass. For two 'point masses', m_1 and m_2, separated by distance r, the force is given by

$$F = \frac{Gm_1m_2}{r^2}$$

in which G is a constant called *Newton's gravitational constant* (or *big G*).

$G = 6.67 \times 10^{-11}$ N m² kg⁻²

4.2.13 ▽ Self-test

Calculate the force between a movable sphere (mass 0.73 kg) and a fixed sphere (mass 158 kg) in Cavendish's experiment. The centres were 305 mm apart. [Assume a value for G!]

4.2.14 ▽ Self-test

For two protons at the same separation, calculate the ratio

$$\frac{\text{electrostatic repulsive force}}{\text{gravitational attractive force}}.$$

[If necessary look up the values for m_p and e – try and remember them!]

(c) The capacitance of a concentric sphere capacitor

A battery connected briefly between two concentric spherical metal shells (hollow spheres) will transfer electrons from one shell to the other. Equal and opposite charges will therefore coat the facing surfaces of the shells, and there will be a field in the gap as shown in Fig. 4.2.14.

The electric field strength in the gap, at distance r from the centre of the spheres will simply be $E = \dfrac{Q}{4\pi\varepsilon_0 a^2}$ radially outwards. This is entirely due to the charged inner sphere; the charged outer sphere contributes nothing, as the gap is part of the space *inside* it.

Now suppose that the gap width d is much less than the radius of the inner sphere, that is $d << a$. We can then take the field as approximately constant across the gap and given by $E = \dfrac{Q}{4\pi\varepsilon_0 a^2}$

Using $\Delta V = -E\Delta x$, the pd between the shells is simply…

$$\left| \Delta V \right| = \frac{Q}{4\pi\varepsilon_0 a^2}\, d = \frac{Q}{\varepsilon_0 A}\, d$$

in which we have put $4\pi a^2 = A$, as it is the surface area of the inner shell, and almost that of the outer shell.

From the definition of *capacitance* (Section 4.1.1) $C = \dfrac{Q}{\left|\Delta V\right|} = \dfrac{\varepsilon_0 A}{d}$.

4.2.6 Newton's law of gravitation

Isaac Newton gave convincing evidence for an inverse square law of forces between bodies with mass (so-called gravitational forces) roughly a century before Coulomb discovered the inverse square law of forces between charges. Newton's law of gravitation is given in modern language in **Terms and definitions**.

- Newton showed that a consequence of the inverse square law would be that a spherical shell of evenly distributed mass would exert the same force on an external body as if all its mass were concentrated at its centre. No force would be exerted by the shell on any body placed *inside* it. We've already met similar results for charges in Section 4.2.4.

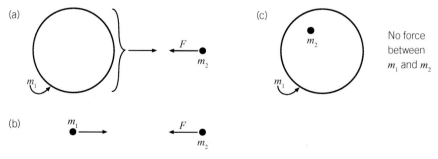

Fig. 4.2.15 Gravitational effect of a spherical shell – in (a) and (b) the forces between m_1 and m_2 are the same

Any spherical body with its mass distributed with spherical symmetry about its centre, can be thought of as an 'onion' of concentric shells, so we can deduce that *it*, too, attracts an external body as if all its mass were at its centre. That's a very useful result, because the bodies that exert appreciable gravitational forces are stars and planets, whose mass *is* distributed with roughly spherical symmetry.

- Newton's main evidence for the inverse square law of gravitation came from the heavens. He showed that an inverse square law of force from the Sun would account for the observed motion of the planets – more about this in Section 4.3.

- Because Newton could only guess at the masses of astronomical bodies such as the Sun, he would not have been able to determine a reliable value for the proportionality constant in the law of gravitation, the constant we now call G. The problem was solved over a century later (in the 1790s) by Henry Cavendish (Fig. 4.2.16) who measured the force between lead spheres here on Earth. He used a torsion balance similar in principle to Coulomb's (Section 4.2.1), but much larger. The experiment was an astonishing feat of design and skill because of the low percentage uncertainty achieved in the result, even though the force to be measured was extremely weak.

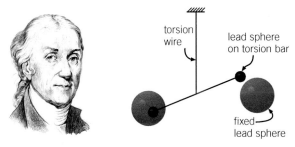

torsion wire

lead sphere on torsion bar

fixed lead sphere

Fig. 4.2.16 Henry Cavendish and a simplified diagram of his apparatus

4.2.7 Gravitational field strength, g and potential, V

(a) Definition of g

Gravitational field strength is defined in a very similar way to electric field strength (see Section 4.1.6), but using a test *mass* rather than a test *charge*. See **Terms and definitions**.

If the only force acting on a body is gravitational, then, using Newton's second law of motion, its acceleration is

$$a_{grav} = \frac{F}{m} = \frac{mg}{m} = g$$

This is a simple but non-trivial result (see **Study point**). It tells us that:

- Measuring a body's gravitational acceleration in a location amounts to measuring the gravitational field strength there. Indeed we shall now use g for both the field strength and the acceleration.

- A body's gravitational acceleration is independent of its mass; all bodies have the same gravitational acceleration in the same place. Galileo confirmed this by experiment, though almost certainly he never dropped cannon balls from the leaning tower of Pisa as the legend goes! The result has since been confirmed to very high accuracy.

(b) g outside a spherically symmetric mass distribution

If the mass of the sphere is M, and m is a test mass outside the sphere we can put $m_1 = M$ and $m_2 = m$ into Newton's law of gravitation, to give

$$F = \frac{GMm}{r^2} \qquad \text{so} \qquad \frac{F}{m} = \frac{GM}{r^2} \qquad \text{that is} \qquad g = (-)\frac{GM}{r^2}$$

in which r is the distance from the test mass to the centre of the sphere.

A minus sign is often included as shown; it signifies that g is directed *inwards* along a radius.

Terms & definitions

The **gravitational field strength,** g, at a point is defined as $g = \dfrac{F}{m}$, that is

$$g = \frac{\text{Force on test mass, } m}{m}$$

at that point. It is a vector.

UNITS: N kg^{-1}

At distance r from the centre of a spherically symmetric mass distribution, M,

$g = (-)\dfrac{GM}{r^2}$ (outside the sphere containing M)

▼ Study point

The independence of a_{grav} from the *mass* of the accelerating body is due, in our derivation of $a_{grav} = g$, to the cancellation of m. On the top line m's role is gravitational: bigger m, bigger gravitational force; on the bottom line m's role is inertial: bigger m, smaller acceleration. In Newtonian physics this dual role of mass is a mystery.

Show that the gravitational field strength *inside* a sphere of uniform density at a distance r from its centre is given by

$$g = g_{surf} \frac{r}{r_{surf}}$$

4.2.15 Self-test

The mean radius of Mars is 3390 km, and its mean density is 3930 kg m⁻³. Calculate a value for the gravitational field strength on its surface.

4.2.16 Self-test

By what factor would the length of the Earth's day have to decrease in order for the observed acceleration of free fall to be zero at the equator?

Example

Calculate a value for the mass of the Earth given that its mean radius is 6370 km and the mean field strength near its surface is 9.81 N kg⁻¹.

Answer

$$M = \frac{R_E^2 g}{G} = \frac{(6.37 \times 10^6 \text{ m})^2 \times 9.81 \text{ N kg}^{-1}}{6.67 \times 10^{-11} \text{ N kg}^2 \text{ m}^{-2}} = 5.97 \times 10^{24} \text{ kg}$$

Dividing this mass by the volume $\frac{4}{3}\pi R_E^2$, gives a value for the mean density of the Earth as approximately 5.5×10^3 kg m⁻³.

It's now time to confess that Cavendish never published a value for G as such. Instead, he used the results of his experiment to calculate a value for the mean density of the Earth, rather as we have just done. This is generally regarded as *equivalent* to determining a value for G.

Observed free fall acceleration on the Earth

On the Earth, at all latitudes except ±90° (the poles), the *observed* acceleration of free fall, g_{obs}, is a little less than the gravitational acceleration, g. This is because g includes the centripetal acceleration, $r\omega^2$, needed to keep the body rotating about the Earth's axis (at a distance r from the axis), as well as g_{obs}. In our rotating frame of reference (the Earth's surface) we don't observe $r\omega^2$ directly.

Fig. 4.2.17 shows the easiest case: when a body is falling at a point **P** on the equator, so $r\omega^2$ like g, is directed towards the centre of the Earth. g_{obs} must therefore also be in this direction and

$$g = g_{obs} + r\omega^2 \qquad \text{that is} \qquad g_{obs} = g - r\omega^2$$

Terms & definitions

The **potential energy** of a test mass, *m*, at a point P in a gravitational field is the work done by the field on *m* if *m* goes from P to infinity.

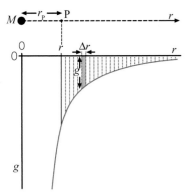

Fig. 4.2.18 Area under *g* vs. *r*

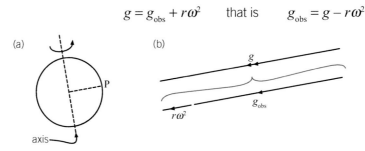

Fig. 4.2.17 *g* at a point P on the equator

At the poles $r\omega^2 = 0$ and $g_{obs} = 9.832$ m s⁻². At the equator, $r\omega^2 = 0.034$ m s⁻² (as you should check!), but $g_{obs} = 9.780$ m s⁻². Clearly the different value of $r\omega^2$ doesn't account for all the 0.052 m s⁻² discrepancy in g_{obs}. The rest is due to the Earth not being a perfect sphere. At the equator, g itself is actually very slightly less than at the poles – which are closer to the centre of the Earth.

Having just dragged out into the open the distinction between real and observed g, we must now admit that for many purposes (including exam questions unless specifically mentioned) we *don't* distinguish – for the simple reason that $r\omega^2 \ll g$.

▼ **Study point**

For those who have been initiated into the mysteries of the integral calculus ...

Shaded area in Fig. 4.2.18 is

$$\int_{r_P}^{\infty} g\, dr = -GM \int_{r_P}^{\infty} \frac{1}{r^2}\, dr$$

$$= -GM \left[-\frac{1}{r}\right]_{r_P}^{\infty} = -GM\left[0 - \left(-\frac{1}{r_P}\right)\right]$$

which, having dropped the 'P' subscript, reduces to the result quoted in the main text.

(c) Gravitational potential, *V*

This concept is the gravitational twin of *electrical* potential, explained in Section 4.2.3(a), so we'll be briefer here.

We start with the definition of *gravitational potential energy* in **Terms and definitions**.

Here, the work done by the field on m is negative because the gravitational force on m will be back towards the body or bodies responsible for the field, that is in essentially the opposite direction to m's journey from P to infinity. This makes gravitational PE, E_p, a negative quantity, approaching zero as P approaches infinity. The zero of E_p at infinity – far enough for the gravitational force to be negligible – is contrived for mathematical convenience.

We now concentrate on the potential energy of m at a point P outside a spherically symmetric mass distribution, M. Fig. 4.2.18 shows how g varies with distance r from the centre of M.

Then, following our method in Section 4.2.3,

$$E_\text{p} \text{ of } m \text{ at P } = m \times \text{area 'under' } g\text{–}r \text{ graph from P to } \infty$$

$$= m \times \left(-\frac{GM}{r}\right) \quad \text{(see \textbf{Study point})}$$

We can express this in terms of the *gravitational potential*, V, due to M at P (see **Terms and definitions**):

$$E_\text{p} \text{ of } m \text{ at P} = mV \quad \text{in which} \quad V = -\frac{GM}{r}$$

Figs 4.2.19 (a) and (b) show how g and V vary with distance r from the centre of a star. See Self-test 4.2.17. Note the different effects of doubling r, in accordance with the inverse square law for g and the inverse law for V.

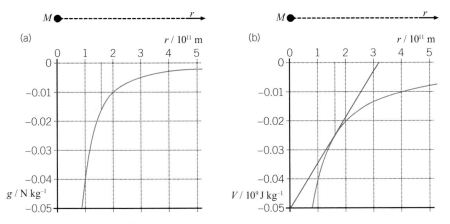

Fig. 4.2.19 (a) g against r (b) V against r, for the same star

(d) Gravitational potential gradient

A test mass, m, moving distance Δx in the direction of a gravitational field will have work $mg \times \Delta x$ done on it by the field,

so   change in gravitational PE, $\Delta E_\text{p} = -mg\Delta x$ (see **Study point**)

and   change in gravitational potential, $\Delta V = -\dfrac{mg\Delta x}{m}$

That is $\Delta V = -g\Delta x$ so $g = -\dfrac{\Delta V}{\Delta x} = -$ potential gradient.

For the field from a sperically symmetric mass distribution,

$$g = -\frac{\Delta V}{\Delta r} = - \text{ gradient of tangent to graph of } V \text{ against } r$$

See Self-test 4.2.18.

Terms & definitions

The **potential**, V, at a point P in a gravitational field is the work done by the field *per unit mass* on a test mass if it goes from P to infinity.

UNIT: J kg^{-1}

So, for test mass m, $E_\text{p} = mV$.

At distance r from centre of M:

$$V = -\frac{GM}{r}.$$

▼ **Study point**

We can calculate ΔE_p from $\Delta E_\text{p} = -mg\Delta x$ only when Δx is small enough for g to be almost constant. We write it as $\Delta E_\text{p} = mg\Delta h$ for a body lifted through a height Δh near the Earth's surface ($\Delta h \ll R_\text{E}$). It won't usually be applicable to space-shots and the like.

Self-test 4.2.17

Determine the mass of the star for which Fig. 4.2.19 (a) and (b) are drawn.

Self-test 4.2.18

Determine the gradient of the tangent that has been drawn to the graph of V against r in Fig. 4.2.19 (b) and compare it with the value of g for the same value of r as read from Fig. 4.2.19 (a).

Self-test 4.2.19

Calculate the speed with which a rocket needs to be launched from the Moon's surface to reach a height of 8.7×10^5 m above the surface.

(Mean $R_\text{Moon} = = 1.74 \times 10^6$ m,

g at surface $= 1.62 \text{ m s}^{-2}$.)

Example: using gravitational potential

The mean radius of Mars is 3.39×10^6 m and the value of g at its surface is 3.71 N kg⁻¹. Neglecting any resistive effects, calculate:

(a) the distance that a rocket will rise if it is launched straight up from the surface of Mars with a speed, u, of 3200 m s⁻¹.

(b) the escape velocity of a rocket from the surface of Mars.

Answer

(a) Since energy is conserved,

initial E_K + initial $E_P = E_K$ at highest point + E_P at highest point.

So $\quad \frac{1}{2}mu^2 + \left(-\frac{GMm}{r_{Mars}}\right) = 0 + \left(-\frac{GMm}{r_{max}}\right)$

in which r_{max} is the distance from the centre of Mars at which the missile runs out of KE. Dividing through by GMm and re-arranging:

$$\frac{1}{r_{max}} = \frac{1}{r_{Mars}} - \frac{u^2}{2GM} \quad \text{therefore} \quad \frac{1}{r_{max}} = \frac{1}{r_{Mars}} - \frac{u^2}{2g_{surf}r^2_{Mars}}$$

Note how we've just overcome a dirty trick: not being given the mass, M, of Mars! See Study point.

So $\quad \dfrac{1}{r_{max}} = \dfrac{1}{3.39 \times 10^6 \text{ m}} - \dfrac{(3200 \text{ m s}^{-1})^2}{2 \times 3.71 \text{ m s}^{-1} \times (3.39 \times 10^6 \text{ m})^2} = 1.75 \times 10^{-7} \text{ m}^{-1}$

Thus $\quad r_{max} = 5.72 \times 10^6$ m $\quad$ and $\quad r_{max} - r_{Mars} = 2.33 \times 10^6$ m

(b) We use energy conservation as in (a), but with less effort, as this time not only the 'final' KE, but also the 'final' PE, is zero ($r = \infty$).

Initial KE + initial PE = KE at infinity + PE at infinity.

So $\quad \frac{1}{2}mu^2 + \left(-\frac{GMm}{r_{Mars}}\right) = 0 + 0$

So $\quad u_{esc} = \sqrt{\dfrac{2GM}{r_{Mars}}} = \sqrt{\dfrac{2g_{surf} \times r^2_{Mars}}{r_{Mars}}} = \sqrt{2g_{surf}\,r_{Mars}} = 5020$ m s⁻¹

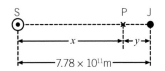

Fig. 4.2.21 Sun and Jupiter

(e) g and V due to more than one body

When the field at a point P arises from more than one body:

- The field strength, g, is the vector sum of the field strengths at P that would arise from each body by itself.

- The potential, V, is the scalar sum of the potentials that would arise from each body by itself.

Example

Fig. 4.2.20 shows the alignment of Sun, Moon and Earth during a total eclipse of the Sun. Calculate the resultant field strength and potential at the Moon due to the Earth and Sun.

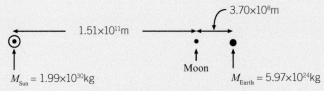

Fig. 4.2.20 Sun, Moon and Earth during a total eclipse

Answers

$$g = \frac{GM_E}{(3.7 \times 10^8\,\text{m})^2} - \frac{GM_S}{(1.51 \times 10^{11}\,\text{m})^2} \text{ towards the Earth}$$

$$= 6.67 \times 10^{-11}\text{N kg}^2\text{m}^{-2}\left\{\frac{5.97 \times 10^{24}\,\text{kg}}{(3.7 \times 10^8\,\text{m})^2} - \frac{1.99 \times 10^{30}\,\text{kg}}{(1.51 \times 10^{11}\,\text{m})^2}\right\}$$

$$= -2.91 \times 10^{-3}\,\text{N kg}^{-1} \text{ towards the Earth}$$

i.e. $2.91 \times 10^{-3}\,\text{N kg}^{-1}$ towards the Sun.

and $V = -\dfrac{GM_E}{3.7 \times 10^8\,\text{m}} + \left\{-\dfrac{GM_S}{1.51 \times 10^{11}\,\text{m}}\right\}$

which similarly leads to $V = -8.80 \times 10^8\,\text{J kg}^{-1}$.

4.2.8 Gravitational field lines and equipotentials

Allowing for the different type of field, gravitational field lines (see **Terms and definitions**) have the same properties as electric field lines (Section 4.2.5) except that they come from infinity and end on anything with mass. Equipotential surfaces are, again, pretty much self-explanatory.

Note that in Fig. 4.2.22 the equipotential surfaces are spheres of less and less negative potential as their radii increase.

Fig. 4.2.23 (a) shows the pattern of field lines and equipotentials for two spherically symmetric bodies of unequal mass. It is identical to the electric field pattern for two negative point charges such that $\dfrac{Q_1}{Q_2} = \dfrac{M_1}{M_2}$.

Figs 4.2.23 (b) and (c) are sketches of the field strength and potential along the line AB (the x direction, say). Note in particular the point between M_1 and M_2 at which the field strength is zero. It corresponds to a local maximum in the potential, in accordance with the relationship

$$g = -\frac{\Delta V}{\Delta x}$$

$$= -\text{ Gradient of tangent of } V\text{–}x \text{ graph}$$

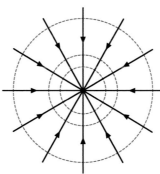

Fig. 4.2.22 Field lines and equipotentials for a small body with mass

Terms & definitions

A **gravitational field line** is a line whose direction at each point along it is the direction of the gravitational field at that point.

An **equipotential surface** in a gravitational field is an imaginary surface on which all points are at the same potential.

4.2.21

Self-test

In Fig. 4.2.23, the centres of M_1 and M_2 are distance D apart. Calculate the ratio $\dfrac{M_1}{M_2}$ from the additional data in the figure.

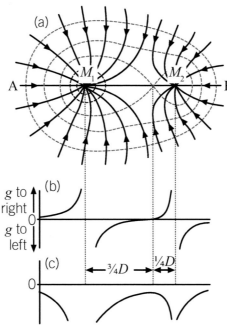

Fig. 4.2.23 Field due to spheres of unequal mass

Summary of properties of electric and gravitational fields

Electric fields	Gravitational fields
The electric field strength, E, at a point is the force per unit charge on a small positive test charge placed at that point.	The gravitational field strength, g, at a point is the force per unit mass on a small test mass placed at that point
The inverse square law for the force between two point electric charges takes the form: $$F = \frac{1}{4\pi\varepsilon_0}\frac{Q_1 Q_2}{r^2}$$ which is Coulomb's law	The inverse square law for the force between two point masses takes the form: $$F = G\frac{M_1 M_2}{r^2}$$ which is Newton's law of gravitation
F can be attractive or repulsive	F is attractive only
$E = \frac{1}{4\pi\varepsilon_0}\frac{Q}{r^2}$ is the field strength due to a point charge in free space (a vacuum) or in air	$g = G\frac{M}{r^2}$ is the field strength due to a point mass
The potential, V, at a point is the work done per unit charge in bringing a small positive test charge from infinity to that point	The potential, V, at a point is the work done per unit mass in bringing a small test mass from infinity to that point
For a point charge: $\quad V_E = \frac{1}{4\pi\varepsilon_0}\frac{Q}{r}$ and for two point charges: $\quad U_P = \frac{1}{4\pi\varepsilon_0}\frac{Q_1 Q_2}{r}$	For a point mass: $\quad V_g = -G\frac{M}{r}$ and for two point masses: $\quad U_P = -G\frac{M_1 M_2}{r}$
Note that all the above equations also apply outside spherically symmetric charges	Note that all the above equations also apply outside spherically symmetric masses
The change of potential energy of a point charge displaced in an electric field is given by: $\Delta U_P = q\Delta V_E$	The change of potential energy of a point mass displaced in an gravitational field is given by: $\Delta U_P = m\Delta V_g$
Field strength at a point is given by: $E = -$ slope of the $V_E - r$ graph at that point	Field strength at a point is given by: $g = -$ slope of the $V_g - r$ graph at that point
$\varepsilon_0 = 8.854\times 10^{-12}\text{ F m}^{-1}$ But $\frac{1}{4\pi\varepsilon_0} \approx 9.0 \times 10^9\text{ F}^{-1}\text{ m}$ is an acceptable approximation	$G = 6.67 \times 10^{-11}\text{ N m}^2\text{ kg}^{-2}$

Exercise 4.2

1 The diagram of electric field lines has several (intentional) mistakes. Identify them. The shaded objects are electrical conductors.

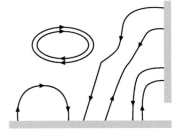

2 Diagrams (a) and (b) are of electrical field lines in different regions of space.

In each case, draw an equipotential surface (line) through the point P and a second equipotential line which is at a higher potential.

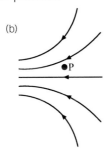

(a)   ●P

(b)   ●P

3 Two points, X and Y, are positioned as shown in a uniform electric field of intensity 150 kV m^{-1}.

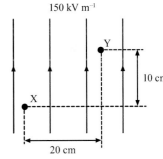

150 kV m⁻¹

(a) Explain which point is at the higher potential.

(b) State the pd between X and Y.

(c) A particle with a charge of $+10 \text{ μC}$ is released at zero speed from X. Its path takes it subsequently through Y (presumably along a frictionless tube!). Assuming no friction, state its kinetic energy at Y.

(d) Calculate the speed a proton would acquire in moving from X to Y.

(e) A $^{7}_{3}\text{Li}$ nucleus moves on a path that takes it through Y and X. Its kinetic energy at Y is 100 keV. Calculate its speed at X. [You may need to use a data book for the mass and charge of a $^{7}_{3}\text{Li}$ nucleus.]

4 A hollow sphere of radius 10 cm carries a positive charge of 1.2 μC. There are no other charges nearby.

(a) Describe the distribution of charge.

(b) Calculate the electric potential of the sphere and the electric field strength at its outer surface.

(c) State the potential and field strength:

    (i) at a point 20 cm from the centre of the sphere

    (ii) at the centre of the sphere

[One of these requires a trivial calculation.]

5 A small conducting sphere of mass 10 mg touches the outside of the charged sphere in Q4 and acquires a charge of $+10 \text{ nC}$. It is observed to move rapidly away.

(a) Explain this observation.

(b) Calculate its initial acceleration.

(c) Assuming no other significant forces act, calculate its final velocity.

6 Three charges are arranged in a right-angled isosceles triangle. The separation of the -20 μC from each of the other charges is 10 cm. The charges are not free to move.

10 μC ●

(a) Calculate the electrical potential energy of the arrangement.

(b) Calculate the resultant electrical force on each charge due to the other two charges.

−20 μC ●   ● 10 μC

7 The -20 μC charge in Q6 is now free to move. Its mass is 1.00 g.

(a) Calculate its initial acceleration.

(b) Describe its subsequent motion assuming no resistance to motion

(c) Calculate the greatest speed.

Questions 8–13 refer to a dwarf planet of radius 1000 km. The gravitational force on an object of mass 1.00 kg, due to the planet, is 1.00 N on the surface of the planet. Question 11 introduces a satellite (moon) of the dwarf planet with a radius of 100 km, of the same density as the planet and with an orbital radius of $20\,000$ km.

8 Calculate:

(a) the mass and mean density of the dwarf planet

(b) the acceleration due to gravity at the surface

(c) the gravitational potential at the surface

(d) the gravitational potential energy of a 100 kg object on its surface.

9 Calculate the gravitational potential and field strength:

(a) $5\,000$ km from the centre and

(b) 1000 km above the surface of the dwarf planet.

10 An object of mass 10 kg is raised 100 km from the surface.

(a) Calculate the increase in gravitational potential energy using the uniform field approximation, $\Delta U = mg\Delta h$.

(b) Calculate the increase in gravitational potential energy using the equation derived from Newton's law of gravitation.

(c) Calculate the percentage error obtained by using the uniform field approximation.

11 For the satellite:

(a) Calculate the gravitational field strength at its surface.

(b) Calculate the gravitational potential at its surface due to:

(i) the satellite alone, and

(ii) the two bodies (i.e. dwarf planet and satellite).

(c) Discuss briefly to what extent the position on the satellite affects the answer to (b)(ii).

12 The satellite is in a circular orbit around the dwarf planet.

(a) Determine the gravitational acceleration due to the dwarf planet at the position of the satellite.

(b) By equating this to the centripetal acceleration of the satellite in its orbit, determine the period of the orbit.

13 Between the bodies on the line joining their centres, the gravitational forces due to the dwarf planet and satellite act in opposite directions.

(a) Determine the distance from the satellite at which these two forces are equal and opposite.

(b) State the significance of this point for the gravitational potential energy variation along the line.

4.3 Orbits and the wider universe

Section 4.2 discussed Newton's law of gravitation. The investigation into the properties of gravity and gravitational mass and the motion of bodies under their influence has led to many important developments in our understanding of the universe. The story of Johannes Kepler's[1] development of these revolutionary laws, which not only debunked the philosophical notion that celestial (and therefore perfect) objects could only travel in perfect (i.e. circular) paths but also dethroned the Earth from its central position in the universe, is well worth looking into. The observational and experimental work of Galileo and mathematical work of Newton would eventually explain them and consolidate the idea that the whole of the universe was subject to the same physical laws.

4.3.1 Kepler's laws of planetary motion

Using the very accurate observations of the Danish astronomer, Tycho Brahe, Kepler was able to promulgate three laws which sum up how planets move relative to the Sun. These were purely empirical laws, i.e. they did not explain the motion but they described it to a much greater precision than Copernicus. In fact, they also apply to comets, dwarf planets, **TNOs** and asteroids, etc., i.e. all objects in orbit around the Sun; they also apply separately to objects in orbit around planets.

(a) Kepler's 1st and 2nd laws

As with Newton's laws we shall refer to Kepler's laws as K1, K2 and (later) K3. The first two laws can be written as:

K1: The orbit of a planet is an ellipse with the Sun at one focus.

K2: The line joining the planet to the Sun sweeps out equal areas in equal intervals of time.

For most planets it is difficult to tell that the orbit is not circular. Mars has the most eccentric of any of the easily observed planets; its path drawn in Fig. 4.3.2 departs from circular by about the thickness of the line!

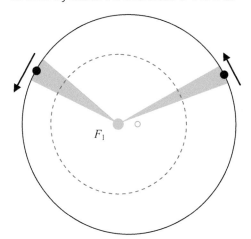

Fig. 4.3.2 Orbit of Mars

In Fig. 4.3.2, the yellow and black discs represent the Sun and Mars (not to scale) and the red pecked line, the orbit of the Earth. The Sun is at, F_1, one of the foci of the ellipse – the other focus is shown by the small grey circle.

Fig. 4.3.1 Kepler's portrait on a German postage stamp

MATHS TIP

See the Maths section for the terminology and mathematical properties of ellipses.

▼ **Study point**

When applying Kepler's laws to moon systems, we replace *planet* with *moon* and *Sun* with *planet*. Thus:

The orbit of a moon is an ellipse with the planet at one focus.

▼ **Study point**

Note that K2 refers to an individual planet (comet, TNO…) in its orbit. It doesn't compare the areas swept out by different bodies. The movement of different objects in orbit around the same central body is the subject of Kepler's third law (K3).

---

1 He was a mathematician, astronomer and astrologer who successfully defended his mother from a charge of witchcraft.

▼ Study point

The comet's orbit shown has a semi-major axis of 10 AU and an eccentricity of 0.866. The orbit of Halley's comet is about the same size but is even more eccentric. The red pecked line represents the Earth's orbit.

4.3.1 **Self-test**

Using your ruler to measure the maximum and minimum distances of the comet from the Sun, compare the gravitational potential energy of the comet at P and A.

NON-SI UNITS

Many calculations in astronomy are carried out in **astronomical units** (AU).

1 AU = semi-major axis of Earth's orbit

$= 1.496 \times 10^{11}$ m

The year, 3.156×10^7 s, is often used as a unit of time.

Another unit of distance is the light-year (l-y) = 9.46×10^{15} m.

4.3.2 **Self-test**

The mass of 67P/Churyumov-Gerasimenko has been estimated as $(1.0 \pm 0.1) \times 10^{13}$ kg. Use this to calculate:

(a) the gravitational potential energy

(b) the kinetic energy

(c) the total mechanical energy of the comet at perihelion.

(Mass of Sun = 1.99×10^{30} kg)

Stretch & Challenge

Assess whether the data in the example are consistent with the Principle of Conservation of Energy.

The elliptical nature of cometary orbits is much more obvious.

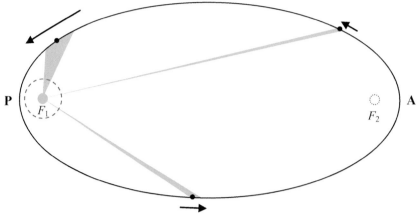

Fig. 4.3.3 Cometary orbit

The grey areas in the two figures illustrate K2. They represent equal time slices in the orbits. The comet in Fig. 4.3.3 travels much faster when it is close to the Sun (near **perihelion**) than near **aphelion**, but the areas swept out are the same.

Example

Comet 67P/Churyumov-Gerasimenko has a speed at perihelion of **38 km s⁻¹** when its distance from the Sun is **1.2432 AU**. The semi-major axis of its orbit is **3.4630 AU**. Using Kepler's 2nd law, calculate:

(a) the area swept out by its radius vector per second

(b) its speed at aphelion.

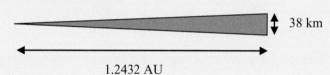

Answer

(a) The diagram shows the area swept out per second at perihelion.

$1.2432 \text{ AU} = 1.2432 \times 1.496 \times 10^8 \text{ km} = 1.86 \times 10^8 \text{ km}$

$\therefore$ Area swept out per second $= \frac{1}{2} \times 38 \times 1.86 \times 10^8 = 3.53 \times 10^9 \text{ km}^2 \text{ s}^{-1}$

(b) Aphelion $= 2 \times$ semi-major axis $-$ perihelion

$= 2 \times 3.4630 - 1.2432 = 5.6828 \text{ AU}$

$= 5.6828 \text{ AU} = 8.50 \times 10^8 \text{ km}$

From K2: Area per second is constant, $\therefore 3.53 \times 10^9 = \frac{1}{2} \times 8.50 \times 10^8 \, v$

$\therefore$ Velocity at aphelion, $v = 8.3 \text{ km s}^{-1}$

In this example, the speed at aphelion could also be calculated without using the area swept out per second. Because of the geometry:

$$v_A A = v_P P,$$

where A = aphelion (distance)   and P = perihelion distance.

(b) Kepler's 3rd law

This law compares the motion of different bodies in orbit around the same central body.

K3: The square of the period, T, of the orbit is proportional to the cube of the semi-major axis, a, of its orbit.

This law can be used to compare the orbital motion of all the planets, dwarf planets, asteroids, comets, TNOs, which orbit the Sun. It can also be used for a family of extra-solar planets around an individual star or the satellites of a planet such as Jupiter. It *cannot* be used to compare objects which orbit around different bodies, e.g. the moons of Saturn and Uranus.

Using T for the period of orbit, K3 can be expressed as $T^2 \propto a^3$, so for orbiting bodies 1 and 2:

$$\frac{T_1^2}{T_2^2} = \frac{a_1^3}{a_2^3} \qquad \text{or} \qquad \frac{T_1}{T_2} = \left(\frac{a_1}{a_2}\right)^{\frac{3}{2}}$$

This tells us that, as we might expect, the period increases with mean distance from the Sun (i.e. the semi-major axis) but more rapidly than a proportional relationship.

Example

The mean distance of the Earth from the Sun (the semi-major axis of its orbit) is 1.496×10^{11} m. Saturn's orbital distance is 1.427×10^{12} m. Show that the orbital period of Saturn is approximately 30 years.

Answer

From K3: $\dfrac{T^2}{a^3}$ = constant. The period of the Earth's orbit is 1 year.

$$\therefore \frac{1^2}{(1.496 \times 10^{11})^3} = \frac{T_s^2}{(1.427 \times 10^{12})^3}, \text{ where } T_s = \text{period of Saturn (in years).}$$

$$\therefore T_s = \left(\frac{14.27}{1.496}\right)^{\frac{3}{2}} = 29.46 \approx 30 \text{ years}$$

4.3.2 Newton's laws and orbital motion

To a very good approximation, bodies in orbit around the Sun move under the influence of just one force; that is the Sun's gravity. The gravitational force of other Solar System objects can usually be ignored because their masses are so much smaller than the Sun; the interplanetary medium exerts a negligible drag and the pressure of sunlight is miniscule.[2] In these circumstances, as Newton showed, objects move in paths which are elliptical, parabolic or hyperbolic. This section explores how Newton explained Kepler's laws in terms of his principles of force, mass and acceleration together with his universal law of gravitation, which we met in Section 4.2. We'll start with the equal areas law.

(a) Kepler's 2nd law – equal areas

We won't do a full derivation here, but we'll point in the direction of the way that Newton explained it in his treatise *Principia Mathematica*. We'll leave some of the work as a Stretch & Challenge exercise.

---

2 In fact all these three forces can exert influences in some circumstances, e.g. when a comet approaches a planet, drag in the early stages of the formation of the Solar System and the effects of sunlight causing outgassing of comets.

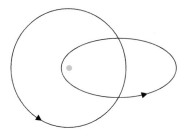
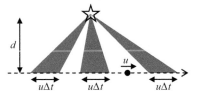

Study point

In Fig. 4.3.5 the area swept out per second is $\frac{1}{2}ud$. We'll use this in the 'derivation' of K2.

Stretch & Challenge

In the derivation of K2 we applied the 'nudge' at the special point of closest approach. The challenge is to repeat the proof for a general point as in this diagram.

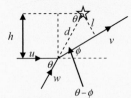

Hints:

1   Express l in terms of h and θ.

2   Use the sine rule.

4.3.5 Self-test

How does this derivation of K2 depend upon Newton's laws of motion?

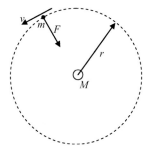

Fig. 4.3.7 Force on an orbiting body

Study point

An alternative derivation uses the angular velocity, $\omega = \frac{2\pi}{T}$.
From Section 3.1, the centripetal force is $mr\omega^2$. Equating this to the gravitational force gives

$mr\omega^2 = G\frac{Mm}{r^2}, \therefore \omega^2 = \frac{GM}{r^3}$.

Hence $T^2 = \frac{4\pi^2 r^3}{GM}$.

This law is a consequence of the fact that gravity is a *central force*, i.e. the force on a planet is always directed towards the Sun and the force on a satellite is directed towards its planet. To give us a run up we'll examine the case of a planet moving past a zero mass star – so the gravitational force is zero! In Fig. 4.3.5, the speed of the planet is a constant u and the closest distance of approach is d. The shaded areas represent the K2 areas in equal times Δt. The area of a triangle is ½ base × height so all the triangles have area $\frac{1}{2}ud\Delta t$, i.e. the areas are all the same.

Now we'll apply Newton's strategy and imagine that, as the planet passes the star, it receives a 'nudge' from the star, which adds a velocity, w, in the direction of the star, changing its velocity from u to v. This is illustrated in Fig. 4.3.6.

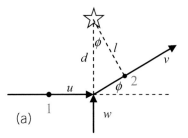 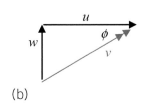

(a)   (b)

Fig. 4.3.6 (a) Space diagram!   (b) Vector diagram

The area swept out per second at position 1 $= \frac{1}{2}ud$

At position 2 the area per second $= \frac{1}{2}vl$

From the vector diagram $v = \frac{u}{\cos\phi}$; from the space diagram, $l = d\cos\phi$

$\therefore \frac{1}{2}vl = \frac{1}{2}\frac{u}{\cos\phi}d\cos\phi = \frac{1}{2}ud$, i.e. the area swept out per second is the same after the nudge as before. Newton argued that the steady force from the star could be thought of as a large number of small nudges so the area swept out per second would be constant as long as the force was always in the direction of the star.

(b) Kepler's 3rd law – distance / period relationship

Once again, we'll simplify the problem (to keep it within the realms of A level Physics and Mathematics) and consider an object in a circular orbit. The various quantities are defined in Fig. 4.3.7 and we'll use Newton's law of gravitation and the concept of centripetal acceleration [see Sections 4.2 and 3.1] as well as N2.

The resultant force, F, on an object moving at constant velocity in a circle is directed towards the centre of the circle and is given by:

$$F = \frac{mv^2}{r} \qquad [1]$$

The only force on the object is the gravitational force $\frac{GMm}{r^2}$, so we can write

$$\frac{mv^2}{r} = \frac{GMm}{r^2}$$

$\therefore$ Simplifying: $v^2 = \frac{GM}{r}$. $[2]$

The velocity is related to the period, T, of the orbit by $v = \frac{2\pi r}{T}$.

Substituting for v in equation [2] and re-arranging gives

$$T^2 = \frac{4\pi^2 r^3}{GM} \qquad [3]$$

A circle is an ellipse with eccentricity zero, i.e. the major and minor axes are equal. Thus we can consider r in equation [3] to be the semi-major axis, a, of the elliptical orbit. The K3 result, that T^2 is proportional to a^3, follows. The significance of equations [2] and [3] cannot be over stated. By measuring the orbital speed or period, we can determine the mass of the central body. This is explored in greater detail in Section 4.3.3.

(c) Kepler's 1st law – elliptical orbits

The derivation of K1 is beyond the scope of this book – indeed beyond A level Mathematics. However, it is worth noting that Newton did indeed show that it was true – for a bound orbit. He showed, further, that objects with sufficient kinetic energy to escape to infinity follow an open hyperbolic orbit.

The orbits in Fig. 4.3.8 illustrate the following cases:

1   Elliptical orbit: the sum of the kinetic and potential energies is negative, i.e. $E_K + E_P < 0$

2   Parabolic orbit: $E_K + E_P = 0$

3   Hyperbolic orbit: $E_K + E_P > 0$

Periodic comets are clearly bound to the Sun and follow elliptical orbits, which are often modified by close encounters with planets. Some comets are one-offs, i.e. they appear from deep space (probably from the Oort cloud) and have parabolic or hyperbolic orbits. Often the number of observations on such comets is insufficient to distinguish between a high-eccentricity elliptical orbit, a parabolic or a hyperbolic orbit.

(d) The energy of orbiting bodies

As Section 4.3.2(c) indicated, the total mechanical energy of bodies which orbit in closed elliptical orbits is negative. The reason for this is that, by definition, the potential energy of an object an infinite distance from a gravitating body is zero. Thus bodies with a positive total mechanical energy would be able to escape to infinity.

First, we'll consider a body in a circular orbit, as in Fig. 4.3.6. The total mechanical energy, E, is given by:

$$E = \text{kinetic energy} + \text{gravitational potential energy}$$
$$= \tfrac{1}{2}mv^2 - \frac{GMm}{r}$$

From equation [2] in Section 4.3.2(c): $v^2 = \dfrac{GMm}{r}$. Substituting for v^2 gives

$$E = \tfrac{1}{2}\frac{GMm}{r} - \frac{GMm}{r}$$

So the total mechanical energy of an object is a circular orbit is $-\tfrac{1}{2}\dfrac{GMm}{r}$.

Now we'll consider elliptical orbits. Remembering that the period of an elliptical orbit depends only on its semi-major axis (i.e. it is independent of the eccentricity – see Fig. 4.3.6 again) it should be no surprise that the total energy of an orbiting object also depends only on the semi-major axis, a. This means that we can write:

$$\text{Energy of an object in an elliptical orbit} = -\tfrac{1}{2}\frac{GMm}{a}.$$

The proof of this requires a bit of algebra. See Stretch & Challenge.

Self-test 4.3.6

The semi-major axis of the orbit of Miranda about Uranus is 129 400 km and the period is 1.413 days. Calculate the mass of Uranus.

$G = 6.67 \times 10^{-11}\ \text{N m}^2\ \text{kg}^{-2}$.

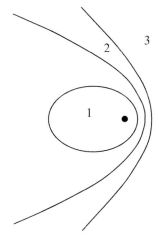

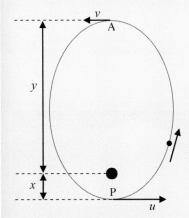

Fig. 4.3.8 Different orbits

Stretch & Challenge

The diagram shows an object in an elliptical orbit. Show that its total energy is given by $E = -\tfrac{1}{2}\dfrac{GMm}{a}$.

Hint: The energies at **A** and **P** are equal. Express the total energy in terms of u and x separately in terms of v and y. Use K2 to eliminate one of the velocities and hence find the total energy in terms of $a\ (= x + y)$.

Self-test 4.3.7

Use the data in S-t 4.3.2 and the value of semi-major axis to calculate the orbital energy of the comet 67P/Churyumov-Gerasimenko.

Compare the two answers.

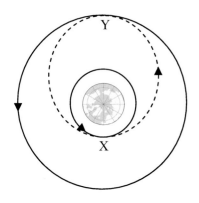

Fig. 4.3.9 Transfer orbit

4.3.8 Self-test

Calculate the energy boost required at **Y** for geosynchronous insertion.

Example

A 1-tonne satellite is to be boosted from its circular low-earth parking orbit (radius 6700 km) to a geosynchronous orbit (42 200 km). It does this via the transfer orbit shown in Fig. 4.3.9 (diagram not to scale). Calculate the energy boost required at the insertion into the transfer orbit at point X.

Mass of Earth = 5.98×10^{24} kg

Answer

Energy of parking orbit $= -\frac{1}{2}\frac{GMm}{r} = -\frac{1}{2}\frac{6.67 \times 10^{-11} \times 5.98 \times 10^{24} \times 1.0 \times 10^3}{6700 \times 10^3}$

$$= -5.95 \times 10^{10} \text{ J}$$

Energy of transfer orbit $= -\frac{1}{2}\frac{GMm}{a} = -\frac{1}{2}\frac{6.67 \times 10^{-11} \times 5.98 \times 10^{24} \times 1.0 \times 10^3}{\frac{1}{2}(6700 + 42200) \times 10^3}$

$$= -1.63 \times 10^{10} \text{ J}$$

∴ Energy boost required $= -1.63 \times 10^{10} - (-5.95 \times 10^{10}) = 4.32 \times 10^{10}$ J

Note: the satellite requires another energy boost at **Y** to be inserted into the geosynchronous orbit. Otherwise it will continue in the pecked orbit and return to **X**. See Self-test 4.3.8. There are lots more calculations that can be carried out on transfer orbits – see Self-test 4.3.8

▼ Study point

Newton's law of gravitation is valid for point objects and also for spherically symmetric objects. Thus the orbit equations work well for stars and planets.

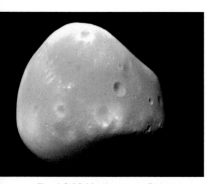

Fig. 4.3.10 Martian moon Deimos

4.3.3 Using orbits to determine the mass of objects

In this section we'll look first at orbits around compact bodies, e.g. planets, stars and black holes. Then we'll examine the orbits of stars in galaxies.

(a) Orbits around compact bodies

In Section 4.3.2(b) we obtained the following equations:

$$T^2 = \frac{4\pi^2 a^3}{GM} \quad [2] \qquad \text{and} \qquad v^2 = \frac{GM}{r} \quad [3]$$

We can apply equation [2] to any elliptical orbit. Equation [3] is only applicable to circular orbits but inserting the semi-major axis, a, instead of r will enable us to calculate an average figure of orbital speed [strictly the rms speed]. Hence, measuring the radius (or semi-major axis) of an orbit and either the orbital period or speed enables astronomers to determine the mass of the central object star or planet. If the diameter of the central body is known, its mean density can also be calculated, as in the following example. This enables planetary scientists to develop models of planetary composition.

Example

Deimos, the smaller of the two moons of Mars, has an orbit with a semi-major axis of 23 460 km and a period of 30.35 hours. The diameter of Mars is 6750 km. Calculate the mean density of Mars.

Answer

The mass, M, of Mars is given by $M = \dfrac{4\pi^2 a^3}{GT^2} = \dfrac{4\pi^2(23460 \times 10^3)^3}{6.67 \times 10^{-11} \times (30.35 \times 3600)^2}$
$$= 6.40 \times 10^{23}\ \text{kg}$$

Density $\rho = \dfrac{M}{V} = \dfrac{M}{\frac{4}{3}\pi r^3} = \dfrac{3 \times 6.40 \times 10^{23}}{4\pi(3375 \times 10^3)^3} = 3970\ \text{kg m}^{-3}$

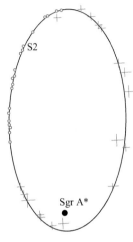

Fig. 4.3.12 The orbit of S2 around Sgr A*

This figure for the density of Mars is much lower than the mean density of the Earth, which is 5520 kg m^{-3}. This suggests that Mars has proportionately much less iron than the Earth does. Also the absence of a significant magnetic field suggests that any iron core is completely solid. This fits in with the fact that small objects (Mars is about half the diameter of the Earth) cool more rapidly than large ones.

A more exciting example of the power of orbits to reveal mass is in the nature of the powerful radio source, known as Sagittarius A*, at the centre of our galaxy – see Fig. 4.3.11. Various stars in the vicinity of **Sgr A*** have been tracked over the past 20 years. The positions of one of them, known as **S2**, are shown schematically in Fig. 4.3.12. The crosses are error bars; the circles have error bars which are too small to plot. Analysis of the data reveals an object of several million solar masses within a space of the order of the size of the solar system.

(b) The rotation curves of the Solar System

A plot of orbital speed for objects circling the Sun is shown in Fig. 4.3.13. This sort of graph is called a **rotation curve**. The equation of the curve is
$$v = \sqrt{\dfrac{GM}{r}}$$
where M is the mass of the Sun, 1.99×10^{30} kg

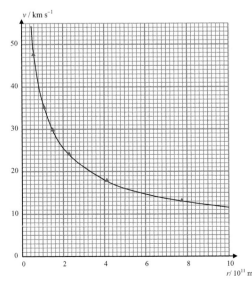

Fig. 4.3.13 Solar system rotation curve

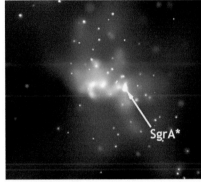

Fig. 4.3.14 The Sun's orbit

4.3.11 Self-test

(a) Express the distance of the Sun from the centre of the galaxy in light-years.

(b) Show that: 1 pc = 3.26 light-years

(c) Express the distance of the Sun from the centre of galaxy in parsecs.

4.3.12 Self-test

Check the 160 km s^{-1} figure for the Sun's orbital speed, for a mass within the orbit of 1.0×10^{41} kg.

The rotation curve has this simple form because:

• the Sun contains the great bulk of the mass in the Solar System,

• the Solar System objects orbit outside the Sun and

• the Sun is almost spherically symmetric.

These features of the Solar System mean that we can treat the motion of any orbiting body as being subject only to the gravitational force due to a point mass in the centre of the Sun.

(c) The rotation curve of spiral galaxies

The stars in our galaxy orbit the centre of the galaxy. The image in Fig. 4.3.14 gives an idea of the appearance of our galaxy seen from 'above' the galactic plane and the position and the circular orbit of the Sun: radius $\sim 2.7 \times 10^{17}$ km, orbital speed ~ 220 km s^{-1}. Measuring the orbital speed of stars and gas clouds within our own galaxy is not easy but it has been done for a large number of other spiral galaxies using the Doppler effect (see Section 4.3.4). The result for a very similar galaxy called NGC[3] 3198 is shown in Fig. 4.3.15.

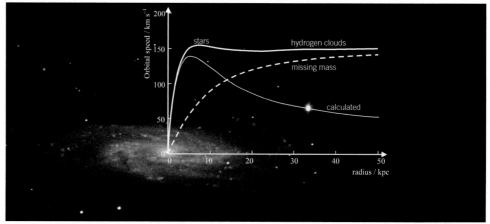

Fig. 4.3.15 Rotation curve of NGC 3198

The upper solid curve is the measured orbital speeds of objects. The inner part of the curve is from star motion and the outer parts from motion of the of hydrogen clouds. The galaxy extends a long way beyond its visible disc: there are large numbers of hydrogen clouds which are invisible but which emit characteristic 21 cm radiation (see AS / Year 12 book page 77) the speed of which can be measured by radio astronomers, again using their Doppler shifts.

We saw in Section 4.2 that the orbital motion of an object is determined by the total mass of the material within the orbit (see Study point). Hence we can use the orbit of the Sun to determine the total mass within the orbit.

Using $v^2 = \dfrac{GM}{r}$, $M = \dfrac{rv^2}{G}$

$\therefore$ Mass within the Sun's orbit $= \dfrac{2.7 \times 10^{20} \times (220 \times 10^3)^2}{6.67 \times 10^{-11}} = 2.0 \times 10^{41}$ kg.

Measurements of the mass of stars and interstellar clouds within the galaxy give a maximum mass within the Sun's orbit of 1.0×10^{41} kg, meaning that the Sun's orbital speed should be about 160 km s^{-1}. So the Sun should fly off into intergalactic space, along with most of the other stars outside the central bulge. This appears to be true of all galaxies which have been investigated. It is also true of clusters of galaxies; there is apparently not enough mass to hold galaxy clusters together.

Fig. 4.3.16 Coma galaxy cluster: gravitationally unstable without dark matter

3 NGC = New General Catalogue of Nebulae and Clusters of Stars

(d) Dark matter holds galaxies together

The lowest curve in Fig. 4.3.15 represents the rotation curve calculated on the basis of the conventional 'baryonic' matter in the galaxy. It drops off roughly as $r^{-0.5}$ beyond the visible galaxy. However, the observed velocity is almost constant out to several times the visible galactic radius. So, assuming Newton's law of gravitation continues to be valid at these distances, there is a lot of 'missing mass' in the galaxy. How much? At 50 kpc, the 'calculated' velocity is about one third of the observed velocity. But $M \propto v^2$, so the total mass within 50 kpc of the centre is about 9 times the observed mass; i.e. we are missing nearly 90% of the mass of the galaxy!

This missing mass only appears to be detectable by its gravitational effect, i.e it doesn't interact by the electromagnetic force, otherwise we would detect radiation from it. It is referred to as **nonbaryonic dark matter**. There is other observational evidence for its existence.

1   Evidence from **Gravitational lensing**.

   One of the successful predictions of Einstein's general theory of relativity was that the path of light should be bent by massive objects. This lensing effect of galaxies and clusters of galaxies sometimes leads to multiple images of objects as in the Chandra image of the 'clover leaf quasar' in Fig. 4.3.17.

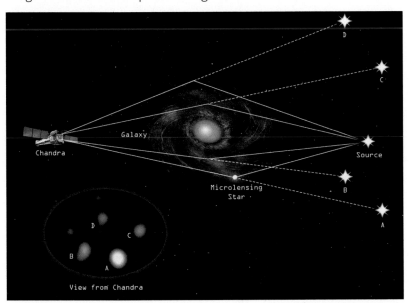

Fig. 4.3.17 Gravitational lensing

Light from the quasar, labelled Source, is diverted into multiple paths, each of which leads to an image. The degree of light bending reveals that the galaxy has much more mass than in the stars and gas clouds.

2   Evidence from the **Cosmic Microwave Background Radiation (CMBR)**.
   The CMBR retains the imprint of density fluctuations in the early universe. Baryonic matter and nonbaryonic matter have different effects. Detailed investigation of the **CMBR** images (see Fig. 1.6.19 in the AS book) leads to the conclusion that 85% of the matter in the universe is nonbaryonic.

Stretch & Challenge

The 'missing mass' curve in Fig. 4.3.15 is the rotational speed due to the dark matter alone:

(a)   Explain why the observed velocity is not equal to the sum of the calculated and dark matter velocities.

(b)   What distribution of matter (dark + baryonic) could account for the constant velocity portion of the rotation curve (see Hints).

Terms & definitions

Nonbaryonic matter is hypothetical material which is not composed of quarks and leptons, which only interacts via the gravitational (and possibly the weak) interactions. It is the main component of **dark matter**.

▼ Study point

The different paths of light due to gravitational lensing often means that the variation of light in the various images of a compact source, such as a quasar, show the same fluctuations but time shifted because of the different path lengths. Time shifts of months to years have been observed.

▼ Study point

It has been suggested that the Higgs boson could be the source of dark matter. The Higgs normally decays into bosons (such as the W or Z particles) or into fermions (such as $\tau^+ + \tau^-$ or bb^-). Swedish theoretician Christopher Petersson has suggested that it could also decay into a photon plus a dark matter particle. No evidence at current energies has been found but at the time of writing (August 2015) the suggestion is being investigated by two teams at CERN.

▼ Study point

See WMAP and Planck mission websites for more information on the evidence for dark matter.

Stretch & Challenge HINTS

1   If $\rho = \rho(r)$, the mass out to r,
   $M(r) = \int_0^r 4\pi x^2 \rho(x)\,dx$

2   $v(r) = \sqrt{\dfrac{GM(r)}{r}} = \text{const.}$

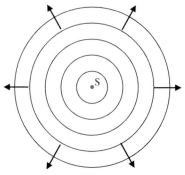

Fig. 4.3.18 Waves from a stationary source

▼ **Study point**

The **radial velocity** of the source in Fig. 4.3.19 is positive. A source moving towards the observer has a negative radial velocity. Such an observer would observe a higher frequency and a shorter wavelength, e.g. if the observer were to the right of the source in Fig. 4.3.19.

Stretch & Challenge

Derive the equation $\frac{\Delta f}{f_0} = -\frac{v}{c}$ from $\lambda = \lambda_0 + \frac{v}{c}\lambda_0$.

Hint: You will need to use the binomial approximation:

$(1 + x)^n \approx 1 + nx$ for $x \ll 1$

▼ **Study point**

These equations are not valid for speeds approaching the speed of light. For more information, search for the 'relativistic Doppler effect'.

4.3.13 Self-test

The radial velocity of the Andromeda galaxy (M31) is given on a website as $-301 \pm 1 \text{ km s}^{-1}$.

The laboratory wavelength of Hα is 656.281 nm. Calculate the measured wavelength of Hα emissions from M31, together with its absolute uncertainty.

4.3.4 Measuring radial velocity using the Doppler effect

The Austrian physicist Christian Doppler (1803–1853) predicted that the observed frequencies of sound waves depend on the motions of the source and observer[4]. He also predicted the same effect with light waves and this effect, together with the Newtonian analysis of Kepler's laws provides a very powerful tool for the investigation of the universe.

Consider the stationary source, **S**, of light waves of in Fig. 4.3.18. The light waves have a frequency f_0 and spread out with the velocity of light, c, producing a set of spherical wavefronts, centred on **S**. A stationary observer in any position will determine the frequency to be f_0 and wavelength to be λ_0, where $c = \lambda_0 f_0$.

Fig. 4.3.19 shows the same source moving away from the observer with a velocity, v. In the context of astronomy, this is called the **radial velocity**.

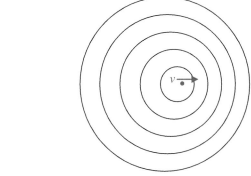

observer

Fig. 4.3.19 Source moving away from observerobserver

The wavefronts are still spherical but their centres move progressively to the right, so the waves which propagate towards the observer have a longer wavelength and therefore a lower frequency. We can derive the magnitude of the effect as follows:

Consider wavefronts which are one cycle apart. The time, T, between the emission of the wavefronts is given by $T = \frac{1}{f_0}$.

In this time the source has moved a distance to the right $= \frac{v}{f_0} = \frac{v}{c}\lambda_0$.

So the observed wavelength, $\lambda = \lambda_0 + \frac{v}{c}\lambda_0$

∴ The fractional change in wavelength, $\frac{\Delta\lambda}{\lambda_0} = \frac{\lambda - \lambda_0}{\lambda_0} = \frac{v}{c}$.

Similarly, the equation for the fractional frequency change, $\frac{\Delta f}{f_0} = -\frac{v}{c}$, which is valid as long as $v \ll c$.

Many astronomical objects have line spectra, either emission or absorption, and the wavelengths of the lines have been accurately measured in the laboratory (see AS book Section 1.6.4). Measuring the wavelengths (or frequencies) of these lines in the spectra of astronomical objects will tell us their radial velocities. In many cases this will enable us to determine a quantity of gravitating material.

Example

A hydrogen gas cloud is observed $8.5 \times 10^{17} \text{ km}$ from the centre of a spiral galaxy which is viewed edge on. The frequency of the 21.1 cm hydrogen line is observed to have a Doppler shift of -0.62 MHz. Estimate the mass of the galaxy within the radius of the gas cloud's orbit.

---

4 This was confirmed in 1845 by the Dutch meteorologist Buys-Ballot in an experiment involving a flat-bed train, a brass band and a group of listeners with perfect pitch! Like all scientists, Doppler didn't always get things right – he tried to explain the different colours of stars by assuming they were moving towards us (blue stars) or away from us (red stars).

Answer

Laboratory frequency of 21.1 cm hydrogen line $= \frac{c}{\lambda} = \frac{3.00 \times 10^8}{0.211} = 1420$ MHz.

$\therefore$ Using $\frac{\Delta f}{f_0} = -\frac{v}{c}$, $v = \frac{-c\Delta f}{f_0} = \frac{3.00 \times 10^8 \times 0.62}{1420} = 131\,000$ m s^{-1}

Using $v^2 = \frac{GM}{r}$, $M = \frac{rv^2}{G} = \frac{8.5 \times 10^{20} \times 131\,000^2}{6.67 \times 10^{-11}} = 2.2 \times 10^{41}$ kg

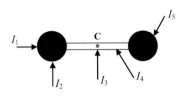

Fig. 4.3.20 Impulses on a barbell

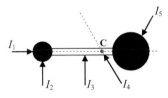

Fig. 4.3.21 Asymmetric barbell

4.3.5 Bodies in mutual orbit

In many cases, one body in a system is overwhelmingly more massive than all the other bodies put together. In these cases we can do as we have done up to now and assume that the massive body is at rest and the light body (or bodies) orbits around it. For example, Mercury and Venus can be considered to orbit around a Sun which is metaphorically nailed to the spot; the stars orbiting SgrA* – see Section 4.3.3(a) – don't tug the black hole noticeably out of position. However, there are many pairs of objects of similar mass, such the Pluto / Charon system and binary stars, which orbit around each other. In order to analyse these we'll first need to develop the concept of the **centre of mass**.

(a) Centre of mass

We'll start with a thought experiment. The barbell in Fig. 4.3.20 is floating in space, away from gravitational influences, and is hit separately by various impulses, $I_1 - I_5$. Three of these, I_2, I_4 and I_5, cause the barbell to rotate as well as to move. Two of them, I_1 and I_3, only cause it to move: I_1 makes it move to the right and I_3 upwards. This is because the lines of action of these two forces pass through the middle of the bar (the red dot). With an asymmetric barbell the 'no-rotation' point, called the **centre of mass (C)**, is shifted towards the more massive ball (Fig. 4.3.21).

The position of the centre of mass for two small objects is calculated using the equation:

$$m_1 r_1 = m_2 r_2$$

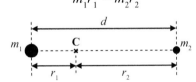

Fig. 4.3.22 Centre of mass

When we investigate binary stars or double planets we shall need to express the position of C in terms of the separation of the bodies. You should be able to show (see Self-test 4.3.14) that:

$$r_1 = \frac{m_2}{m_1 + m_2}d \qquad \text{and} \qquad r_2 = \frac{m_1}{m_1 + m_2}d$$

When we study the motion of binary systems, e.g. the orbit of a binary star system, we can consider the motion in two parts:

1. For the system as a whole, we consider the whole mass to be concentrated at the centre of mass.

2. For the motions within the system we consider the motions relative to the centre of mass.

So: the Earth and Moon orbit around their common centre of mass (see Self-test 4.3.14) and (strictly) it is this centre of mass that orbits in an elliptical orbit around the centre of mass of the Earth-Moon-Sun system rather than the Earth orbiting the Sun. We'll now see how to apply this.

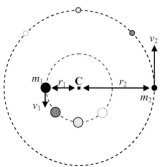

Fig. 4.3.23 Mutual orbit

▼ **Study point**

$(m_1 + m_2)$ is the total mass of the binary system, so we could rewrite the formula as $T = 2\pi\sqrt{\dfrac{d^3}{GM_{tot}}}$.

▼ **Study point**

Notice the relationship between the formulae:

$T^2 = \dfrac{4\pi^2 a^3}{GM}$ and $T = 2\pi\sqrt{\dfrac{d^3}{GM_{tot}}}$.

4.3.16 ⌄ Self-test

The speed of one component of the binary pair in the example is 58 km s^{-1}. Calculate:

(a) the radius of its orbit, and

(b) the masses of the two stars.

┌─ *Terms & definitions* ─┐

An **exoplanet** is a planet which orbits a star other than the Sun.

A **brown** dwarf is a stellar-type object with too low a mass for hydrogen fusion.

▼ **Study point**

If the plane of the orbit is tilted through θ, the radial velocity fluctuates by $v \cos \theta$. A level exam questions will usually assume that the observer is in the plane of the orbit.

4.3.17 ⌄ Self-test

The plane of a star's orbit is at $15°$ to the line of sight to the Earth. The star's orbital speed is 150 ± 10 km s^{-1}. Compare any error caused by ignoring the angle of inclination with the uncertainty in the speed.

(b) Orbits of binary objects

We'll consider circular orbits: as usual, the results also apply to elliptical orbits. The two objects from above are in mutual orbit – see Fig. 4.3.23. We are only considering motion within the system, so the centre of mass remains stationary. Starting off with the object in the black position, the velocities v_1 and v_2 must be such that $m_1 v_1 = m_2 v_2$, otherwise the centre of mass would move. Hence each orbits around the centre of mass (fading into the future!) with the same period.

To calculate the period of the orbit, consider the motion of m_1:

The centripetal force on m_1 is provided by the gravitational force from the other mass, i.e.

$$m_1 r_1 \frac{4\pi^2}{T^2} = \frac{G m_1 m_2}{d^2} \qquad \text{where we have used} \qquad \omega = \frac{2\pi}{T}.$$

Substituting $r_1 = \dfrac{m_2}{m_1 + m_2} d, \rightarrow \dfrac{m_1 m_2}{m_1 + m_2} d \dfrac{4\pi^2}{T^2} = \dfrac{G m_1 m_2}{d^2}$

Dividing by $m_1 m_2$ and rearranging $\rightarrow T = 2\pi\sqrt{\dfrac{d^3}{G(m_1 + m_2)}}$

Example

A close pair of stars has a separation of **5 million km** and orbit each other with a period of **2.5 days**. Calculate the total mass of the system.

Answer

$$T^2 = \frac{4\pi^2 d^3}{GM_{tot}}, \text{ so} \qquad M_{tot} = \frac{4\pi^2 d^3}{GT^2} = \frac{4\pi^2 (5 \times 10^9)^3}{6.67 \times 10^{-11} \times (2.5 \times 86400)^2}$$
$$= 1.6 \times 10^{30} \text{ kg}$$

More information is needed to find the masses of the individual components of the binary system, e.g. the orbital speed of one of them (see Self-test 4.3.16).

If the separation of the bodies is not known, other information such as the orbital speeds of both objects also enables the system to be solved. This is explored below.

4.3.6 Exoplanets and binary stars

An observer on an exoplanet would not be able to image the Earth directly with current Earth technology because it would be lost in the glare of the Sun. However, there are two ways in which the existence of the planets or invisible low mass companions, such as red or **brown dwarfs**, could be inferred and their properties investigated.

(a) The Doppler technique

The star and the unseen companion are in mutual orbit. The radial velocity of the star changes throughout its orbit; if our line of sight is in the plane of the orbit, the amplitude of the variation or radial velocity is the orbital speed of the star. This can be detected by Doppler shift in the wavelength of the star's Fraunhofer lines.

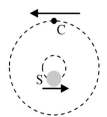

Fig. 4.3.24 Doppler method

In Fig. 4.3.24 the radial velocity of the star, **S**, is negative, giving a negative $\Delta\lambda$. Half an orbit later $\Delta\lambda$ is positive; the companion, **C**, is detected only by this oscillating Doppler shift.

If the period and amplitude of the wavelength variation are determined, the mass of the planet can be estimated – even without using the *mutual orbit* formula in Section 4.3.5(b). This is illustrated by an example – the details of the calculations are left for self-test questions.

Example

The wavelength of the sodium D-line (laboratory value 589 nm) in the spectrum of a star of mass 1.0×10^{30} kg was found to vary by ± 0.20 pm with a period of 1.0×10^{6} s, due to the effect of an unseen orbiting companion. Estimate the orbital distance and mass of the companion.

Answer method

Start by assuming that the mass of the star is much greater than that of the companion, so we can use the equations in Section 4.3.3, then:

1. Use $T^2 = \dfrac{4\pi^2 r^3}{GM}$ to find the orbital distance of the companion, then calculate its orbital speed using $v = \dfrac{2\pi r}{T}$.

2. Find the speed of the star using the Doppler shift.

3. Finally use $M_s v_s = m_c v_c$ to estimate the mass of the unseen companion.

4. Check the initial assumption.

Now work through these steps – see Self-test 4.3.18.

You should have found that the mass of the companion to be about 1.4×10^{27} kg, making the star ~700× as massive. This makes the original assumption reasonable. It is worth checking that, putting in this value of m_C and using the given values of M_S and T give a value of $\Delta\lambda$ which is indistinguishable from the measured value.

If the two objects in a binary system, such as Pluto and Charon, have comparable masses, it is helpful to have extra information such as the orbital velocity of the two objects. The graphs in Fig. 4.3.25 are typical.

The whole system is moving away from us with a velocity v, derived from the Doppler shift. The two components, A and B, have orbital velocities given by $v_A = v_1 - v$ and $v_B = v_2 - v$ (assuming we are in the orbital plane). The masses of the two components can be determined from these and the orbital period, T – see the Study point and Self-test 4.3.20.

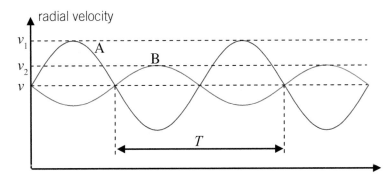

Fig. 4.3.25 Objects in mutual orbit

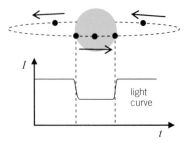

Fig. 4.3.26 Transit method

─── Terms & definitions ───

Hubble's law states that the **recession velocity**, v, of deep-space objects is proportional to their proper distance.

The **recession velocity** of a deep space object is the Hubble shift interpretation of their red shift.

A correct definition of the **proper distance,** D, of a deep-space object is out of the scope of this book but it is, in essence, the distance from us to the current position of the object.

The **Hubble constant,** H_0, is the gradient of the v-D graph for deep-space objects.

(b) The Transit method

If the observer is more or less in the plane of the orbit of the companion, the starlight will dim periodically as the planet passes in front of the stellar disk. This is shown schematically in Fig. 4.3.26. Search *Kepler mission* for more information. The duration of the eclipse and its depth provide information about the speed of the companion and its diameter. Both these properties depend upon knowledge of the diameter of the star itself. Up to January 2015, the Kepler spacecraft had found over 1000 planets using this technique. In conjunction with Doppler measurements, which deliver the mass of the planet, the mean density and therefore its nature can be investigated.

A close binary star system which we happen to view edge on, such as MY Camelopardalis [MY Cam] is called an *eclipsing binary.* Its components have 31 and 38 solar masses and have an orbital period of 1.17545 days. In fact, they are so close that their outer atmospheres overlap. Such pairs are also called *contact binaries.* The details of the light curve and Doppler measurements reveal a wealth of data. This system is explored in Exercise 4.3. The variation in brightness shown in the light curve in Fig. 4.3.27 is due to the fact that at some periods in the orbit, one of the components is obscured by the other.

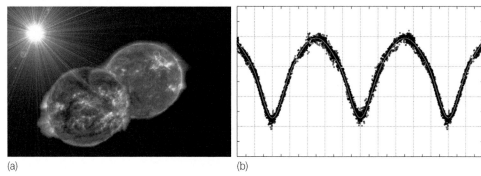

(a)   (b)

Fig. 4.3.27 MY Cam (a) artists impression and (b) light curve

4.3.7 The expansion of the universe

(a) Hubble's law

Astronomers have discovered that all objects outside our local cluster of galaxies have a spectrum which is red-shifted, which can be interpreted as a Doppler shift showing a relative velocity away from our galaxy. This velocity, v, is found to be approximately proportional to their distance, up to a distance of several hundred megaparsecs. This relationship, known as **Hubble's law**, can be expressed as:
$$v = H_0 D,$$

where H_0 is the **Hubble constant** and D is the distance (strictly the **proper distance**) to the object.

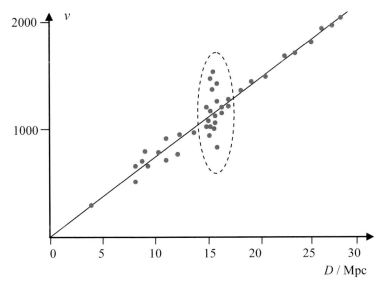

Fig. 4.3.28 Recession velocity – distance relationship

The value of the Hubble constant is not easy to determine. The recession velocity emerges from the red shift, but the distance to objects beyond about 100 kpc cannot be determined directly by parallax (see the AS book p73). A common method is to use the apparent brightness of **standard candles** such as type 1a supernovae. The details of this method are beyond the scope of this book but are well worth investigating.

The red dots in Fig. 4.3.28 represent nearby galaxies. The scatter about the line of best fit is due to the motion of the galaxies <u>through</u> space. The line of best fit itself represents the expansion <u>of</u> space. The galaxies within the pecked ellipse are members of the Virgo cluster of galaxies.

At the time of writing, the consensus value of H_0 is $67.9 \text{ km s}^{-1} \text{ Mpc}^{-1}$, which means that a galaxy with a proper distance of 20 Mpc has a recession velocity of $20 \times 67.9 = 1360 \text{ km s}^{-1}$.

Why are all galaxies moving away from us? Are we the 'centre of the universe'? No. The whole of space is expanding and the galaxies are embedded in this expanding space. Fig. 4.3.29 is a schematic illustration of a region of space, the dots being clusters of galaxies. As space expands, the clusters recede from one another: an observer in *any* galaxy cluster observes all the other clusters receding, with a red-shift proportional to distance.

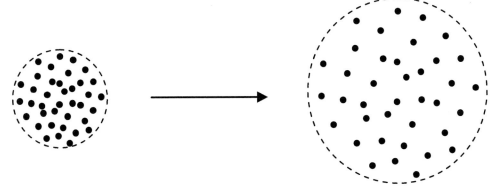

Fig. 4.3.29 The expansion of the universe

UNITS

From its definition, the SI unit of the **Hubble constant** is s^{-1}. Cosmologists normally express it in $\text{km s}^{-1} \text{ Mpc}^{-1}$.

Terms & definitions

A **standard candle** in astronomy is an object of known (or assumed!) luminosity, such as a Type 1a supernova or a Cepheid variable star. Measuring the apparent brightness of such an object in a distant galaxy allows the distance to be calculated, using the inverse square law.

Self-test 4.3.22

Show that the value of the Hubble constant given in the main text $(67.9 \text{ km s}^{-1} \text{ Mpc}^{-1})$ is equivalent to $2.20 \times 10^{-18} \text{ s}^{-1}$.

$[1 \text{ pc} = 3.08568 \times 10^{16} \text{ m}]$

Self-test 4.3.23

An Hγ emission line ($\lambda_{lab} = 434$ nm) from a distant galaxy has a measured wavelength of 456 nm. Use the Hubble constant and the red shift to estimate the distance of the galaxy (a) in **Mpc** and (b) in **m**.

Stretch & Challenge

A very neat way of considering the Hubble red-shift is to think of the light from a distance object as being 'embedded' in space: as space expands, so does the wavelength of the light. Calculate by what factor the universe has expanded since:

(a) the light in Self-test 4.3.23 was emitted

(b) the CMBR was emitted.

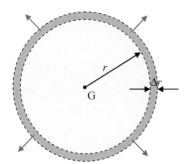

Fig. 4.3.30 Expanding shell

▼ **Study point**

According to the General Theory of Relativity, the geometry of space itself relates to its energy density. If the energy density is negative, space curves in on itself; a positive density gives an outward curvature. If the energy density is zero then space is said to be *flat*, which means that Pythagoras' theorem is obeyed and angles in a triangle add up to 180°.

— **Terms & definitions** —

The **Hubble time** is an estimate of the age of the universe and is defined as the reciprocal of the **Hubble constant**.

4.3.24 Self-test

Show that $t_H = 14.4 \times 10^9$ years.

(b) Will the universe expand for ever?

Treating the Doppler shift, for the moment, as representing the speed of recession of galaxies *through space* rather than with space, we can use Newton's law of gravitation to investigate this question. We'll consider the matter in a shell of space at a radius, r, from our galaxy, G. Assuming that the universe is homogenous:[5]

The mass of the shell, $m_S = 4\pi r^2 \rho \Delta r$, and

The mass within the shell, $M_S = \frac{4}{3}\pi r^3 \rho$

where $\rho =$ the mean density of the universe.

The recession speed of the shell, $v_S = H_0 r$.

The KE and PE$_{grav}$ of the shell are: $E_k = \frac{1}{2}m_S v_S^2$ and $E_p = -\dfrac{GM_S m_S}{r}$.

Substituting for M_S, m_S and v_S from above and simplifying gives:

$E_{Tot} = E_k + E_p = 2\pi r^4 \rho H_0^2 \Delta r - \frac{16}{3}G\pi^2 r^4 \rho^2 \Delta r$ – you should check this.

If $E_{Tot} > 0$, the shell will expand for ever; if $E_{Tot} < 0$, it will reach a maximum distance and then collapse; If $E_{Tot} = 0$ it will just come to a halt at an infinite distance and the universe has a *flat geometry* (see Study point). For a flat universe, then, setting the expression for E_{Tot} to zero and rearranging leads to the following expression for density:

$$\rho_c = \frac{3H_0^2}{8\pi G},$$ where ρ_c is called the *critical density*.

Current measurements of baryonic matter show that its density is much less than this. But the universe appears to have a flat geometry – more evidence for the existence of dark matter.

(c) How old is the universe

We have seen that the accepted value of the Hubble constant at the current time is 2.20×10^{-18} s^{-1} (see Section 4.3.7(a) and Self-test 4.3.22). In other words, for every metre an object is away its recession speed is 2.20×10^{-18} m s^{-1}. *If this rate of expansion were constant* then the time, t_H, since all matter was gathered in the same infinitesimal spot is given by:

$$t_H = \frac{1}{2.20 \times 10^{-18} \text{ s}^{-1}} = 4.55 \times 10^{17} \text{ s}.$$

And more generally: $t_H = \dfrac{1}{H_0}$, where t_H is called the **Hubble time.**

---

5 .... meaning that matter is spread evenly through space. This is obviously not correct on the small scale [the Earth is a lot denser than interplanetary space, to say nothing of neutron stars] but it is a pretty good approximation on the scale of hundreds of Mpc

We might expect the expansion of the universe to be slowing down because of the effect of gravity. If this is true, then it was expanding more quickly in the past and therefore the time from the big bang is less than the Hubble time. This is illustrated in Fig. 4.3.31.

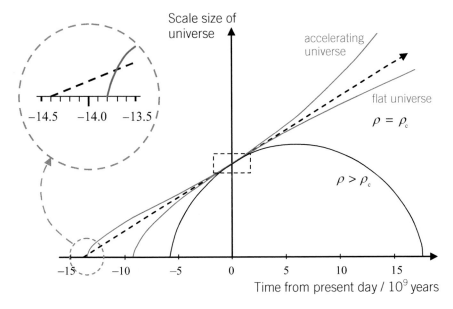

Fig. 4.3.31 Expansion models of the universe

The pecked arrow represents a universe with a constant rate of expansion. This is a universe without any matter in it, so no gravitational deceleration. The red line shows a 'flat' universe, with the same observed value of the Hubble constant: the gradient within the pecked box is the same. Notice that this universe is 'only' 9 billion years old. A 'closed' universe, i.e. with enough gravity to stop and reverse the expansion would be even younger. However, our universe is none of these: recent measurements show that it has been *accelerating* for the last 5 billion years (blue line). Co-incidentally the resulting best age estimate for the universe (13.8 billion years) is very close to the **Hubble time** (14.4 billion years).

▼ **Study point**

The 'scale size' plotted in Fig. 4.3.31 is any representative length which expands with the universe, e.g. the mean distance between clusters of galaxies.

Self-test 4.3.25

The wavelength of the Hδ line (laboratory $\lambda = 410$ nm) of the radiation from a distant source is measured and found to be 820 nm. Estimate the distance of the sourse according to the four graphs in Fig. 4.3.31.

[Hint: see the Stretch & Challenge at the end of Section 4.3.7(a)].

Stretch & Challenge

The first indications that the universe is accelerating came from measurements of the brightness and red-shift of distant type 1a supernovae; for a particular red-shift the supernovae are fainter than expected. Explain this.

[Hint: Consider your answer to Self-test 4.3.25]

Exercise 4.3

1 A planet has two moons, A and B. Moon A has a period of 15 days; the period of moon B is 30 days. Find the ratio of the semi-major axes of the orbits.

2 The ratio of the orbital radii of two moons of a planet in a circular orbit is 2.25. Calculate the ratio of the orbital periods.

3 The semi-major axis of the orbit of Io around Jupiter is 421 700 km. Its period is 1.769 days.

(a) Calculate the mass of Jupiter.

(b) The orbital distance of Callisto (another Jovian moon) is about 4.5 times that of Io. Estimate its orbital period.

4 The periods and orbital radii of the five major satellites of Uranus are given in the table to two significant figures. You will use these to investigate Kepler's 3rd law.

Satellite name	Miranda	Ariel	Umbriel	Titania	Oberon
Orbital radius / 10^3 km	130	190	270	440	580
Orbital period / day	1.4	2.5	4.1	8.7	13.5

(a) In view of the spread of data, explain why a plot of T^2 against r^3 is an inappropriate graph.

(b) What is the gradient and intercept of a graph of $\ln T$ against $\ln r$?

(c) What values of m and n should give a straight line graph in a plot of T^m against r^n?

(d) Select a suitable pair of values of m and n from part (c), plot a graph of T^m against r^n.

[Hint: the values of r are approximately equally placed], use it to test Kepler's 3rd law and estimate the mass of Uranus.

5 According to a website, the star S2 orbits the radio source SgrA* with a period of 15.8 ± 0.11 years [see Section 4.3.3(a)]. The semi-major axis of the orbit has an apparent angular size of $(5.832 \pm 0.13) \times 10^{-7}$ rad and recent measurements give the distance of SgrA* from the Earth as 7.94 ± 0.42 kpc. Newspaper reports identify the radio source as a black hole with a mass of 4 million Suns ($M_{Sun} = 1.99 \times 10^{30}$ kg).

(a) Ignoring the uncertainty values, calculate the semi-major axis of the orbit and hence the mass of the radio source. Does it agree with the newspaper reports?

(b) The perinigron [closest approach to the black hole] is estimated at 17 light-hours. Calculate the speed of S2 at this point. [Hint: consider the total orbital energy.] What fraction of the speed of light is this?

(c) Use the uncertainty values to estimate the uncertainty in your value of the mass of SgrA*.

6 A neutron star of mass $1.4M_\odot$ and a white dwarf of mass $0.6M_\odot$ are in circular orbits about their centre of mass with a period of 15 hours. Calculate:

(a) the separation of the stars and hence the radii of their orbits

(b) the orbital speed of each of the two bodies

(c) the observed Doppler shift of the Hα spectral line (laboratory wavelength 656.281 nm).

4.4 The magnetic field

We say that there is a magnetic field, or an electric field (Section 4.2), in any region where electric charges experience forces. The key difference between the fields is that only moving charges are affected by magnetic fields. With electric fields, both moving and stationary charges experience forces. See **Terms and definitions** and **Study point**.

We'll be looking at the magnetic forces (forces due to applied magnetic fields) on free electrons moving in wires that are carrying electric currents, and on charged particles moving in empty space in particle accelerators. Later we'll study how currents in wires, straight or coiled, act as *sources* of magnetic fields.

Considering the name 'magnetic field', you might well ask: where do magnets fit into all this? The moving charges are orbiting and spinning electrons inside the magnets' atoms. The electron spins of neighbouring atoms in regions called 'domains' (Fig 4.4.1) align so that their effects re-inforce. In magnets the domains themselves are predominantly aligned roughly in one direction. All this is clearly quite complicated, or at any rate not exactly basic – which is why magnets aren't studied in depth at A-level. Nevertheless, by their very familiarity, magnets do give us an easy way into our work on magnetic fields…

4.4.1 The direction of a magnetic field

A useful way to explore magnetic fields is with a small magnet, pivoted so it can point in any direction. A magnetic 'plotting compass' is usually good enough, though its magnet can point only in horizontal directions.

If we find that a freely-pivoted magnet points in a particular direction we know that it's in a magnetic field. We can use this for our second definition: the **direction of a magnetic field** (**Terms and definitions**).

(a) Magnetic field lines (lines of magnetic flux)

By moving a plotting compass so that its South pole is successively where its North pole was previously, we can trace or 'plot' a magnetic field line. If we start in different places we can plot (in theory) as many lines as we like in a region.

Fig. 4.4.2 shows what emerges when we investigate the region around a bar magnet in this way. The compass magnet's **North pole** points in the direction of the resultant magnetic field, in this case the resultant of the magnet's field and the Earth's field. As long as we stay near the magnet, its field is much stronger than the Earth's, so the pattern of field lines is almost entirely due to the magnet.

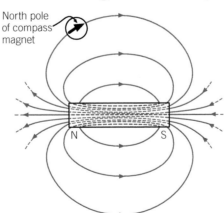

North pole of compass magnet

Fig. 4.4.2 Magnetic field lines of a bar magnet (horizontal section)

Terms & definitions

We say that a **magnetic field** exists in a region in which a moving electric charge experiences a force, but a stationary charge does not.

The direction of a magnetic field in a certain place is the direction in which the North pole of a small, freely pivoted magnet tends to point.

The **North pole (or North-seeking pole) of a magnet** is the end that tends to point roughly geographically North if only the Earth's magnetic field is present.

▼ Study point

A charge that is stationary in our lab will be moving if viewed from a passing vehicle. Doesn't that make nonsense of the first paragraph in 4.4? No: it tells us that magnetic (and electric) fields depend on our frame of reference. At A-level we stick with one frame. Phew!

Fig. 4.4.1 False colour image of oppositely magnetised domains – image diameter ~ 0.2 mm

4.4.1

Self-test

Magnetic field lines *emerge* from the magnet's North pole. Deduce this from the 'like poles repel, unlike poles attract' rule for magnets. [Remember how the lines are plotted!]

┌─ *Terms & definitions* ─┐

A magnetic field line (or line of magnetic flux) is a line whose direction at every point is the direction of the magnetic field at that point.

▼ **Study point**

The properties of magnetic field lines are analogous to those of electric field lines, except that electric field lines due to stationary charges are not closed loops, but start on positive charges and end on negative. Arguably, the concept of electric charge is indispensible, but we could do without that of a magnetic pole. See later…

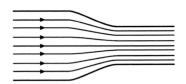

Fig. 4.4.3 Streamlines for liquid flowing in a narrowing pipe

4.4.2   Self-test

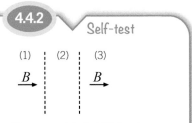

The arrows labelled '*B*' denote magnetic field directions in two regions, (1) and (3). The field in (3) is stronger. Use the properties of magnetic field lines to show that the direction of the field can't be the same all over the transition region (2).

(b) Properties of magnetic field lines

So what *is* a magnetic field line? The definition is very straightforward (see **Terms & definitions**). Note the following properties of the lines, using Fig. 4.4.2 as an illustration:

1   Magnetic field lines never cross or meet. If they could, which way would a compass point at their intersection? In fact, it points in the direction of the *resultant* field, which is everywhere well defined.

2   Magnetic field lines are closed (end-less) loops. Although we can't show this with a plotting compass, the lines that arrive at the South pole of a magnet continue (from S to N) through the magnet, emerging at the North pole.

3   When field lines become closer together, the field is becoming stronger – as near the poles of a magnet.

Compare the lines with streamlines in a smoothly flowing liquid… The direction of a streamline shows the direction of liquid velocity at a point, and where streamlines become closer together, as when the pipe in Fig. 4.4.3 narrows, the magnitude of the velocity becomes greater. Indeed, magnetic field lines are sometimes called *lines of magnetic flux*, the primary meaning of the word 'flux' being 'flow'. Don't take the analogy too far: we don't believe anything is actually flowing along a magnetic field line.

4.4.2 The motor effect: force on a current-carrying conductor

We're now ready to move on to the central topic of this chapter: the forces experienced by charges moving in magnetic fields.

We find that a wire, carrying a current, placed in a magnetic field, at an angle to the field, experiences a force. This is 'the motor effect'. Fig. 4.4.4(a) shows one way of demonstrating it.

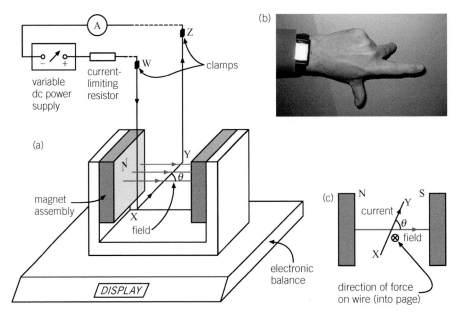

Fig. 4.4.4 The motor effect and Fleming's left hand motor rule

(a) The direction of the motor effect force

In the apparatus of Fig. 4.4.4(a) we are concerned with the force on the horizontal portion, XY, of the current-carrying wire, WXYZ. Note the position of XY, at an angle to the magnetic field direction.

With a current of a few ampère in the XY direction, XY experiences an appreciable downward force. The magnet assembly producing the field experiences an upward force of equal magnitude. This shows up on the electronic balance supporting the assembly. See **Study point**.

The force on the wire isn't an attraction or repulsion, as it is at right angles to both the magnetic field and the current. **Fleming's left hand motor rule** (FLHMR) has been verified by countless observations and we will be using it to *predict* the direction of motor effect forces. Note the alliterative association of **F**irst finger with **F**ield, and so on. Fig. 4.4.4(b) shows how the rule works for XY in Fig. 4.4.4(a). Do try it with your own left hand; no injuries have been reported over more than a century of the rule's use ...

The motor effect, then, has to be understood in three dimensions, but there's often a simple alternative to drawing a diagram that tries to project a third dimension on to a piece of paper. We draw a section, as in Fig. 4.4.4(c), adding dot or cross symbols:

- ⊙ based on an arrow coming towards us, means a current, field or force – we have to label it with which it is – coming out of the page;

- ⊗ based on the tail-feathers of an arrow, means something going into the page. Again it must be labelled.

The magnetic field direction is usually labelled 'B', the current, I, as in **Self-test 4.4.3**.

(b) The magnitude of the motor effect force

Again investigating this with the apparatus in Fig. 4.4.4, we find, when the other variables are held constant, the force is proportional to:

- the length, ℓ, of the wire (XY),

- the current, I,

- sin θ, in which θ is the angle between current and field.

The force must surely also increase with the strength of the field. But we can't even attempt to discover an exact relationship by experiment, as we haven't defined in a quantifiable way what we *mean* by the strength of a magnetic field! We turn the situation on its head and use the motor effect itself to provide us with a definition...

We define the magnitude, B, of the **magnetic flux density** (the official name for magnetic field strength) as

$$B = \frac{F}{I\ell\sin\theta}$$

in which F is the magnitude of the force on a short wire of length ℓ, carrying a current, I, at angle θ to the field direction.

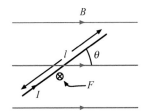

Fig. 4.4.5 Motor effect

[Note how, if we use twice the current, the force will be twice as much, so we'll get the same value for B, which is as it should be, because the field can't depend on what we're using to measure it. Similar points can be made about sin θ and ℓ, but see **Study point**.]

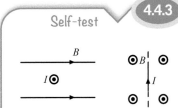

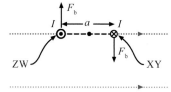

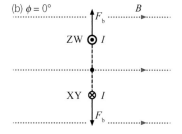

Fig. 4.4.8 (a) maximum torque and (b) zero torque

The equation tells us that the SI unit of the magnetic flux density, B, must be $\text{N A}^{-1} \text{m}^{-1}$. We give this combination the name 'tesla', abbreviated to 'T', so $\text{N A}^{-1} \text{m}^{-1} = \text{T}$.

Nicola Tesla (1856–1943) was an American pioneer of alternating current technology, who revelled in high voltages and high frequencies.

Re-arranging the defining equation for B,

$$F = BI\ell\sin\theta$$

Fig. 4.4.6 Nicola Tesla

(c) The magnetic flux density vector

Magnetic flux density is a vector. We can find its direction, as already explained, using a pivoted magnet. In the equation $F = BI\ell\sin\theta$, $B \sin\theta$ is the component of the vector perpendicular to the wire, sometimes written in one lump as $B_\perp$. See Fig. 4.4.5.

It may seem messy to define the direction of the vector in terms of a pivoted magnet, and its magnitude in terms of a current-carrying wire. It's fine to stick with these definitions, but see if you prefer the one-stop-shop approach given in **Terms & definitions**.

(d) Example of the motor effect: the current-carrying coil

We examine the forces on a rectangular current-carrying coil (of N turns) orientated with the normal to its plane at an angle ϕ to a uniform magnetic field.

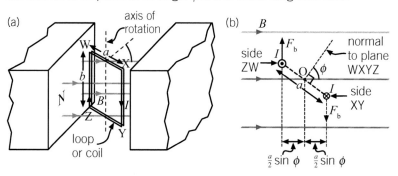

Fig. 4.4.7 Torque on a coil: (a) general view (b) horizontal section

First we note that, according to FLHMR, the force on side WX is upwards (towards the top of the page) and the force on YZ is downwards. These forces are not only equal and opposite, but in the same vertical plane – that of the coil. Their only effect is to stretch the coil – unnoticeably.

The forces (of magnitude F_b) on XY and ZW are also equal and opposite (see Fig. 4.4.7(b)), but this time they are 'staggered' apart by a distance of $a \sin\phi$, making up what is called a 'couple'. Adding the moments of these forces about the 'axis of rotation' (which goes through point O) …

Total torque, $\tau = 2F_b \times \dfrac{a}{2}\sin\phi = 2NBIb \sin 90° \times \dfrac{a}{2}\sin\phi$

Tidying up, $\tau = NBIA \sin\phi$

Here we have put $ab = A$, since ab is the area of the coil. There's more to this substitution than mathematical short-hand: it turns out that the last equation applies for a flat coil of *any* shape, A being the coil area.

Note that the effect of having N turns on the coil is to make the effective current N times larger.

The torque on a coil is exploited in an electric motor. The current through the coil has to be reversed as the coil turns through the positions where the field through it changes direction, to prevent the torque direction reversing. Most motors have several coils at different angles so the total torque never even drops to zero. The torque is increased by winding the coils on iron 'cores'. See Fig. 4.4.9

Fig. 4.4.9 Simple electric motor: coils, iron cores (green), magnets (blue and white)

Self-test   **4.4.6**

Insulated wire is wound around the edges of a door to make a flat, 250-turn coil, 0.75 m × 2.00 m. The open door is in the magnetic North-South plane. Taking the horizontal component of the Earth's field to be 20 µT, calculate the current needed to start the door closing, against a frictional torque of 1.2 N m.

4.4.3 The effect of magnetic fields on moving charges

(a) The force on a moving charge

Since a current in a metal wire is made up of moving free electrons, it seems likely that the motor effect force on a current-carrying wire is simply equal to the sum of forces acting on the individual moving electrons. See second **Study point**. If so we can easily find the magnetic force on an electron with drift velocity v …

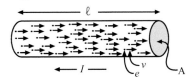

Fig. 4.4.10 Current in wire

First, recall that if the wire has cross-sectional area A and has n free electrons per unit volume, then the current will be $I = nAve$

We can therefore write the motor effect force on length ℓ of the wire as

$$F_{\text{wire}} = BI\ell \sin \theta = BnAve\ell \sin \theta$$

But the volume of the wire is $A\ell$, so there are $nA\ell$ free electrons in it.

Therefore force on each free electron $= \dfrac{BnAve\ell \sin \theta}{nA\ell} = Bev \sin\theta$

We now make the leap that this applies to any particle carrying charge q, provided we replace e by q. See **Terms and definitions.**

Terms & definitions

The magnetic force on a charge, q, moving in a magnetic field as shown has magnitude
$$F = Bqv \sin \theta$$

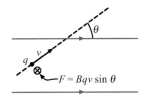

Fig. 4.4.11 Charge moving in magnetic field

The direction of the force is given by FLHMR.

▼ Study point

In $F = Bqv \sin \theta$, if $q < 0$, then $F < 0$, i.e. F acts in the opposite direction.

▼ Study point

The magnetic forces don't push drifting electrons out of the sides of a wire, because opposing electric field forces come into play. Strictly, the rest of the wire experiences the Newton 3 electric field 'partner' forces, rather than the magnetic forces themselves.

(b) Path of a charged particle moving in a uniform magnetic field

Suppose that a charge, q, moving at speed v is 'injected' into a uniform magnetic field, B, with angle θ between the directions of the field and the particle velocity. If no forces other than the magnetic force act on the particle, how will it move?

First note that the magnetic (motor effect) force is always at right angles to the particle's direction of motion. Therefore the field can't do work on the particle. Its kinetic energy, and hence its speed, remain constant.

Instead, the force causes the particle's *direction* of motion to change at a constant rate. That's unless $\theta = 0°$, in which case there will be no force so the particle will carry on moving parallel to the field lines, at speed v.

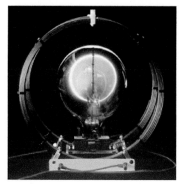

Fig. 4.4.13 Electrons' circular path in a uniform magnetic field

Stretch & Challenge

Show that for a helical or circular path (that is for any θ except 0° and 180°), the time for one turn is $\frac{2\pi m}{Bq}$.

Hence derive an equation giving the pitch of the helix (see Fig. 4.4.14) in terms of B, m, q, v, and θ.

Fig. 4.4.15 Particles trapped by Earth's magnetic field

Fig. 4.4.16 Aurora borealis photographed in Manitoba

4.4.7

Self-test

Electrons accelerated through a pd of 970 V are injected at right angles to the field direction into a magnetic field of strength 1.50 mT. The path radius is measured as 71 ±2 mm. Show whether or not this is consistent with the equations just developed.

[$e = 1.60 \times 10^{-19}$ C, $m_e = 9.11 \times 10^{-31}$ kg]

$\theta = 90°$

If, initially, the particle is moving at right angles to the field, it will continue to do so – in a circular path.

The motor effect force supplies the centripetal acceleration, so it's easy to find the radius of the circle, using $F = ma$…

Force on particle = mass × acceleration

So $Bqv = \frac{mv^2}{r}$. Hence $r = \frac{mv}{Bq}$

Fig. 4.4.13 shows electrons moving in a circular path – the bright ring. The light comes from gas molecules excited by hits from electrons and de-excited by photon emission. [The gas pressure in the glass vessel is very low so that some electrons can complete a circle without hitting molecules!] The fast electrons are injected by an electron gun at the bottom of the ring. The magnetic field, almost uniform in the region of the ring, is made by passing current through the large (red) coils of wire.

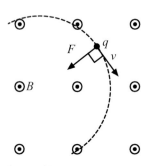

Fig. 4.4.12 Force on a + charge moving in a uniform magnetic field

θ neither 0° nor 90°

We can resolve the particle's velocity into a component $v_{\parallel} = v\cos\theta$ parallel to the field direction and a component $v_{\perp} = v\sin\theta$ at right angles to it. By themselves, $v_{\perp}$ would give circular motion, and $v_{\parallel}$, unaffected by the field, would give straight line motion parallel to the field lines.

A combination of these two motions results: the particle traces out, at constant speed, a spiral staircase path called a helix (adjective: helical). It may help to imagine yourself viewing the particle from a platform moving with velocity $v_{\parallel}$. From this viewpoint the particle would be moving in a circle with velocity, $v_{\perp}$

Fig. 4.4.14 Helical path of particle in uniform magnetic field

Paths resembling helices are traced out by charged particles, mainly protons and electrons emitted by the Sun, that are trapped by the Earth's magnetic field and condemned to gyrate around field lines.

The Earth's magnetic field is believed to be generated by the motion of molten iron-rich material in the Earth's core. On a large scale the pattern of the Earth's field lines is quite similar to the pattern that would arise from a small but very strong magnet near the centre of the Earth and aligned at a few degrees to the Earth's axis of rotation. Clearly, then, the field is not uniform on a large scale, so the paths of the gyrating particles are not true helices.

Fig. 4.4.15 shows what happens in a stylised way. At near-polar latitudes the field lines are close together. They also pass through the upper atmosphere, where the fast-moving charged particles that they have captured may collide with molecules, exciting them. The light emitted during de-excitation is a major cause of the spectacular displays in the night sky that are the Northern and Southern 'lights' (aurora borealis and aurora australis).

(c) Particle accelerators

From early in the 20th century, physicists have used high-speed particles as missiles to 'fire' at nuclei and, later, at smaller targets, such as nucleons. By noting how the missile particles are deflected, and what comes out of the target particles and in which direction(s), physicists have learned much about the structure of sub-atomic particles.

At first, α particles from radioactive sources were used as missile particles. This is how the very existence of the atomic nucleus was revealed. But for more detailed work, α particles (helium nuclei) have too much internal structure of their own, and not enough KE (usually a few MeV).

In the 1920s, physicists started to develop ways of accelerating particles such as electrons and protons. The electron gun (Section 4.1.7 (c)) is a simple form of accelerator – one in which the acceleration occurs in a single step. Unfeasibly enormous pds would be needed to give the particles enough KE for use as missiles in modern particle physics. The problem is overcome by accelerating the particle over and over again, through a large (but managable) pd …

The linear accelerator

Fig 4.4.17 shows a simplified linear accelerator (or linac). Suppose the particles to be accelerated are positive. They will be accelerated across the first gap into drift tube (1), when this is at a negative potential with respect to the particle production chamber (0). The particles will travel through the first drift tube at constant speed.

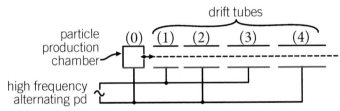

Fig. 4.4.17 Linear accelerator (simplified)

The pd needs to alternate at such a frequency that (2) is negative with respect to (1) when the particles reach the gap between (1) and (2), but the pd has reversed again so that (3) is negative with respect to (2) when the particles reach the gap between (2) and (3) – and so on. The increasing length of the drift tubes is designed to keep pace with the increasing speed of the particles, so that the frequency of the alternating pd can be roughly constant.

The longest linear accelerator in the world (3.2 km) forms a main part of the SLAC laboratory in California. It accelerates electrons or positrons up to kinetic energies of $50\,\text{GeV}$.

The cyclotron

The cyclotron is a particle accelerator that can take up much less space than the equivalent linear accelerator. This is because a cyclotron uses a magnetic field (at right angles to the particle velocity) to confine the charged particles to (semi)circular paths between the times when they acquire kinetic energy.

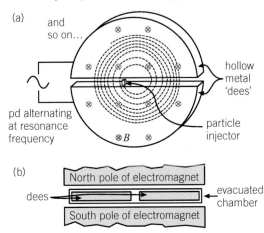

Fig. 4.4.18 A cyclotron: (a) path of particles (b) vertical section

Self-test

A mass spectrometer enables the masses of ions to be determined. Positive ions are accelerated by a known pd and a magnetic field is adjusted until the ions reach a detector by travelling round an arc of known radius.

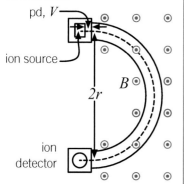

Singly ionised magnesium atoms, accelerated through 4500 V, are found to need a magnetic flux density of 0.296 T to travel a semicircular path of diameter 0.320 m. Identify the nucleon number of the magnesium isotope.

[$e = 1.60 \times 10^{-19}$ C, 1 kg $= 6.02 \times 10^{26}$ u]

▼ **Study point**

In a synchrotron we are usually dealing with particles near the speed of light, where Newtonian formulae for momentum and kinetic energy have to be modified. Particles can still go on acquiring kinetic energy from the electric fields, even though their speed increases very little!

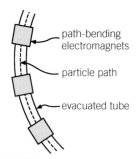

Fig. 4.4.19 Particle path in synchrotron

Fig. 4.4.20 Part of the Super Proton Synchrotron at CERN

A particle sent out from the injector gets faster each time it crosses the gaps between the metal 'dees', because a high voltage is applied between them. In Fig. 4.4.18 a positive particle will travel anticlockwise – use that left hand rule! – so the top dee will need to be negative and the bottom positive when the particle crosses the gap on the right, but the other way round when the particle reaches the left hand gap. Therefore the pd must alternate with a periodic time equal to the time, T_C, for the particle to make one revolution.

We now calculate T_C for when the particle has attained a speed v and a path radius r…

$$T_C = \frac{2\pi r}{v} = \frac{2\pi}{v} \times \frac{mv}{Bq} = \frac{2\pi m}{Bq}$$

The reciprocal of T_C gives us the frequency of revolution, the number of circles per second! It is sometimes called *the cyclotron resonance frequency*, f_C, but it applies for any charged particle moving at right angles to a uniform magnetic field, provided $v \ll c$ (the speed of light).

We have then: $f_C = \dfrac{Bq}{2\pi m}$

The really neat thing is that T_C and f_C are independent of v and r! The faster the particle the wider its circular path, and these factors cancel.

Therefore we should be able to stick with one frequency of alternating voltage for all particles of the same type anywhere in the cyclotron…

If only life were that simple… The cyclotron was invented in 1932 and was brilliant for producing fast protons to use as missiles in nuclear research. Cyclotrons are still used in hospitals to produce proton beams for cancer therapy and for bombarding nuclei to produce short-lived isotopes. Unfortunately, for cutting-edge experiments in Particle Physics the cyclotron simply can't give particles enough kinetic energy.

For one thing, as a particle's speed approaches the speed of light, its path radius increases faster than its speed, and the time per revolution increases. A cyclotron can be modified to deal with this, but a more serious issue is the very large path radii of high energy particles, which would require the magnetic field to be applied over a very wide disc-shaped region. The electromagnet would have to be vast.

The synchrotron

In a synchrotron the particles go round and round one circular path. Thus no magnetic field is needed at other radii. Even better, the field can be applied at regular stations along the path (Fig. 4.4.19) rather than all along it. All the same, the Super Proton Synchrotron (SPS) at CERN (Fig. 4.4.20) has 744 path-bending electromagnets along its 7 km circumference! In the SPS, the particles are given kinetic energy by electric fields in 'radio frequency cavities' at two places in the circular path, every time they go round it.

What is sacrificed in a synchrotron is the ability to produce high energy particles in a continuous stream. For the particles to remain in the same circular path as they gain KE, the magnetic field has to be increased continuously. The magnetic flux density that is needed for particles that have done many circuits of the synchrotron would be too large for particles at earlier stages in their ride. The only thing to do is to inject particles into the ring a batch at a time; in the SPS, typically around 10^{13} per batch, occupying less than the ring's circumference.

A by-product of a synchrotron's operation is 'synchrotron radiation': photons of up to X-ray frequencies emitted by the charged particles as they are deflected by the electromagnets and undergo radial acceleration. Although extra power has to be supplied to the beam by the electric fields to compensate for this energy loss, the synchrotron radiation may be used for other experiments.

4.4.4 'Crossed' electric and magnetic fields

(a) In a vacuum

A charged particle in a uniform electric field moves in a straight or parabolic path with changing speed (See Section 4.1.7), but, a charged particle injected into a uniform magnetic field moves in a straight, helical or circular path at constant speed. If we apply *both* E and B fields over the same region, a particle's trajectory is likely to be complicated, but we consider just one simple and useful case…

Suppose the E and B fields are at right angles to each other. A set-up for producing such fields is shown in Fig. 4.4.21. A charged particle is sent into the 'crossed fields' region travelling at speed v in a direction at right angles to both these fields (along the dotted line). If the fields are as shown (or if *both* fields are reversed!) the electric and magnetic forces on the particle will be in opposite directions. For zero resultant their magnitudes must be equal…

That is $\qquad Bqv = Eq \quad$ so $\quad Bv = E$

This is therefore the – delightfully simple – condition for the particle to continue moving in a straight line at constant speed, even though there are fields present.

This gives us a neat way of measuring the speed of the particles in a beam of charged particles: send the beam normally through crossed fields and adjust the field strengths until the beam is undeflected. Then $v = E/B$. A very similar use is as a 'velocity filter': the ratio E/B is set at a desired value, so if particles of various speeds are sent normally through the crossed fields, only those with speed E/B will go straight through. [In one arrangement these can pass through a hole in a metal electrode while others, with the 'wrong' speed, are obstructed.]

In the next section we meet crossed fields inside a solid…

Self-test 4.4.11

What happens to a particle initially at rest in:

(a) a uniform E field,

(b) a uniform B field?

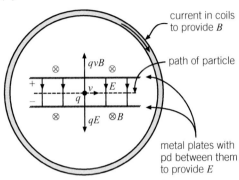

Fig. 4.4.21 Particle in crossed E and B fields

current in coils to provide B

path of particle

metal plates with pd between them to provide E

Self-test 4.4.12

Calculate the speed of a particle that passes undeflected through crossed E and B fields, if $B = 0.37$ T and E is provided by placing a pd of 3300 V between plates separated by 8.0 mm.

(b) In a conductor – the Hall effect

The Hall effect (discovered by Edwin Hall in 1879 – before the discovery of the electron) gave the first indication that the motor effect force on a current-carrying metal conductor arises from forces on negative charge-carriers moving inside it. The Hall effect results from the magnetic field exerting a force towards one side of the conductor on the carriers.

Although the effect occurs in metals, it is much greater in semiconductors, such as silicon. The thinner the specimen, the larger the effect. Thin pieces of semiconductor used to show the Hall effect are often called 'Hall wafers'. We shall first consider a wafer made of silicon doped (alloyed) with a small amount of phosphorus This material is called an 'n type' semiconductor for reasons we shall see.

▼ **Study point**

Connections to the external current-supplying circuit are made right across the small end-faces of the wafer, rather than to single points.

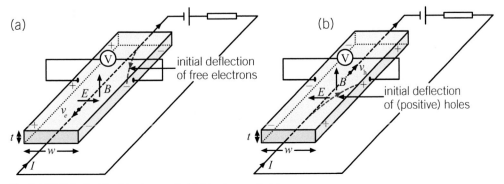

(a)

initial deflection of free electrons

(b)

initial deflection of (positive) holes

Fig. 4.4.22 Hall effect in: (a) n-type material (b) p type material

▼ Study point

There is a longitudinal electric field in the wafer, which causes the current. This means that the voltmeter will show a pd even with no magnetic field, if the pick-up points are not exactly opposite each other. This zero-error can be compensated for.

▼ Study point

Pure silicon has equal numbers of electrons and holes, but the drift speed of the holes is lower, so there is still a resultant Hall voltage.

Fig. 4.4.23 Hand-held Hall probe (the wafer is near the end of the stick)

4.4.13 Self-test

In copper, $n = 8.5 \times 10^{28}$ m^{-3}; in quite heavily doped n-type silicon n is typically ~ 1×10^{22} m^{-3}.

Explain in physical terms why the Hall effect is much greater in the silicon than the copper.

Stretch & Challenge

Discuss how you would seek to maximise the Hall voltage across a copper specimen. Consider both the size and shape of specimen, and the circuit in which it is included.

A circuit is set up to send a current (usually a few **mA**) lengthways through the wafer (Fig. 4.4.22 (a)). See **Study point**. With no applied magnetic field there should be no pd between points exactly opposite each other on the side faces, so the voltmeter should read zero.

When a magnetic field is applied in the direction shown, across the whole of the wafer – not difficult, because the area of its largest face may well be less than that enclosed by this letter 'O' – the voltmeter indicates a small pd, V_H with the left-hand face at a higher potential.

This is just what we'd expect if free electrons in the wafer had been deflected, as shown in Fig. 4.4.22 (a), in accordance with Fleming's Left Hand Motor rule (FLHMR), remembering their negative charge!

If we repeat the experiment with a wafer made of a p-type semiconductor, e.g. silicon doped with a smallish proportion of indium atoms (see Section 2.7.5 in the AS textbook), we find that the Hall voltage is the other way round. The right-hand face is at a more positive potential than the left hand. We infer that the mobile charge carriers in this wafer are predominantly the positive 'holes', and are deflected as shown in Fig. 4.4.22 (b), again in accordance with FLHMR.

The magnitude of the Hall voltage

Within a small fraction of a second of applying B, the side faces of the wafer acquire charge, and there will be a transverse (width-wise) electric field, E, inside the wafer. This will present further build-up of charge by producing a force to the left on the free electrons (or holes), equal and opposite to the magnetic force.

Thus $\qquad (\pm e)E = (\pm e)vB \qquad$ that is $\quad E = Bv$

This is the 'crossed fields' condition for the charge carriers, with drift speed v, to pass undeflected through the wafer. The magnitude of the Hall voltage, V_H, measured between points separated by the width, w, of the wafer with width-ways electric field strength, E, is (see Section 4.1.6)

$$V_H = Ew \qquad \text{so} \qquad V_H = Bvw$$

It helps to re-express v in terms of the current I through the wafer, using our old friend $I = nAve$. A is the cross-sectional area of the conductor normal to the current direction, that is the area, wt, of the small face.

So $\qquad\qquad v = \dfrac{I}{nwte}$

Thus $\qquad\qquad V_H = \dfrac{BIw}{nwte} \qquad$ that is $\qquad V_H = \dfrac{BI}{nte}$

in which n is the free electron or hole concentration (number per unit volume) and t is the thickness of the wafer (Fig. 4.4.22).

Note these features of the equation:

- V_H is proportional to B, the magnetic flux density. A Hall wafer, mounted in a 'probe' (Fig. 4.4.24), with a connecting cable to external current supply and voltmeter (usually calibrated directly in **mT**), provides a practical way of measuring B, even when it varies significantly over distances of a few mm. Measuring the force on a current-carrying wire isn't practicable in such cases (nor, indeed, we now confess, in most others)!

- The smaller the thickness, t, of the wafer and the smaller the free electron concentration, n, the larger V_H. This is because, for a given current, I, decreasing either of these factors increases the drift speed, v, and therefore the magnetic force on each charge carrier.

- The factor $\dfrac{1}{ne}$ in the equation for V_H is called the 'Hall coefficient'. It contains the quantities determined by the wafer *material*. A minus sign is sometimes put in front of it if the predominant charge carriers are negative.

4.4.5 The magnetic field due to a steady current

Almost all of this chapter so far has concerned the *effects* of magnetic fields. We've not discussed in any systematic way how the fields are related to their *sources*, though we've described the magnetic field around a bar magnet and mentioned the use of current-carrying coils to provide magnetic fields. It's time to put this right – by looking in more detail at the magnetic fields due to steady currents.

(a) The magnetic field due to a current in a circuit

A steady current requires a complete conducting circuit (including a cell or a power supply). We find that the circuit produces a magnetic field and that the lines of flux are linked with the circuit in the same topological way that handles of keys are linked with a key ring.

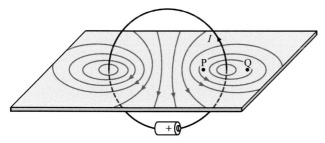

Fig. 4.4.24 Magnetic field due to current in a circuit

The direction of the field is given, as we can show with a compass, by the *right hand grip rule*: see **Terms and definitions**. You need to be able to use it confidently.

The rule can be applied all round the circuit; it's natural to suppose that each part of the circuit is contributing to the magnetic field. The flux density at any point in the vicinity of the circuit is the vector sum of flux densities due to all the little bits of the circuit.

(b) The magnetic field due to a current in a long straight wire

The wire will have to be part of a circuit, of course, but we shall consider the field only in a region which is close to the wire and far from (i) the **ends** of the straight part, (ii) the rest of the circuit.

We can plot the field with a compass (if the wire is vertical). The current, I, needs to be large enough for the wire's field to be much greater than the Earth's. The field lines are circles around the wire (Fig. 4.4.26).

The magnitude of the field strength (flux density), B, can be investigated with a Hall probe, orientated at each point to give maximum Hall voltage (See Section 4.4.8). We find that for points at various distances, r, from the centre of the wire, $B \propto \frac{1}{r}$, so doubling r halves B, and so on: inverse proportionality. We also find that at any given point $B \propto I$.

Self-test 4.4.14

Give one example of a circuit which carries a current, but not a steady current.

—— *Terms & definitions* ——

The right hand grip rule
Grasp the conductor in your right hand with your thumb sticking out in the direction of the current. The fingers will curl round in the direction of the field. See Fig. 4.4.25.

Fig. 4.4.25 Right hand grip rule

Self-test 4.4.15

Remembering that magnetic flux density is a vector, explain why we should expect the field lines to be closer at P than at Q in Fig. 4.4.24.

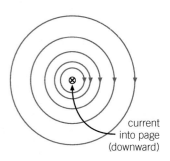

current into page (downward)

Fig. 4.4.26 Selected field lines due to current in long straight wire

4.4.16 Self-test

Using the data in the example:

(a) determine the magnetic flux density at point Y, 35 mm due North of the wire in Fig. 4.4.26,

(b) determine the force on 0.20 m of the wire due to the Earth's horizontal field.

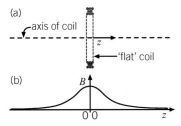

Fig. 4.4.28 (a) Section through a coil
(b) Magnitude of B along coil axis

The dependences of B on both r and I are contained in the equation

$$B = \frac{\mu_0 I}{2\pi r} \text{ in which } \frac{\mu_0}{2\pi} \text{ is the constant of proportionality}$$

μ_0 is a constant officially called *the permeability of free space*. Blame the 19th-century pioneers! Most of us call it 'mu nought'. Its value is

$$\mu_0 = 4\pi \times 10^{-7} \text{ T m A}^{-1} = 4\pi \times 10^{-7} \text{ H m}^{-1} \text{ (henry metre}^{-1})$$

This value seems suspiciously exact. It is, in fact, an artefact of the SI, arising from the definition of the ampère – see Section 4.4.6. The henry (**H**), is the unit of self inductance which is introduced in Section 4.5.6.

Example

If there is only a small current in a vertical wire, its field strength a few centimetres away will be comparable with the horizontal component, B_{EH}, of the Earth's magnetic field. B_{EH} varies from place to place and over time, but assume a value of 20 μT due North. The resultant magnetic field is shown in Fig. 4.4.27. Point X (due East of the wire) is a null point – where the (horizontal) field strength is zero.

How large a current must there be in the wire for X to be 35 mm from the wire?

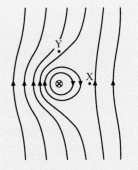

Fig. 4.4.27 Resultant of Earth's field and field due to wire.

Answer

The magnitude of the wire's field at X is equal to that of the Earth's horizontal field.

$$\frac{\mu_0 I}{2\pi r} = B_{EH}, \quad \text{so} \quad I = \frac{2\pi r B_{EH}}{\mu_0} = \frac{2\pi \times 0.035 \text{ m} \times 20 \times 10^{-6} \text{ T}}{4\pi \times 10^{-7} \text{ H m}^{-1}} = 3.5 \text{ A}.$$

(c) The magnetic field due to a current in a coil

Where the field lines pass through the plane of the loop in Fig. 4.4.24 the fields due to all parts of the loop are in the same direction, and vector addition leads to reinforcement. We can make the field here (and at every point) N times stronger still, by replacing the loop with a coil of N turns of insulated wire, giving, effectively, N times the loop current.

If we use a Hall probe to measure the field along the axis of a 'flat' circular current-carrying coil (Fig. 4.4.28) we find that it varies with distance z from the centre of the coil as shown in the graph.

We now concentrate on the **solenoid**, a coil whose turns are wound in a close-packed helix, usually around a 'former' in the shape of a tube. [The word 'solenoid' means tube-like.] Field lines due to current in a **long solenoid** (length >> diameter) are shown in Fig. 4.4.29. Note these points:

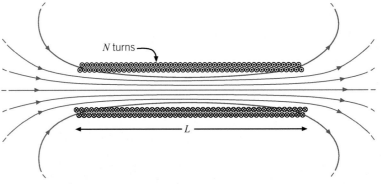

Fig. 4.4.29 Field due to current in long solenoid

- The field lines are closed loops, and the pattern resembles that of a bar magnet with its North pole near the solenoid's right hand end. North because, with the current as shown, the field lines emerge from the right, according to the right hand grip rule.

- Inside the long solenoid, except near the ends, the field is almost uniform over the cross-section and along the solenoid. See first **Study point**. Its magnitude is given by

$$B = \mu_0 nI \quad \text{in which} \quad n = \frac{\text{number of turns}}{\text{axial length}} = \frac{N}{L}$$

- The same equation gives the flux density inside a toroidal coil (Fig. 4.4.30), provided that the 'tube' radius is much less than the overall radius of the torus. (A torus is an object shaped like an inflated inner tube.) A toroidal coil can be thought of as a solenoid with no ends – and no poles.

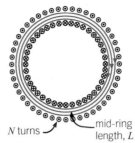

N turns — mid-ring length, L

Fig. 4.4.30 Toroidal coil

4.4.6 Magnetic flux, Φ

A striking feature of the flux density inside a long solenoid is that, for given n and I, it doesn't depend on the cross-sectional area, A. But, in a way, with a larger A you do get more field for your money, even if the field is no stronger! We can make this idea precise by defining a 'new' quantity, the **magnetic flux**, Φ. See **Terms and definitions**.

More flux is generated by the solenoid with larger A.

We can now see the sense in calling B the 'flux density', because, writing $B_\perp = B \cos \phi$ for the component of flux density normal to the area,

$$B_\perp = \frac{\Phi}{A} = \text{normal flux per unit area}$$

The properties of magnetic field lines (Section 4.4.1) are connected with the 'conservation of magnetic flux'. An example is given in Fig. 4.4.32... All the flux that passes to the right through the disc-shaped area A_{int} 'returns' leftward through area A_{ext}, an imaginary annulus (washer shape) surrounding the solenoid and extending indefinitely. The flux *density*, on the other hand, varies over A_{ext} and is everywhere *much* less than over A_{int}.

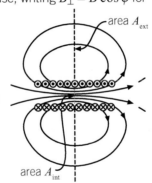

area A_{ext}

area A_{int}

Fig. 4.4.32 Solenoid flux

Use of an iron core in a coil

(a)

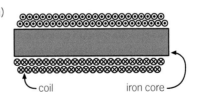

coil   iron core

We can greatly increase the flux density (and the flux) in a solenoid or toroidal coil by including a soft iron core (Fig. 4.4.34(a)).

Soft iron is relatively pure iron. Its magnetic domains (see start of section) are usually orientated so that their effects cancel, but the magnetic field produced by the current expands the domains that reinforce the field and shrinks those that oppose it. When the current is switched off the iron loses almost all its magnetism. (The domains are no longer aligned.)

(b)

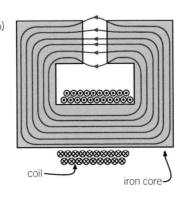

An **electromagnet** is an iron-cored coil designed to produce a strong field beyond the ends of the core (the poles). In the electromagnet of Fig. 4.4.34(b) the field will be roughly uniform in the gap between the poles, though there is some bulging out of field lines near the edges. The flux crossing the gap will be the same as that passing through a cross-section of the core.

coil — iron core

Fig. 4.4.34(a) Sections through (a) iron-cored coil and (b) electromagnet (one type)

---Terms & definitions---

The magnetic flux, Φ, through an area, A is defined by $\Phi = BA \cos \phi$, in which ϕ is the angle between the normal to the area and the flux density (Fig. 4.4.31). It is a scalar.

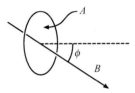

Fig. 4.4.31 Magnetic flux

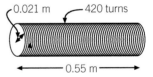

0.021 m — 420 turns

0.55 m

Fig. 4.4.33 Solenoid

▼ **Study point**

Putting an iron core into a current-carrying coil may increase the flux density inside it by a factor of several hundred!

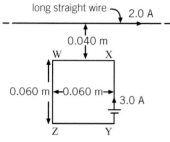

4.4.7 Forces between current-carrying wires

Knowing the magnetic flux density around a long straight wire carrying a current I_1, we can calculate the motor effect force on a nearby parallel wire of length ℓ carrying current I_2.

The force on ℓ will be the same whether the set-up is that shown in Fig. 4.4.35 (a) or (b).; we're not concerned with the forces on the rest of the circuit that carries I_2.

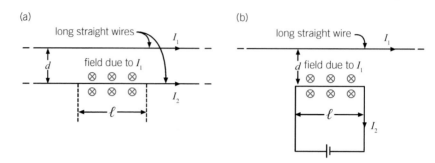

Fig. 4.4.35 Current-carrying wires that exert forces on each other

In the vicinity of ℓ, the field due to I_1 is into the page, as shown (using the right hand grip rule). So ℓ experiences a force towards I_1 (using Fleming's left hand motor rule). The magnitude of the force on ℓ is

$$F = BI_2\ell \quad \text{so} \quad F = \frac{\mu_0 I_1}{2\pi d} \times I_2\ell \quad \text{that is} \quad F = \frac{\mu_0 I_1 I_2 \ell}{2\pi d}$$

Newton's third law tells us that ℓ exerts an equal and opposite force on the I_1 wire *as a whole* (because the field that ℓ sits in is due to the current in this wire as a whole).

We see that 'like currents attract'. Check that unlike currents repel.

The ampère and μ_0

We can now deal swiftly with a loose end: the curiously exact value of μ_0. It is entailed by the definition of the ampère, which is as follows...

'The ampère is that current which, in two long parallel wires exactly 1 metre apart in a vacuum, would produce a force between the wires of exactly 2×10^{-7} newton per metre of their length.'

Putting the values from this definition into the equation just derived...

$$2 \times 10^{-7}\,\text{N} = \frac{\mu_0 \times 1\,\text{A} \times 1\,\text{A} \times 1\,\text{m}}{2\pi \times 1\,\text{m}} \quad \text{so} \quad \mu_0 = 4\pi \times 10^{-7}\,\text{N A}^{-2}\ (\equiv\text{H m}^{-1})$$

4.4.8 Specified practical work

(a) Investigation of magnetic flux density using a Hall probe

A Hall probe contains a wafer connected as in Fig. 4.4.22 and often mounted in an enclosure at the end of a stick with a handle. The Hall pd (most likely amplified and with a zero-adjusting facility) is fed to a meter that may have been calibrated directly in mT or μT.

Here are two possible investigations...

The field inside a long solenoid

The solenoid is included in a circuit to provide a known current. Probably less than an ampère will be needed. The probe needs to be of a type with wafer orientated to measure fields that are *parallel* to its stick. The reason should be clear from Figs 4.4.29 and 4.4.37.

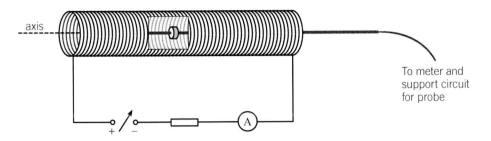

Fig. 4.4.37 Investigating B in a long solenoid with a Hall probe

(a) The flux density at the centre of a solenoid whose length is many times its diameter is given – see Section 4.4.5(c) – by

$$B = \mu_0 nI$$

Knowing n and having read the current I, we can calculate B. This enables us to check the calibration of a probe that gives a direct reading of B, or to calibrate a voltmeter wired to a special 'Hall chip'. We need to check for zero error, and having eliminated or allowed for this, to test for linearity: doubling the solenoid current should give twice the reading.

(b) With the probe near the centre, lengthways, of the solenoid we measure the field at three or four places across a diameter, to see if the field is uniform over a cross-section of the interior.

(c) The main investigation would be of the flux density, B, at different distances, z, along the solenoid axis from the centre ($z = 0$) to a little way beyond one end of the solenoid, with the goal of plotting a graph of B against z. Near the end, B decreases rapidly with z, so readings need to be at closer z-intervals than are needed at smaller z. Readings can also, of course, be taken for negative z values.

The field due to a long straight wire

Fig. 4.4.38 shows a practical set-up (not the only one possible) for doing the investigation. The main problem it addresses is that a long, straight, current-carrying wire can't exist in isolation!

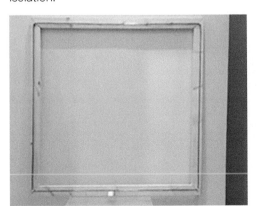 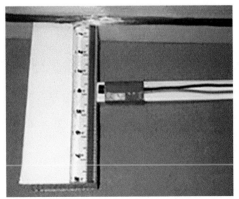

Fig. 4.4.38 (a) Set-up for investigating B due to long straight wire (b) enlarged detail

Stretch & Challenge

For a truly long solenoid (length >> diameter), the field at the point on the axis at the very end of the solenoid is expected to be

$$B_{end} = \frac{1}{2} B_{middle}.$$

Prove this without intricate mathematics by using symmetry. Hint: half a long solenoid is still a long solenoid!

▼ Study point

'In the region of the probe, the other three sides of the coil produce almost no field *in this direction*.' (fourth bullet in main text). Check this claim using the *right hand grip rule*.

Self-test 4.4.23

What is the theoretical gradient of a graph of B against $\frac{1}{r}$ for a long straight wire?

Stretch & Challenge

Because of the thickness of the bundle of wires forming the coil, it may be difficult to decide where to position the zero of the scale used to measure r. For a zero error in r simply to produce a non-zero intercept we need to plot a different straight line graph from the one recommended in the main text. What should be plotted against what?

A circuit is set up (as in Fig. 4.4.36) to drive a measured current through the big square coil (about 80 cm × 80 cm) in Fig. 4.4.38(a). Note that:

- We study the field due to the bottom side of the coil, keeping the probe within about 5 cm of the centre of this side.

- We measure the field at points along a radial line from the midpoint of this side. This radial line (the line marked by the edge of the scale in Fig. 4.4.38 (b)) is at right angles to the plane of the coil.

- The Hall probe chip is seen as a small black square to the left of the red band in Fig. 4.4.38(b).

- In the region of the probe, the other three sides of the coil produce almost no field *in this direction*.

- If the coil carries a current of (say) 2.0 A and has 20 turns, each side behaves as a single wire carrying 20 A. With a *single* turn, the field at, say, 5 cm from the wire might be too weak to be reliably measured.

Having used a long solenoid to calibrate the probe if necessary, and having noted or eliminated zero errors, we take readings of B for values of r ranging from a few millimetres up to about 50 mm. B should be plotted (on the vertical axis) against $\frac{1}{r}$, and the gradient compared with the theoretical value (Self-test 4.4.22).

4.4.9 Investigating the force on a current-carrying conductor

The apparatus is shown in Fig. 4.4.4, and indeed the investigation itself is outlined in Section 4.4.2. Here are a few practical points:

- The digital balance will no doubt be calibrated in gram, but it actually responds to force. A balance made for use in the UK registers 1.000 gram for a force of 9.81×10^{-3} N on its top-pan.

- The main investigation is of how force, F, varies with current, I, for a fixed length, ℓ, of wire (the length of the horizontal portion, XY) at a fixed angle θ (the obvious choice is 90°). Currents of up to a few ampère will probably be needed to produce forces large enough to measure reliably (see **Study point**).

  We plot F against I. A value for B can be found from the gradient, if ℓ has been measured with a ruler. See Self-test 4.4.24. Note that the value of B that emerges will be an average over XY; the field in the gap is unlikely to be very uniform.

- Another investigation could be of the dependency of F on θ. Some ingenuity is needed to measure θ reliably with a protractor. F should be plotted against sin θ, hoping for a straight line through the origin.

- It should be possible to check the dependency of F on ℓ for constant I and θ, but this may not be easy.

4.4.24 ∨ Self-test

The lower parts of the vertical portions, WX and YZ, of the wire shown in Fig. 4.4.4 experience magnetic forces. Why don't these forces affect the balance reading?

▼ **Study point**

To increase the forces we are trying to measure, we could take the wire several times round the path WXYZW. This will be equivalent to increasing the current in a single wire this number of times.

4.4.25 ∨ Self-test

If the gradient of a graph of F against I for $\theta = 90°$ is $(4.2 \pm 0.1) \times 10^{-4}$ N A^{-1}, and ℓ is measured to be 20 ± 1 mm, calculate a value for B in the region of the wire, together with its absolute uncertainty.

Exercise 4.4

1 State the direction of the force on each red wire. The dot and cross symbols relate to the direction of the current or magnetic field as indicated. The circle in part (e) is a plane current-carrying coil.

2 A horizontal wire of length 10 m is oriented W-E and carries a steady current of 25 A in the eastward direction. The flux density of the Earth's magnetic field at the position of the wire is 50 μT and its direction is downwards at an angle of $70°$ to the horizontal in a NS plane. Calculate:

(a) the magnitude and direction of the force on the wire due to the Earth's magnetic field;

(b) the horizontal and vertical components of the force on the wire.

3 The wire in question 2 now carries an alternating current given by $I = 50 \cos 100\pi t$, where I is in A and t in s. Sketch a graph to show the variation in the horizontal component of the force on the wire over a period of 40 ms. [Hint: first find the period of the AC.]

4 An electric motor contains a coil, with 500 turns of area 50 cm², carrying a current of 3.0 A which rotates in a magnetic field of flux density 0.20 T. Calculate the torque on the coil when it is (a) parallel to the magnetic field and (b) at $60°$ to the magnetic field.

5 (a) An improved version of the motor in question 4 has two coils at right angles. Each of the coils carries a current of 1.5 A. The rotation rate of the motor is 20 Hz. On the same axes, sketch graphs the variation of the torque for each of the coils over a period of 0.1 s and also the variation of the total torque.

(b) Explain why it would be even better to design the motor so as to have three coils arranged $60°$ apart.

6 Electrons of kinetic energy 1.0 keV are travelling at right angles to a uniform magnetic field. They are observed to travel in circular paths of radius 5.0 cm. Calculate:

(a) the momentum of the electrons,

(b) the flux density of the magnetic field and

(c) the gyrofrequency of the electrons (i.e. the number of orbits per unit time).

[Required data: electron charge and mass]

7 Electrons in the upper atmosphere have a gyrofrequency of 1.7 MHz. This results in radio wave absorption due to resonance. Calculate:

(a) the flux density of the Earth's magnetic field and

(b) the gyrofrequency of protons.

[Required data: electron and proton charge and mass]

8 A 2 keV electron in a region of the magnetosphere, where the flux density is 20 μT, travels at an angle of $30°$ to the magnetic field lines. Calculate the radius and pitch of its helical path.

9 A cyclotron operates with a magnetic flux density of 0.50 T. It is used to accelerate protons. When the protons cross the space between the dees, the pd is 1000 V.

(a) Calculate the frequency of the pd required.

(b) State the energy of the protons after 2000 complete orbits.

(c) Calculate the radius of protons' path after 2000 complete orbits.

(d) State why a cyclotron could not be used in this way to accelerate electrons to this energy.

10 Monoenergetic charged metal ions are passed through a voltage selector. The magnetic field has a flux density of 0.10 T. The electric field is provided by a pair of parallel plates 4.0 cm apart. The ions are observed to pass undeviated through the selector when the pd between them is 116 V.

(a) Calculate the speed of the ions.

(b) The ions (assumed to be singly ionised) are produced by acceleration through a pd of 100 V. Identify them by calculating their mass.

[Data required: atomic masses]

11 The diagram shows the principle of a simple mass spectrometer, a device for separating atoms of different masses. Singly ionised atoms, which are initially of negligible energy, are accelerated through a pd pass through a slit, S, and enter a region with a uniform magnetic field.

The ions with travel in circular paths, and those with different masses are detected in different places, **X** and **Y**.

(a) State the direction of the magnetic field.

(b) There are two isotopes of masses, m_X and m_Y.

Give an expressions for the velocities of the two isotopes as they enter the magnetic field.

(c) Use your answer to part (b) to derive expressions for the radii of the circular paths, r_X and r_Y.

(d) A mass spectrometer uses a magnetic field of flux density 0.050 T. It is used to separate isotopes of copper. The $^{63}Cu^+$ ions were detected at a distance of 25.0 cm from S.

(i) Calculate the potential, V.

(ii) Calculate the separation of $^{63}Cu^+$ and $^{65}Cu^+$ ions.

12 A Hall probe uses a p-type Silicon chip with a conductor concentration of 1.0×10^{15} cm^{-3}. It has a resistivity of 13.5 Ω cm. The dimensions, $w \times t \times l$ of the chip are 2 mm $\times 0.1$ mm $\times 5$ mm. It is supplied with a current of 5.0 mA.

(a) Calculate the resistance of the chip.

(b) Calculate the longitudinal pd required.

(c) Calculate the ratio of the Hall voltage, V_H, to the magnetic flux density, B, when the chip is placed at right angles to a magnetic field. [Care units]

(d) The contact points for measuring the hall voltage are not exactly opposite each other on the chip. The difference is 0.01 cm. Estimate the pd between these points with zero magnetic field.

In order to achieve the (almost) uniform magnetic field in the deflection tube shown in Fig. 4.4.13 a Helmholtz coil arrangement is used. This is two plane coils of the same radius, a, and number of turns, N, arranged on the same axis and separated by a. The coils are connected in series with the currents in the same direction. For a single plane coil, the variation of axial flux density with position, shown in Fig. 4.4.27 has the formula:

$$B = \frac{\mu_0}{2} \frac{NIa^2}{(a^2 + x^2)^{\frac{3}{2}}}$$

1. Plot a graph of this function from $x = -2a$ to $x = 2a$. We are only interested in the shape of the graph so you can set μ_0 to 2 and N,

    I and a to 1. So you should plot the function $\dfrac{1}{(1 + x^2)^{\frac{3}{2}}}$ between $x = -1$ and $x = 2$. [Hint: Use a spreadsheet]

2. Superimpose a second graph shifted to the right by a [i.e. 1], to represent the field of the second coil.

3. Add the two fields of the two coils together and comment on the value of the field between the coils.

4. The field is roughly constant because, in the middle region, the field from one coil decreases at the same rate as the other increases. This is achieved when the points of inflection of the two graphs coincide. By differentiating the flux density formula twice, show that the points of inflection are at $x = \pm\frac{1}{2}a$.

167

4.5 Electromagnetic induction

4.5.1 Self-test

The rectangular loop, PQRS in Fig. 4.5.2 is moved steadily to the right. The loop is partly in a magnetic field that is negligible outside the dotted boundary.

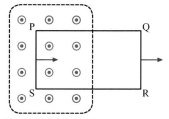

Fig. 4.5.2 Cutting of flux

(a) Explain in terms of cutting of flux why no emf is induced in sides PQ, QR or RS.

(b) State the direction of the emf induced in SP.

(c) State the direction of current in QR.

▼ Study point

If the circuit is not complete, electrons urged towards X will pile up there, and there'll be an electron deficit at Y. The process is quickly self-limiting, stopping when the pd due to the charge separation is equal to the emf.

This is the physical effect made use of in electrical generators (for example, in power stations and in cars – to charge their batteries). Transformers (as used in electrical energy distribution and to produce low voltages from 'mains' supplies) also depend for their operation on electromagnetic induction, though the effect involved is really rather different. We explain how one equation covers both cases.

4.5.1 The emf induced in a moving conductor

An emf is 'induced' in a conductor when it moves in such a way that it cuts (moves across) lines of magnetic flux. Fig. 4.5.1 shows one way of demonstrating this: we pull the horizontal wire XY sharply upwards and observe that the voltmeter gives a reading (usually a few millivolts) while the wire is moving.

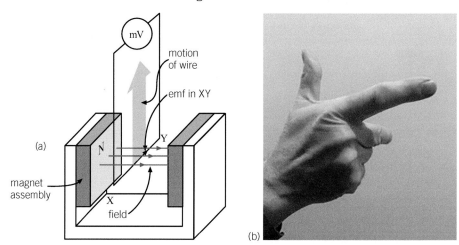

Fig. 4.5.1 (a) Wire cutting flux (b) Fleming's right hand generator rule

Note these points:

- There is no reading if we move the wire XY in Fig. 4.5.1(a) from left to right; it is not *cutting* lines of flux, but slipping between them.

- If the moving conductor is part of a complete circuit, the emf will give rise to a current. An emf in the X to Y direction in the wire XY urges a current from X to Y in the wire and therefore from Y to X in the rest of the circuit! From the voltmeter's point of view, Y acts like the positive terminal of a battery. If the circuit is *not* complete there will still be a pd between X and Y equal in magnitude to the emf. See **Study point**.

- The direction of the emf is given by Fleming's right hand generator rule (FRHGR): see **Terms and definitions**. It's the third and last 'hand rule'!

- Strengthening the magnetic field, lengthening the wire within the field, and moving the wire faster all increase the emf. This much can be discovered using apparatus like that in Fig. 4.5.1 (a). We can derive a simple equation for the emf using a thought-experiment...

How the emf arises when a conductor cuts magnetic flux

We consider a metal rod of length *l* moving at constant velocity, v, cutting a uniform magnetic field, B. The set-up is shown in Fig. 4.5.3.

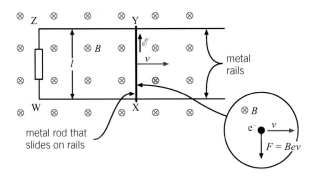

Fig. 4.5.3 Rod on rails: thought-experiment

Each free electron in the rod is carried along with the rod and so has the rod's velocity, v. It therefore experiences a magnetic force Bev as shown in the bubble. See Section 4.4. Free electrons in the rod are, then, urged to drift in the direction YX, making a conventional current in direction XY, (so XYZWX around the circuit). This is just as predicted by FRHGR.

We can go further… The emf, $\mathscr{E}$, is the energy per unit charge given to electrons passing through the rod, that is from Y to X. Therefore

$$\mathscr{E} = \frac{\text{work done on electron}}{e} = \frac{\text{force on electron} \times \text{distance moved in force direction}}{e}$$

So $\quad \mathscr{E} = \dfrac{Bev \times \ell}{e} \quad$ that is $\quad \mathscr{E} = B\ell v$

This equation applies to the rather limited case of a straight wire, when wire, field and wire's velocity are all at right angles to each other.

4.5.2 Faraday's law

The law – see **Terms and definitions** – gives the emf induced in a circuit in the *general* case. The concept of magnetic flux, needed to make sense of the law, was introduced in Section 4.4.6, where the defining equation, $\Phi = BA \cos \phi$, is explained.

How does the law work for the rod on rails? In time Δt the rod sweeps across an area $\ell \times v\Delta t$ (shaded in Fig. 4.5.4), so, because the flux density is normal to the area ($\phi = 0$) the flux linked with the circuit has increased by an amount

$$\Delta\Phi = B \times \ell \times v\Delta t.$$

This much more flux goes *through* WXYZ than when it was W[X][Y]Z. Since $N = 1$, Faraday's law gives the induced emf magnitude as

$$\mathscr{E} = \frac{\Delta\Phi}{\Delta t} = \frac{B\ell v\Delta t}{\Delta t} = B\ell v$$

We now apply Faraday's law to some cases where using $\mathscr{E} = B\ell v$ is either not wholly straightforward – or quite impossible.

Terms & definitions

Faraday's law

When the magnetic flux, Φ, through a circuit changes, an emf is induced in the circuit, proportional to the rate of change of flux linkage. For a circuit with N identical turns

$\mathscr{E} = -\dfrac{\Delta(N\Phi)}{\Delta t}$.

$N\Phi$ is the **flux linkage**. If the circuit is a coil of N turns, with the same flux, Φ, through each, the emf is the same as if there were N times the flux through a single turn!

Self-test 4.5.2

Show that the equation $\mathscr{E} = B\ell v$ is homogeneous in terms of units.

▼ **Study point**

The minus sign in the Faraday's law equation relates to the direction of the emf: see last **Study point** of 4.5.3. It's no great sin to leave out the minus sign, as we shall be using the equation to find only the magnitude of the emf.

Self-test 4.5.3

The wire XY in Fig. 4.5.1(a) is 0.020 m long and is at right angles to the magnetic field. When it is moved upwards at 3.0 m s^{-1} the meter reads 1.3 mV. Calculate B.

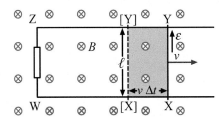

Fig. 4.5.4 Rod on rails again

▼ Study point

$\frac{\Delta(N\Phi)}{\Delta t}$ really gives us the mean emf induced over the time interval Δt. If the flux linkage is changing at a constant rate (as in the rod on rails), the emf is constant and we can drop the word 'mean'. It's not seriously misleading to do so even when the rate of change isn't constant, provided that we understand $\frac{\Delta(N\Phi)}{\Delta t}$ to be applied to a small enough time interval Δt.

4.5.4   Self-test

The wingspan of the Airbus A380 is 80 m.

(a) Calculate the emf generated between the wingtips when the plane is flying horizontally at 260 m s⁻¹ in a vertical magnetic field of 45 μT.

(b) Explain why you would not expect to read this voltage on a voltmeter inside the cabin, connected to the wingtips by wires.

4.5.5   Self-test

Take the 'loop' in Fig. 4.5.7 to be a ring of *diameter* 30.0 cm, initially with the normal to its plane parallel to a *B* field of 25 μT (Fig. 4.5.7(a)). Calculate the mean emf induced in the ring when it is turned through 30° (see Fig. 4.5.7(b)) in a time of 0.24 s.

Example 1: Emf in a loop moving in a uniform magnetic field

In Fig. 4.5.5 side ZW of the wire loop WXYZ has an induced emf, Bbv, in the direction WZ and XY has an emf, Bbv, in the direction XY. These emfs are in opposite 'senses', meaning that, as we go (either way) round the circuit the emfs encountered are equal and opposite. The resultant emf is zero.

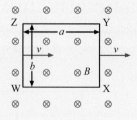

Fig. 4.5.5 Translating loop in uniform *B*

Faraday's law makes even lighter work of things:

As the whole loop is moving, and the field is uniform, the same amount of flux is always passing through it, so $\mathscr{E} = \dfrac{\Delta\Phi}{\Delta t} = 0$

Example 2: Shrinking band cutting flux

Suppose a round balloon with a band of metallic paint around its equator deflates from a radius of 0.16 m to a radius of 0.11 m in a time of 0.18 s. What would be the mean emf induced in the band if the Earth's field component field at right angles to the balloon's equatorial plane is 30 μT?

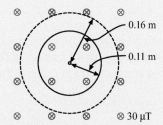

Fig. 4.5.6 Balloon deflating in uniform *B*

The area enclosed by the band decreases, so the flux linked with it decreases. $N = 1$ and the flux density is always normal to the area, so $\phi = 0$. We therefore have

$$\mathscr{E} = \frac{\Delta\Phi}{\Delta t} = \frac{\Delta(BA)}{\Delta t} = \frac{B\Delta A}{\Delta t} = \frac{30 \times 10^{-6}\,\text{T} \times \pi((0.16\,\text{m})^2 - (0.11\,\text{m})^2)}{0.18\,\text{s}} = 7.1\,\mu\text{V}$$

Note that this is the mean emf. See **Study point**.

Example 3: Mean emf in a loop turning in a uniform field

Suppose that, in a time Δt, we turn a metal loop through 180° in a magnetic field so that the normal to its plane shown in Fig. 4.5.7(b) changes from being parallel to a uniform magnetic field as in Fig. 4.5.7(a)) to being 'antiparallel' as in Fig. 4.5.7(c).

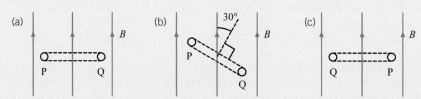

Fig. 4.5.7 Loop flipped through 180° in uniform *B* field (section)

Initially the flux through the loop is BA, in which A is the area enclosed by the ring. Finally, flux of the same magnitude, BA, threads through the ring in the opposite direction. This counts as a flux change of $2BA$.

So the mean emf is ..... $\mathscr{E} = \dfrac{\Delta\Phi}{\Delta t} = \dfrac{2BA}{\Delta t}$

Example 4: Mutual induction

This occurs when a changing current in one circuit causes an emf to be induced in another circuit, because some or all of the (changing) magnetic flux produced by the first is linked with the second.

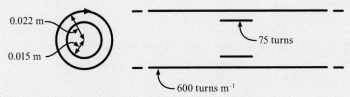

Fig. 4.5.8 'Small' coil inside long solenoid. Sectional diagrams.

In Fig. 4.5.8, suppose that (by some deft action with a variable power supply?) the current in the long solenoid is reduced at a constant rate from 3.0 A to zero in 0.25 s. What emf would be induced as a result in the 75-turn coil inside the solenoid?

In the central region of the solenoid the flux density is uniform and parallel to the axis of the solenoid. Its magnitude is $B = \mu_0 n I_{sol}$.

So the initial flux linkage for the 75 turn coil is $75\Phi = 75\, BA = 75\,(\mu_0 n I_{sol})A$

in which A = area of small coil = $\pi(0.015\text{ m})^2$

So the emf in the 75 turn coil is

$$\mathscr{E} = \frac{\Delta(N\Phi)}{\Delta t} = \frac{75 \times (4\pi \times 10^{-7} \times 600 \times 3.0\text{ T}) \times \pi(0.015\text{m})^2 - 0}{0.25\text{ s}} = 0.48\text{ mV}$$

Self-test 4.5.6

Calculate the emf induced in a 75-turn coil wound tightly around the outside of the middle of the solenoid in example 4, for the same variation of solenoid current.

> **▼ Study point**
>
> Mutual induction is an example of when the induced emf is *not* due to magnetic forces on free electrons in a moving conductor. There *is* no moving conductor! In this case the emf arises from forces on the electrons due to an *electric* field around a closed path through which there is changing magnetic flux. Faraday's law covers both kinds of e-m induction!

4.5.3 Lenz's law

The law, as stated in **Terms and definitions**, gives the direction of the induced emf in the *general* case. The wording of the law is deliberately rather open, as it can often be applied in more than one way. Examples will show this. We first re-visit an old friend (Sections 4.5.1, 4.5.2)… pushing it!

(a) Lenz's law applied to the rod on rails

What counts, in this case, as the *change* producing the emf? Is it the rod moving to the right (due to some external agency pushing it)? Or is it the consequent increase in flux linked with the circuit? Relax: either will do…

What are the *effects* of the emf? The emf drives a current round the circuit and this in turn has two effects of interest…

There will be a motor effect force on the rod XY. Lenz's law demands that this must be to the left to oppose the rod's motion. Using Fleming's left hand motor rule we see that the current in the rod is in the direction XY, so the emf must be in the direction XY (see Fig. 4.5.4).

The current in the loop will set up its own magnetic field. Across the area WXYZ this must be out of the page, opposing the increase of 'original' flux into the page due to the increasing area. According to the Right hand grip rule, to produce flux out of the page the loop current has to be in the sense WXYZW, so the emf in the rod is in the direction XY. The same conclusion as before! Choose whichever method you prefer.

> **Terms & definitions**
>
> **Lenz's law** = The direction of an induced emf is such that its effects oppose the change producing it.

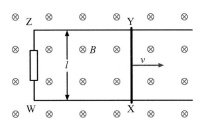

Fig. 4.5.9 Rod on rails (for the last time!)

> **▼ Study point**
>
> If there's a break in the circuit WXYZ, there is still an induced emf in XY even though it can't drive a current round. We use Lenz's law to find the direction of the emf by *imagining* the circuit to be closed!

(b) Lenz's law and energy conservation

Why *should* the effects of the emf oppose the change producing it? Suppose the emf in the rod XY were in the other direction: then, once the rod were given an initial push to the right, the motor effect force, now to the right, would take over the rod's propulsion, while at the same time we'd be getting free heat from the resistor! Marvellous, but a violation of the Principle of Conservation of Energy.

In the real world, with the direction of emf required by Lenz's law, the external agency has to do work against the opposing motor effect force, and it is this work input which supplies the resistor's thermal energy (via the action of an electric current). In other words, Lenz's law is required by the Principle of Conservation of Energy.

(c) More applications of Lenz's law

We go back to Example 2 in Section 4.5.2. The emf in the band must be in the clockwise sense in Fig. 4.5.6 so that the current is in this sense and – according to the right hand grip rule – produces flux directed into the page through the area bounded by the band. This opposes the decrease, due to the shrinking of the bounded area, in the original flux.

In Example 4 in Section 4.5.2 (Fig. 4.5.8) we can similarly deduce that a decrease in the 'clockwise' current in the solenoid will induce an emf in the clockwise sense in the 75-turn coil. Check this!

Next we look at a favourite way of demonstrating electromagnetic induction: plunging a bar magnet into a coil connected to a voltmeter. Fig. 4.5.10 (a) – (c) are 'snapshots' of a magnet moving at constant speed through a coil.

In (a) the sense of the emf and current in the coil is such that flux produced by the coil goes from right to left through the coil, opposing the increase in left to right flux from the magnet. And the coil's left-hand end acts as a North pole repelling the advancing magnet!

Check that you understand the emf in (c) and the lack of one in (b).

By studying the magnet's field lines in relation to the coil in (a) – (c), observe how the sketch-graph in (d) shows the time variation of the flux linkage ($N\Phi$). The gradient of this graph gives us (minus) the induced emf, in accordance with Faraday's law. See graph (e) and **Study point**.

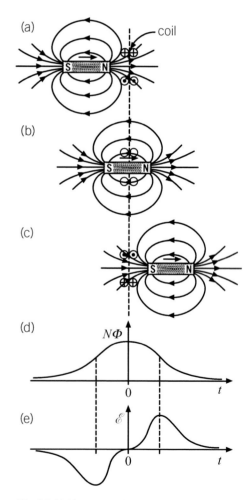

Fig. 4.5.10 Magnet passing through coil

4.5.4 The simple alternating current generator

This is a 'flat' coil that is turned by some external means in a uniform magnetic field. As it rotates, the flux linkage changes, so an emf is induced. For simplicity, we consider a rectangular coil, WXYZ, represented as a single turn in Fig. 4.5.11. The first **Study point** deals briefly with a practical matter.

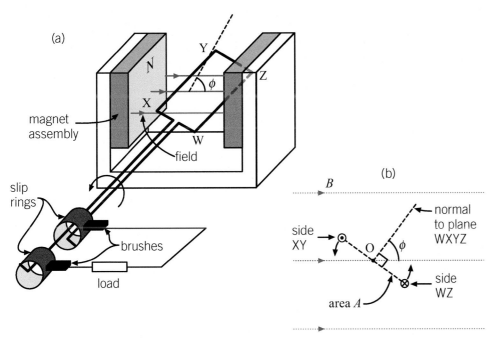

Fig. 4.5.11 (a) Simple ac generator (b) Section through its coil

When the normal to the coil (see second **Study point**) is at angle ϕ to the field, the magnetic flux, Φ, (Section 4.4.5) through the coil is

$$\Phi = BA \cos \phi \qquad \text{[}A = \text{side length XY} \times \text{side length YZ]}$$

For a coil of N turns, the flux *linkage* is

$$N\Phi = NBA \cos \phi$$

$A \cos \phi$ is the area that the coil presents normally to the flux. (For example, when $\phi = 0$, the coil presents 'full frontally' and $A \cos \phi = A$, but when $\phi = 90°$, the coil presents 'edge-on' and $A \cos \phi = 0$.)

To find out how the emf changes with time, we note that, if the coil is being turned at a steady angular speed ω, then at time t,

$$\phi = \omega t + \phi_0$$

Here, ϕ_0 is the angle that the normal to the coil makes with the field direction at time $t = 0$. [If we choose to start timing from an instant when the normal to the coil is in the field direction, then $\phi_0 = 0$.]

In general,

$$N\Phi = NBA \cos (\omega t + \phi_0)$$

In Fig. 4.5.12 the first graph shows how $N\Phi$ varies over one revolution, for $\phi_0 = 0$. The coil orientations are shown every quarter of a revolution. The time scale on the graphs is given in terms of T, the periodic time, or time for one revolution.

▼ **Study point**

How do we connect a rotating coil to a stationary 'load' without twisting the connecting wires to destruction? One solution, shown in Fig. 4.5.11 (a), is to connect the coil to metal 'slip rings' which rotate with the coil, and the load to carbon brushes pressing against the slip rings, but allowing slipping.

In a serious ac generator (or 'alternator'), brushes and slip rings are used to supply a rotating electromagnet. The coil in which the emf is generated (the *stator*) is stationary and wound on a soft iron core.

▼ **Study point**

It was an arbitrary choice for our normal to stick out as shown, rather than in the opposite direction. But we stay with this choice. As the coil rotates, the normal rotates with it.

▼ **Study point**

With the coil as shown in Fig. 4.5.11 and turning so that ϕ is increasing, $N\Phi$ in the direction of the chosen normal is decreasing. According to Lenz's law, the sense of the emf is as shown by ⊙ and ⊗, so any current will produce flux through the coil in the direction of this normal, opposing change in flux.

▼ **Study point**

Recall from Section 3.2.3 that periodic time, T, and frequency, f, are related to ω by:

$$T = \frac{2\pi}{\omega} \text{ and } f = \frac{1}{T} = \frac{\omega}{2\pi}$$

Remember that ω must be in radians per second.

4.5.9 Self-test

Refer to the instant shown in Fig 4.5.11.

(a) By considering the rate of cutting of flux by XY and WZ explain whether $|\mathscr{E}|$ is increasing or decreasing.

(b) By considering the directions of forces on XY and WZ dictated by Lenz's law, explain the sense of the emf in the coil.

The emf at any instant is, from Faraday's law, minus the gradient of the graph of $N\Phi$ at that instant. Hence the bottom graph.

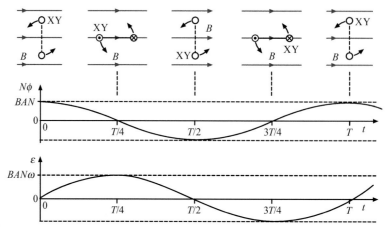

Fig. 4.5.12 Flux linkage and emf for coil turning in uniform field

Mathematically, because $N\Phi = NBA \cos (\omega t + \phi_0)$ and $\mathscr{E} = -\dfrac{\Delta(N\Phi)}{\Delta t}$

we have $\mathscr{E} = BAN\omega \sin (\omega t + \phi_0)$

This equation for $\mathscr{E}$ follows from that for $N\Phi$ in just the same way that $v = -A\omega\sin(\omega t + \phi_0)$ follows from $x = A\cos(\omega t + \phi_0)$ for a body doing *simple harmonic motion*. In *that* case,

$v = \dfrac{\Delta x}{\Delta t}$

Commonly seen versions of the equation for $\mathscr{E}$ (for $\phi_0 = 0$ and $\phi_0 = \frac{\pi}{2}$), are

$$\mathscr{E} = BAN\omega \sin \omega t \quad \text{and} \quad \mathscr{E} = BAN\omega \cos \omega t$$

Peak emf

4.5.10 Self-test

A 30-turn, circular coil of diameter 0.025 m is rotated at 20 Hz in a uniform magnetic field of 0.20 T at right angles to the rotation axis. At time $t = 0$ the normal to the coil is in the direction of the field. Calculate:

(a) the peak flux linkage,

(b) the peak emf,

(c) the flux linkage at $t = 0.010$ s,

(d) the emf at $t = 0.010$ s.

The maximum value of $\sin(\omega t + \phi_0)$ is exactly 1, so the *peak* or maximum value, $\mathscr{E}_{max}$, of $\mathscr{E}$ is given by

$$\mathscr{E}_{max} = BAN\omega$$

This equation makes sense because the greater the maximum flux linkage, BAN, and the greater the coil's angular velocity, ω, the greater the *rate of change* of flux at any instant, including at the peak.

Example

Suppose that the coil in Fig. 4.5.11 has 24 turns, and that WX = 0.030 m, XY = 0.040 m. The rotation rate is 50 revolutions per second and the magnetic flux density is 0.25 T. Calculate the maximum emf and the emf at a time of 3.0 ms after the normal to the coil was parallel to the field.

Answer

$\mathscr{E}_{max} = BAN\omega = BAN2\pi f$
$= 24 \times 0.25 \text{ T} \times 0.030 \text{ m} \times 0.040 \text{ m} \times 100\pi \text{ s}^{-1} = 2.26 \text{ V} = 2.3 \text{ V (2 s.f.)}$

Let t be 0 when $\phi = 0$, so $\phi_0 = 0$. Then, when $t = 0.0030$ s,

$\mathscr{E} = BAN\omega \sin \omega t = 2.26 \text{ V} \times \sin(100\pi \times 0.0030) = 1.8 \text{ V}$

4.5.5 Eddy currents

Fig. 4.5.13 (a) shows a magnet moving towards a suspended aluminium ring. An emf is induced in the ring, as in the coil in Fig. 4.5.10 (a). The ring is effectively a single turn coil with its ends connected together ('short-circuited'), so the emf drives a current round it. The ring is repelled by the magnet in accordance with Lenz's law.

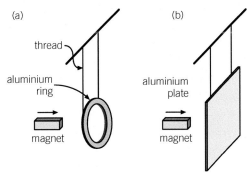

The plate in Fig. 4.5.13 (b) is also repelled. We can think of the plate as *containing* a ring like that in (a) – and

Fig. 4.5.13 Repulsion of ring and plate by approaching magnet

other rings besides. The constraint of an *actual* ring or loop is not needed for the current to take such a path. Currents that follow closed loops in conducting material in bulk are called *eddy currents* (after the eddies that can occur in fluids).

A striking demonstration of the effect of eddy currents uses the apparatus in Fig. 4.5.14 (a).

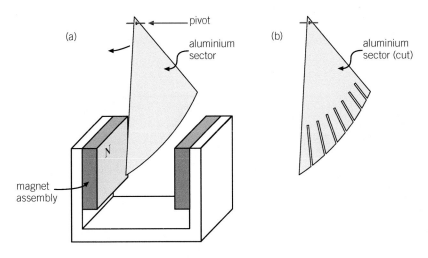

Fig. 4.5.14 Aluminium sector swinging in a magnetic field

We displace the sector through an angle from its lowest position, and release it. As it swings through the magnetic field its motion is very noticeably damped. Eddy currents in the metal experience motor effect forces that oppose the motion: *electromagnetic damping*. A convincing control experiment is to substitute a sector with radial slots (Fig. 4.5.14 (b)). The electromagnetic damping is greatly reduced, as there can no longer be large eddy current loops.

Energy dissipation through eddy currents

Eddy currents heat conductive material, just as a current heats a wire. The atoms of the material gain extra random energy of vibration – and the temperature rises. We say that energy is *dissipated*.

'Induction hobs' use eddy currents in the bottoms of suitable saucepans to heat them and hence, by thermal conduction, and (usually) convection, the contents of the saucepan. The eddy currents arise from emfs generated by rapidly alternating (and therefore changing) currents in coils underneath the pans. See **Study point**.

> **Terms & definitions**
>
> **Eddy currents** are currents that take closed loops in conducting material in bulk. They arise from electromagnetic induction.
>
> **Electromagnetic damping**
> This is the slowing of motion due to forces experienced by induced currents, in accordance with Lenz's law.

> **Self-test** 4.5.11
>
> When a bar magnet is dropped down a copper pipe, held vertically, it soon reaches a terminal velocity. The magnet takes much longer to reach the bottom than if dropped through a plastic pipe of the same length and diameter.
>
> Explain how an electromagnetic force arises opposing the magnet's motion.

> ▼ **Study point**
>
> Energy is also dissipated (by domain realignment) in iron if its magnetisation is constantly changing. This is called 'hysteresis loss'. It contributes to the heating of suitable saucepans on an induction hob. It would be a nuisance in transformer cores, which are therefore made of 'soft' (low hysteresis) iron.

Eddy currents are often unwanted. For example, if the cores of coils (Section 4.4.5) were lumps of solid iron, they would heat up if the coils carried alternating currents. Changing magnetic flux in the core would be linked with closed paths in the core (as in Fig. 4.5.15).

Energy dissipation in the iron cores of transformers and of ac motors is greatly reduced by *laminating* the cores, that is making them out of thin sheets of iron, insulated from each other, so preventing large eddy current loops.

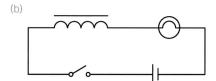

Fig. 4.5.15 Eddy currents in iron core

— Terms & definitions —

Self-induction is the induction of an emf in a coil (or circuit) because of a changing current in that coil (or circuit).

4.5.6 Self-induction

A changing current in one coil induces an emf in another coil, if flux from the first is linked with the second. See **Example 4**, Section 4.5.2. But flux produced by any coil is linked with the very coil producing it (Fig. 4.4.28). So we must have self-induction, as defined in **Terms and definitions**.

A striking demonstration uses an *inductor* made of many turns of wire wound around an iron core. (In Fig. 4.5.16 (a) the two coils can be connected in series so that their fields are in the same sense round the core.) The inductor is included in the circuit shown in Fig. 4.5.16 (b), which features the circuit symbol for an inductor. (The straight line next to the coil symbol represents an iron core.)

(a)

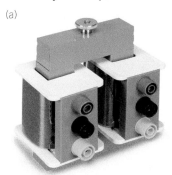

(b)

Fig. 4.5.16 (a) An inductor
(b) Circuit to demonstrate effects of self-induction

When the switch is closed the lamp is slow to light, because of the emf in the inductor (often called a *back-emf*) opposing the battery emf, in accordance with Lenz's law. **When the switch is opened it is important for no-one to be touching any part of the circuit**. This is because the current, and therefore the magnetic flux, decreases to zero in a very short time, and a large emf (maybe several hundred volts) is induced in the coil. This produces a fat spark or 'arc' at the switch contacts (prolonging the current through motion of ions in the arc, thereby temporarily opposing the decrease of current!).

A very nice control experiment is possible with the two-coil inductor of Fig. 4.5.16 (a). We simply reverse the connections to one of the coils so that the fields of the coils oppose, but the resistance stays the same. The lamp now comes now with little delay when the switch is closed, and the spark is much less noticeable when the switch is opened.

Self-inductance

We can calculate the magnitude of the emf when the current changes in a long solenoid with no iron core.

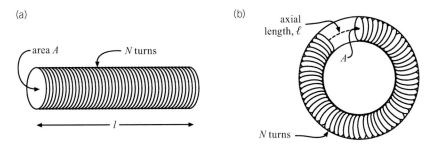

Fig. 4.5.17 Coil parameters (a) long solenoid (b) toroidal coil

Referring to Fig. 4.5.17 (a), the flux through the solenoid, except near the ends, when it carries current I is

$$\Phi = BA = \mu_0 nIA = \mu_0 \frac{N}{\ell}IA$$

Neglecting the decrease of flux near the ends (a good approximation if $\ell \gg$ diameter),

$$\text{Flux linkage} = N\Phi = \frac{\mu_0 N^2 A}{\ell}I$$

If the current changes by ΔI, the change $\Delta(N\Phi)$ in flux linkage is

$$\Delta(N\Phi) = \frac{\mu_0 N^2 A}{\ell}\Delta I$$

We call the lump $\frac{\mu_0 N^2 A}{\ell}$ the inductance, or self-inductance, L, of the coil.

So we can write $\quad \Delta(N\Phi) = L\Delta I$

And, using Faraday's law, $\quad \mathscr{E}_L = -\frac{\Delta(N\Phi)}{\Delta t} \quad$ so $\quad \mathscr{E}_L = -L\frac{\Delta I}{\Delta t}$

From this equation, we see that the unit of L must be $V\,A^{-1}\,s = \Omega\,s$. This unit is given the name: henry (H), after Joseph Henry, an American pioneer of electromagnetism.

The last three equations hold for coils (indeed for circuits) of *any* shape, but L can be calculated from the simple formula, $L = \frac{\mu_0 N^2 A}{\ell}$, only for the cases of a long solenoid or a toroidal coil (Fig. 4.5.17).

An iron core increases the inductance of a coil (maybe by a factor of many hundred if the core is a closed loop) because of the extra flux due to the alignment of magnetic domains – see Section 4.4.5. This extra flux is not strictly proportional to the current, so the inductance of an iron-cored coil is not really a constant, but depends upon the current, upon whether it is increasing or decreasing, and from what previous values! We can often make do with some sort of mean value of inductance.

─── Terms & definitions ───

When the current in a coil (or circuit) changes by ΔI the flux linkage of that same coil or circuit changes by $\Delta(N\Phi)$, given by $\Delta(N\Phi) = L\,\Delta I$

L is a called the **Self-inductance** of the coil or circuit. It is a constant if there is no iron core.

UNIT: Ω s = henry (H)

The emf induced in a coil or circuit due to changing current in that coil or circuit is given by

$$\mathscr{E} = -L\frac{\Delta I}{\Delta t}$$

This equation may also be used to define self-inductance.

Self-test 4.5.13

Calculate the time in which a current of 2.0 A through an inductor of 4.2 H must fall to zero in order to induce an emf of 280 V

▼ Study point

A zero resistance inductor, i.e. one which is superconducting, must have zero electric field inside it – otherwise the current will be infinite. If we apply a pd, V, the current grows at such a rate that the induced emf is equal and opposite to V. So $V = L\frac{\Delta I}{\Delta t}$

Self-test 4.5.14

A pd of 3.0 V is applied to a 2.0 H inductor (assumed to have zero resistance). What happens to the current?

Exercise **4.5**

Questions 1–4 relate to the diagram of the magnet and copper loop. The magnet is moving to the **left**. We'll apply Lenz's law in two ways to predict the direction of the induced current in the loop.

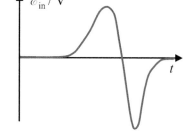

1 In terms of Faraday's law, why is there an induced current in the loop?

2 (a) What is happening to the flux density of the magnetic field in the loop?

(b) What is happening to the flux linkage in the loop?

(c) What direction does Lenz's law predict for the direction of the magnetic flux produced by the induced current? Explain your answer.

(d) Apply the right hand grip rule to predict the direction of the induced current in the loop.

3 Given that there is an induced current in the loop...

(a) What does Lenz's law predict about the direction of the motor effect force on the coil due to the interaction between the magnet's field and the induced current? Explain your answer.

(b) Use FLHR to predict the direction of the current in the top of the loop so that there is a component of the force in the direction you predicted in part (a).

(c) How does this apply to the whole loop?

4 Explain why the direction of the induced current would be the same as you have predicted in questions 2 and 3 if the magnet were the other way around **and** moving towards the loop.

5 A small upright magnet is dropped through a horizontal coil of wire, which is connected to a data logger. A graph of the induced emf against with time is shown.

(a) Briefly explain the overall shape of the graph.

(b) Give a reason why the negative part of the graph is shorter and taller.

(c) [**Stretch & Challenge**] Compare the areas between the positive and negative parts of the graph and explain your answer.

6 A Boeing 787 Dreamliner, with a wingspan of 60 m is cruising at 900 km h^{-1} through the Earth's magnetic field which is 60 μT at an angle of $70°$ to the horizontal.

(a) Calculate pd between the wingtips because of the induced emf.

(b) Give a reason why this pd does not give rise to a current.

Questions 7–10 relate to a square loop of 100 turns of copper wire, which has a length of 50 cm and a width of 40 cm. It is placed so that its 50 cm side is perpendicular to a magnetic field of flux density 0.05 T.

7 The loop is slowly rotated about the dashed line. Calculate the flux linkage when the loop is (a) at 90° to, (b) at 30° to and (c) parallel to the magnetic field.

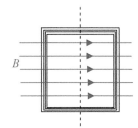

8 (a) The loop is now rotated at a steady speed of 2 revolutions per second. At time, $t = 0$ the loop is parallel to the magnetic field. Give expressions for:

   (i) the flux linkage, $N\Phi$, as a function of time, and

   (ii) the induced emf, $\mathcal{E}_{IN}$, as a function of time.

9 The wire of the coil has a diameter of 0.50 mm and is made from copper of resistivity $1.68 \times 10^{-8}\ \Omega$ m. The ends of the coil are connected together. Sketch a graph of the induced current against time over a period of 2 seconds.

10 Use your answers to questions 8 and 9 to determine the mean power dissipated in the coil.

11 An alternator has 100 turn coil of area 10 cm², which is spun in a uniform magnetic field. The induced emf has a peak value of 6.0 V and a period of 25 ms. Calculate:

   (a) the frequency, f, and angular frequency, ω, of the coil,

   (b) the flux density of the magnetic field,

   (c) the period and peak emf if the angular frequency is doubled.

Stretch & Challenge

An inductor has both inductance and resistance. It can be considered to be a series combination of a pure inductor (i.e. one with zero resistance) and a resistor. This is rather like the emf and internal resistance in a power supply.

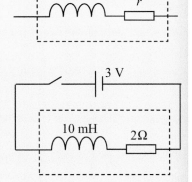

We are going to use this idea to derive the I-t relationship when an inductor is switched into a DC circuit. We'll start by considering a 10 mH inductor which has a resistance of 2 Ω. The cell has a negligible internal resistance.

1 At the instant the switch is closed, the current is zero, so the pd across the 2 Ω 'resistor' is zero and the pd across the 10 mH 'pure' inductance is 3 V. Use $V = L\frac{dI}{dt}$ to calculate the rate of change of current.

2 A (very) short time later, the current is 0.1 A. Calculate the new rate of change of current.

3 Repeat part 2 for currents of 0.2 A, 0.3 A, 0.4 and 0.5 A

4 What can you say about time intervals between the 0, 0.1, 0.2, 0.3, 0.4 and 0.5 A currents?

5 When the current is 1.5 A, state the pd across the inductor. What is the rate of change of current?

6 Estimate the time taken for the current to increase from 0 to 0.5 A.

7 Sketch a graph I against t.

Now we'll move on to the general case, with a power supply V_0, an inductance L and a resistance R.

8 When the current is I, show that $\frac{dI}{dt} = \frac{1}{L}(V_0 - IR)$

9 By rearranging to give $\int \frac{dI}{V_0 - IR} = \int \frac{dt}{L}$, integrate and apply the initial condition $I(0) = 0$ to show that the relationship between I and t is $I = \frac{V_0}{R}\left(1 - e^{\frac{-Rt}{L}}\right)$.

10 Show that the ratio $\frac{L}{R}$ has units of time and work out the value of this 'time constant' for the 10 mH, 2 Ω inductor. Discuss how this relates to your answers to 6 and 7.

Overview: Unit 4 Options

Option A: Alternating currents p181

The principles of DC electrical circuits as well as electric and magnetic fields are applied to alternating currents, which can be used to transmit energy and information.

- The application of Faraday's laws to the quantitative understanding of emf generation by rotating coils in a uniform magnetic field.
- The characteristics of alternating potential differences and currents.
- The definition and use of rms values of current and potential difference in calculating power dissipation.
- Measuring voltages, currents and frequencies using an oscilloscope.
- Current / pd relationships for resistors, capacitors and inductors; resistance, reactance and impedance.
- Lack of power dissipation in inductors and capacitors.
- Phasor analysis of series RCL circuits.
- Resonance in RCL circuits; the Q factor.

Option B: Medical physics p204

The focus of this option is on the principles and practice of medical imaging by a variety of techniques, including X rays, ultrasound, MRI and radiotracers.

- The nature and production of X-rays; their use in diagnosis and therapy.
- X-ray imaging techniques: fluoroscopy, digital imaging, CT scans.
- X-ray attenuation, $I = I_0 \exp(-\mu x)$.
- Generation and detection of ultrasound; ultrasound A and B scans.
- Acoustic impedance and the need for a coupling medium.
- Doppler scans to investigate blood flow.
- The physical principles and the use of MRI in diagnosis.
- Comparison of ultrasound, X-ray and magnetic resonance imaging.
- The effects of α, β and γ radiation on living tissues.
- Absorbed dose, equivalent dose and effective dose and their units.
- The use of radiotracers for imaging, including PET scanning.
- The use of the gamma camera.

Option C: The physics of sports p220

The physics of rotation is extended to cover spinning objects and the principles of linear and rotational motion applied in sporting contexts.

- The application of the principle of moments and centre of gravity in sporting contexts and human skeletomuscular system.
- Newton's second law and the coefficient of restitution.
- Calculations of moment of inertial, torque, angular acceleration and angular momentum.
- Conservation of angular momentum and energy.
- Application of projectile theory.
- Application of Bernoulli's equation $p_1 = p_0 - \frac{1}{2}\rho v^2$.
- Drag force and the drag coefficient, $F_D = \frac{1}{2}\rho v^2 A C_D$.

Option D: Energy and the environment p235

Techniques of energy generation are examined in the context of the changing atmospheric influence on climate as well as the physics of insulation.

- Analysis of the radiant energy balance of the Earth including the changing effect of the atmosphere due to the increase in the levels of greenhouse gases, such as carbon dioxide and methane.
- Comparison of the effect on sea levels of melting land-based and floating ice.
- Analysis and comparison of the following renewable energy sources: solar radiant power, photovoltaic cells, wind power, tidal, hydroelectric and pumped-storage power.
- Principles and practical problems of nuclear fission and fusion, including breeding and the fusion triple product.
- The principles and benefits of fuel cells.
- Thermal conduction, the coefficient of thermal conductivity.
- Heat loss from buildings; U values.

Option A: Alternating currents

Electrical currents and potential differences in circuits are very often alternating. They are generally referred to as AC [or ac] which stands for *alternating current*, even when referring to voltages! Energy is transmitted from power stations to homes by alternating currents because the voltages can easily and efficiently be stepped up and down by transformers to save energy and for safety reasons. Radio aerials convert em radiation into very high frequency electrical signals, which are controlled using resistors, capacitors and inductors. This optional unit investigates the generation, properties and control of sinusoidal electrical oscillations.

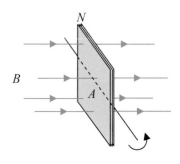

Fig. A1 Plane coil in a magnetic field

A.1 The generation of sinusoidal voltages and currents.

Fig. A1 shows a rectangular coil with N turns and area A, placed at right angles to a magnetic field of flux density B. The flux linkage, $N\Phi$, is defined by:

$$N\Phi = BAN$$

Self-test
A1

For the coil in Fig. A1, state the flux linkage if:

$B = 2.0$ T; $A = 150$ cm²;

$N = 1000$ turns.

If the coil is made to rotate about the axis shown, then the area the coil presents to the field changes and so the flux linkage changes, which induces an emf in the coil.

The coil in Fig. A2 is shown edge on. It is rotating with angular velocity ω.

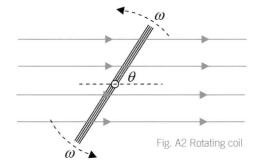

Fig. A2 Rotating coil

The area the coil presents to the magnetic field is reduced by a factor of $\sin\theta$.

$$\therefore N\Phi = BAN\sin\theta$$

If $\theta = 0$ when $t = 0$, then $\theta = \omega t$

$$\therefore N\Phi = BAN\sin\omega t$$

Faraday's and Lenz's laws state that the induced emf, $\mathcal{E}_{in}$, is equal to and opposed to, the rate of change of the flux linkage.

$$\mathcal{E}_{in} = -\frac{\Delta(N\Phi)}{t} = -\frac{d}{dt}(N\Phi)$$

We saw in Section 3.2, Vibrations, that

if $\qquad x = A\sin\omega t \qquad$ then $\qquad v = A\omega\cos\omega t$.

Remembering that v is the rate of change of x we deduce that the induced emf is given by

$$\mathcal{E}_{in} = -\omega BAN\cos\omega t.$$

If an opening is made in the coil, this induced emf results in an output pd across this opening. If no current is taken, or the resistance of the coil is negligible, the output pd, $V_{out} = \mathcal{E}_{in}$, so we can write:

$$V_{out} = -\omega BAN\cos\omega t.$$

▼ **Study point**

Lenz's law: the direction of the induced emf is such as to tend to oppose the change that produces it.

This is the '−' sign in

$\mathcal{E}_n = -\dfrac{\Delta(N\Phi)}{t}$.

▼ **Study point**

Note that $\mathcal{E}_{in}$, V_{out} and I are all proportional to all the significant quantities: B, A, N and ω (and remember that $\omega = 2\pi f$).

So, if the output is connected to an external resistor, R, a current, I, is induced given by:

$$I = \frac{V_{out}}{R} = -\frac{\omega BAN \cos \omega t}{R}$$

For a practical generator, the problem is to connect the coil to the external circuit without the connecting wires becoming increasingly tangled. In low-power dynamos, the output pd is connected to an external circuit by the use of slip-rings, which make sliding contacts with carbon brushes. This is shown in Fig. A3.

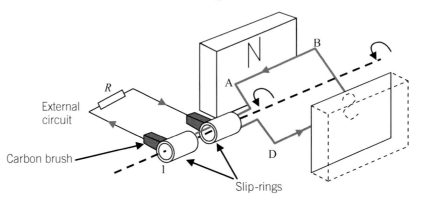

Fig. A3 A simple dynamo

▼ **Study point**

Slip-rings

In Fig. A3 we see that side AB of the coil (see label) is permanently connected to slip-ring 1. When coil side AB is moving downwards the current flows as shown; half a cycle later, the current in the external circuit is reversed.

We use Lenz's law to predict the direction of the emf and hence the current.

A2

Self-test

What emf is produced by a 1000-turn coil of area 10 cm² rotating at 3000 rpm in a magnetic field of flux density 1.5 T?

▼ **Study point**

In bicycle dynamos, the coil is fixed and the changing field is provided by a cylindrical magnet which rotates inside the coil. The advantage of this is that a permanent fixed connection to the coil is possible – no need for slip-rings!

The red arrows in Fig. A3 show the direction of the induced current. We can predict this direction using Lenz's law (see Study point). How do we apply this law to the rotating coil in Fig. A3? The following three ways are suggested – don't worry, they all give the same answer!

1   We know that the flux linkage in the coil is changing so an emf is induced which causes a current because there is a complete circuit.

→   The magnetic field exerts a force on the coil by the motor effect.

→   The direction of the force is to oppose the motion, i.e. upwards on side AB.

→   Applying Flemings Left hand rule with a field from left to right and an upward force gives a current from B to A.

2   →   We know that the flux linkage in the coil is changing so an emf is produced which causes a current because there is a complete circuit.

→   At the instant of Fig. A3, the flux in the coil is decreasing with the field lines from left to right (N to S).

→   So the induced current produces supporting flux in the coil, i.e. a field from left to right.

→   Applying the Right hand grip rule, the induced current in the coil is anti-clockwise as seen from the right.

3   Using the Right hand [generator] rule:[1]
→   The wire AB is moving downwards in a magnetic field that is from left to right.

→   by the Right hand generator rule the current in AB is from B→ A.

Example

At the instant of the diagram in Fig. A3, determine whether the emf is increasing or decreasing in magnitude.

---

1 Many people do not like the Right hand rule because it is **only** applicable in cases where an emf is generated by a wire moving in a magnetic field, and not, for example, where the field is rotating or the flux density is changing.

Answer

The wire AB's motion is becoming closer to vertical, so it is cutting field lines more and more rapidly so the emf is increasing. [Alternatively: the rate of change of flux is greatest when the coil is horizontal because the gradient of $\sin\theta$ is greatest when $\theta = 0$ or $\pi -$ giving the same answer.]

Self-test **A3**

Sketch a labelled graph of $\mathscr{E}_{in}(t)$ from Self-test A2. Add a second curve for $\mathscr{E}_{in}(t)$ if the rotational frequency is halved.

The relation between the coil orientation, $N\Phi$ and $\mathscr{E}_{in}$ is shown in Fig. A4.

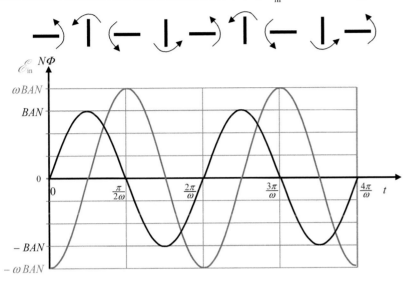

Fig. A4 Coil position and emf

▼ **Study point**

In power station generators, the field is produced by rotating electromagnets. The coils in these field magnets have a much smaller current than in the fixed coil, so maintaining an electrical connection is easier.

A.2 Basic AC terms

If you connect a 12 V AC supply to an oscilloscope[2] you'll see a trace which appears something like the one in Fig. A5 [but without the axes and scales!]. It is a typical voltage-time graph for AC.

Self-test **A4**

From Fig. A5, find the first time for which the pd is 0 and increasing.

Self-test **A5**

State 3 values of θ for which $\sin\theta = 0$ [and increasing as θ gets larger].

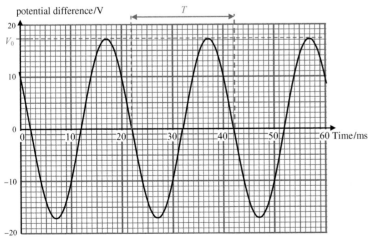

Fig. A5 AC V-t graph

▼ **Study point**

Most of the time the phase angle ϕ can be taken as 0 or $\pm\frac{\pi}{2}$, so the sinusoids will be of the form

$V = V_0 \cos \omega t$,
$V = V_0 \cos (\omega t \pm \frac{\pi}{2})$,
$V = V_0 \sin \omega t$, etc.

This graph has just the same form as those in simple harmonic motion. It is a sinusoid. In other words we can represent it by an equation of the form:

$$V = V_0 \cos (\omega t + \phi)$$

where V_0 is the maximum voltage, which we call the *peak voltage*.

2 See Section A9 for oscilloscopes.

▼ Study point

When taking oscilloscope measurements it is often useful to measure the *peak-to-peak* voltage, which is from top to bottom of the graph, i.e. $2V_0$.

A6 Self-test

Use the answers to Self-tests A4 and A5 to write the equation of the graph in Fig. A.5 in the form:
$V = V_0 \sin(\omega t + \varepsilon)$.

GRAPH SKETCHING

$\langle \cos^2\omega t \rangle = \frac{1}{2}$. Why?

Graphs of $\cos \omega t$ and $\cos^2\omega t$:

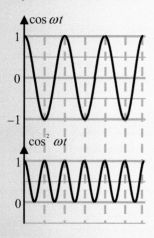

Using the trig relationship

$$\cos 2\theta = 2 \cos^2 \theta - 1$$

you should be able to show that the $\cos^2\omega t$ graph is as shown. You need to be able to sketch this graph.

But $\quad \langle \cos 2\theta \rangle = 0$

∴ $\quad 2\langle \cos^2\theta \rangle - 1 = 0$

∴ $\quad \langle \cos^2\theta \rangle = \frac{1}{2}$.

A7 Self-test

Show that the rms voltage of the supply graph shown in Fig. A5 is 12.0 V.

Similarly the relationship between current, I, against time is expressed as:

$$I = I_0 \cos (\omega t + \phi)$$

and I_0 is called the *peak current*.

The quantity ω is called the *angular frequency* or *pulsatance* and is related to the frequency, f, and the period, T, by the equations:

$$\omega = 2\pi f = \frac{2\pi}{T}.$$

The angle ϕ is a *phase angle*. The graph in Fig. A5 corresponds to the equation:

$$V = 17.0 \cos (100\pi t + 0.3\pi).$$

Example

Find the equation for the graph in Fig. A5 in the form
$$V = V_0 \cos (\omega t + \phi).$$

Answer

1 The peak voltage = 17.0 V, ∴ V_0 = 17.0 V

2 **2 cycles** of the graph take 42.0 − 2.0 ms = 40.0 ms. ∴ T = 20 ms

$$\therefore \omega = \frac{2\pi}{20 \times 10^{-3}} = 100\pi \text{ s}^{-1} \ [= 314 \text{ s}^{-1}]$$

3 V_{max} occurs when $t = -3$ ms. This is $\frac{3}{20}$ of a cycle

$$\therefore \phi = \frac{3}{20} \times 2\pi = 0.3\pi$$

$$\therefore V = 17.0 \cos (100\pi t + 0.3\pi)$$

A.3 rms values

Question: How does the graph in Fig. A5 relate to a 12 V power supply?

The answer is to do with power dissipation. The 12 V figure is the voltage of the constant voltage DC power supply that would dissipate the same power through a resistor. Let's do the maths. Consider the circuit in Fig. A6:

The power supply dissipated in the resistor at any instant is given by:

$$P = \frac{V^2}{R}$$

If V varies with time the mean value of the power, $\langle P \rangle$ is given by

$$\langle P \rangle = \frac{\langle V^2 \rangle}{R}.$$

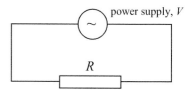

Fig. A6 AC power supply

We **define** the *root mean square* voltage V_{rms} as $\sqrt{\langle V^2 \rangle}$.

Then the mean power dissipated is given by: $\quad \langle P \rangle = \frac{V_{rms}^2}{R}$

So the rms voltage V_{rms} is the voltage of a smooth DC supply that would produce the same power dissipation. In the case of a sinusoidal variation, $V = V_0 \cos \omega t$, so
$$V_{rms} = V_0 \sqrt{\langle \cos^2\omega t \rangle}$$

The value of $\cos \omega t$ oscillates between -1 and $+1$ so its mean value is 0, but the value of $\cos^2 \omega t$ is always positive (e.g. $(-1)^2 = +1$) and it oscillates between 0 and 1: its mean value is $\frac{1}{2}$.

$$\therefore V_{rms} = \frac{V_0}{\sqrt{2}}$$

Example

A student in Britain uses a CRO and measures the peak mains voltage, V_0 as 340 V. Calculate the value of V_{rms}.

Answer

$$V_{rms} = \frac{V_0}{\sqrt{2}} = \frac{340}{\sqrt{2}} = 240 \text{ V}.$$

Note: Historically the domestic mains (rms) voltage has been 240 V in Britain and 220 V in mainland Europe. This has recently been 'harmonised' to 230 V with a tolerance of $+10\% / -6\%$. Both 220 V and 240 V are thus within the allowed range, and following the harmonisation, the actual domestic voltages remain as they were!

Similarly, the rms current, I_{rms}, defined as $\sqrt{\langle I^2 \rangle}$ can be used to calculate the mean power dissipated using:

$$\langle P \rangle = I_{rms}^2 R = I_{rms} V_{rms} = \frac{V_{rms}^2}{R}$$

Because it is so useful, a quoted value of voltage or current is almost always the rms value unless otherwise specified. For power and current calculations, in circuits containing only resistors (i.e. no capacitors or inductors), the rms values can be used just like as DC voltage and current values.

Example

A 15 Ω resistor is connected across a 6 V (rms) power supply. Calculate: (a) the rms current, (b) the peak current and (c) the mean power dissipated.

Answer

(a) $I_{rms} = \dfrac{V_{rms}}{R} = \dfrac{6.0}{15} = 0.4\text{A}$ (b) $I_0 = \sqrt{2}I_{rms} = \sqrt{2} \times 0.40 = 0.57 \text{ A}$

(c) $\langle P \rangle = I_{rms}^2 R = 0.4^2 \times 15 = 2.4 \text{ W}$

Note: for part (c) we could also use $\langle P \rangle = I_{rms} V_{rms}$, $\dfrac{V_{rms}^2}{R}$ or even $\langle P \rangle = \dfrac{I_0^2 R}{2}$, $\dfrac{I_0 V_0}{2}$, or $\dfrac{V_0^2}{2R}$ (as long as it is a sinusoidal AC supply).

Self-test A8

(a) Sketch graphs of $\sin \omega t$ and $\sin^2 \omega t$.

(b) Show that $\langle \sin^2 \omega t \rangle = \frac{1}{2}$.

Hint: See the **Graph sketching** box above.

Self-test A9

An AC supply delivers an rms current of 0.2 A through a 47 Ω resistor. Calculate:

(a) the rms pd of the supply.

(b) the mean power dissipated.

Self-test A10

For the graphs in Fig. A.7 calculate:

(a) the value of the resistance

(b) the frequency and angular frequency (pulsatance)

(c) the rms voltage and current

(d) the mean power dissipated.

▼ **Study point**

The power dissipation at any instant is the product IV at that moment.

A.4 Alternating currents through resistors, inductors and capacitors[3]

(a) Resistors

The current through a **resistor** at any instant is directly proportional to the potential difference at that instant. Because of this, the current and voltage oscillate **in phase** with each other. Fig. A7 gives an example of this.

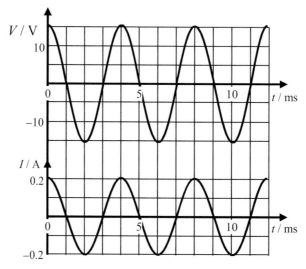

Fig. A7 Current and pd variation for a resistor

For inductors and capacitors, there is a phase difference between V and I, so the situation is more complicated. We'll examine these phase differences qualitatively before looking at it mathematically.

(b) Capacitors

Consider a **capacitor**, C, which is initially uncharged. Suppose it is being charged up by a current, as shown in Fig. A8.

Positive charge builds up on the left-hand plate and negative charge on the Right plate, so a pd becomes established.

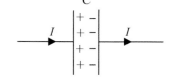

Fig. A8 Charging capacitor

We see from this description that, for a capacitor, current causes a change in voltage – the current leads the voltage.

Taking the conclusions from the Stretch & Challenge box, the current and voltage graphs for a capacitor in an AC circuit are as in the example in Fig. A9. The current graph is ¼ of a cycle ahead of the voltage graph, i.e. it is ahead in phase by $\frac{\pi}{2}$.

A11 Self-test

For the graphs in Fig. A7 calculate the power at:

(a) t = (a) 0

(b) t = 2 ms

(c) t = 5 ms.

A12 Self-test

Sketch a graph of the variation of power with time for the graphs in Fig. A7 and state the mean power.

Stretch & Challenge

For a capacitor $Q = CV$.

The current, I, is the rate of change of charge, $I = \frac{\mathrm{d}Q}{\mathrm{d}t}$.

(a) By differentiating the first equation write I in terms of V.

(b) Let $V = V_0 \sin \omega t$. Use your answer to (a) to show that:

$I = \omega C V_0 \cos \omega t$

$I = \omega C V_0 \sin \left(\omega t + \frac{\pi}{2}\right)$.

Conclusions

(1) The current is $\frac{\pi}{2}$ **ahead** of the voltage and

(2) the peak current and voltage are related by $I = \omega C V_0$.

---

[3] It is slightly strange to talk about the current **through** a capacitor – charge cannot flow through the dielectric, which is an insulator. The current is in the wires on either side of the capacitor!

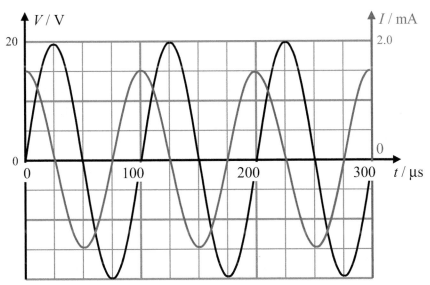

Fig. A9 Current and pd variation for a capacitor

A13

Self-test

For the graphs in Fig. A9:

(a) Determine the values of V_0, I_0, V_{rms}, I_{rms}, T, f and ω.

(b) Use the equation $I_0 = \omega C V_0$ to calculate the capacitance, C.

(c) Confirm that the rms pd and current are related by $I = \omega C V$.

(c) Inductors

Another component that is often found in an AC circuit is an **inductor**. It consists of a coil of wire often wound on a soft iron core. Fig. A10 shows some images of inductors and the circuit symbol. L is the **inductance** of the inductor, defined in the box.

 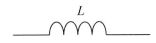

Fig. A10 Inductors

The coil of wire usually has a very low resistance and its effect on the circuit is due to its magnetic properties. A current in an inductor produces a magnetic field and so there is a magnetic flux linking the circuit. If a pd is applied across the inductor so the current starts to grow, this produces a change in the flux linkage which produces an emf which, by Lenz's law, opposes the growth in current. The inductor therefore acts as a brake on a change of current: a reduction in current induces a supporting emf, which slows down the drop in current.

So, for a pure inductor (i.e. one with zero resistance) the **rate of change of current** is proportional to the voltage applied across it. For example, if a voltage of 10 V is applied across a 2 H inductor, the current grows at a rate of 5 A s^{-1}.

From the above discussion, it should be clear that an inductor behaves in just the opposite manner to a capacitor. A pd across an inductor produces a changing current, so in an AC circuit the voltage leads the current for an inductor – also by $\frac{\pi}{2}$ for a pure inductor (see Stretch & Challenge).

Terms & definitions

The **inductance** of an inductor is defined by the equation:

$V = L\frac{\Delta I}{\Delta t}$ or, in calculus, $V = L\frac{dI}{dt}$.

The unit is the henry (H).

Stretch & Challenge

For an inductor $V = L\frac{dI}{dt}$.

Let $I = I_0\cos \omega t$.

By differentiating the first equation, show that:

$V = \omega L I_0 \cos (\omega t + \frac{\pi}{2})$

and hence that:

(1) the voltage is $\frac{\pi}{2}$ **ahead** of the current and

(2) the peak current and voltage are related by

$V_0 = I_0\omega L$.

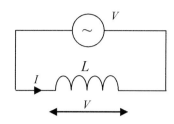

Fig. A11 Inductor in a circuit

Consider an inductor connected across an AC power supply as shown in Fig. A11. If the current varies as $I = I_0 \cos \omega t$, i.e. is at a maximum when $t = 0$, the voltage maximum is ¼ of a cycle before this, i.e. at $-\frac{T}{4}$. An example of this is shown in Fig. A12.

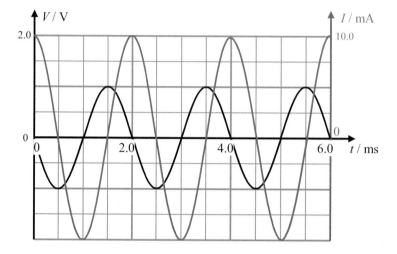

Fig. A12 Current and pd variation for an inductor

A14   Self-test

For the graphs in Fig. A12:

(a) Determine the values of V_0, I_0, V_{rms}, I_{rms}, T, f and ω.

(b) Use the equation $V_0 = I_0 \omega L$ to calculate the inductance capacitance, L.

(c) Confirm that the rms pd and current are related by $V = I\omega L$.

A useful mnemonic

CIVIL: In a **C**apacitor, the current (**I**) leads the **V**oltage; the **V**oltage leads the current (**I**) in an inductor (**L**).

Exercise 4A.1

1 An alternating pd with $V_{rms} = 9.0$ V and $f = 50$ Hz is applied across a $10\ \Omega$ resistor.

   (a) Calculate (i) the rms current and (ii) the mean power dissipation.

   (b) If the pd is at a maximum at $t = 0$, write the variation of pd with time in the form $V = V_0 \cos(\omega t + \varepsilon)$.

   (c) Write the variation of current with time in the form $I = I_0 \cos(\omega t + \phi)$.

   (d) Sketch graphs of the variation of V, I and the power P with time between $t = 0$ and $t = 50$ ms.

2 A 12 V (rms), 1 kHz power supply is connected separately across (a) a 1 μF capacitor and (b) a 100 mH inductor. Calculate the rms current in each case. [Hint: see Self-tests A8(c) and A9(c).]

3 A 9 V (rms), 50 Hz supply is connected across a $4.7\ \Omega$ resistor and an $8.2\ \Omega$ resistor in series. Calculate the **peak** voltage across the $4.7\ \Omega$ resistor. [Hint: potential divider.]

4 For the circuit in Q3 (a) calculate the rms current and (b) write the instantaneous current in the form $I = I_0 \cos \omega t$.

5 A 32 μF capacitor and a $47\ \Omega$ resistor are connected in series. A 0.50 A, 50 Hz current is in the two components.

   (a) Calculate the pd across each component. [Hint: the current is the same for components in series.]

   (b) Without doing detailed calculations state and explain what would happen to the pd across each component if the frequency of the current increased to 100 Hz but its magnitude remained 0.50 A.

6 Three components, **A**, **B** and **C**, are connected in series. The graphs show the current (red) through the components and the pd across each. Identify the components in as much detail as possible.

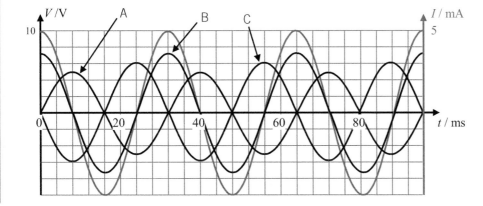

A.5 Capacitors and inductors dissipate no power

We have seen that resistors in AC circuits dissipate a mean power, $\langle P\rangle$, given by

$$\langle P\rangle = IV = I^2R = \frac{V^2}{R}$$

where V and I are the rms values. Let us examine the $V(t)$ and $I(t)$ graphs over one complete cycle to explain why this is not the case for capacitors and inductors. The graphs in Fig. A13 are the current (red) and voltage graphs for a capacitor.

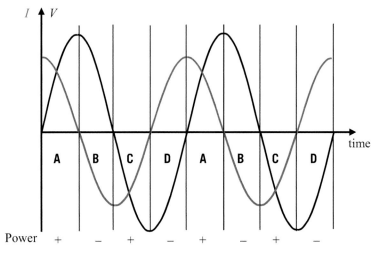

Fig. A13 $I(t)$ and $V(t)$ graphs for a capacitor

Consider the energy exchange between the power supply and the capacitor. The instantaneous energy transfer, P, **from** the power supply is given by $P = IV$.

In the quarter-cycle periods labelled **A**, the current and pd are both positive, so P is positive and the capacitor takes energy from the supply. In the periods labelled **C**, both I and V are negative, so again P is positive. In **B** and **D**, one of I and V is positive and the other negative so, in these two periods, P is negative, i.e. the capacitor returns energy to the supply. From the symmetry of the curves energy given by the power supply in **A** and **C** is the same as that returned in **B** and **D**. Thus, over a complete cycle, the mean energy taken is zero.

Let us now consider the energy balance over 1 cycle for an **inductor**. From the previous discussion, it should be clear that, because of the $\frac{\pi}{2}$ phase difference between I and V, the mean power taken from the power supply over a complete cycle is zero. It doesn't matter whether the voltage or current leads, the instantaneous power will be sometimes positive and sometimes negative.

In order to behave like this, the inductor must store energy in its magnetic field. It can be shown that the energy density is $\dfrac{B^2}{2\mu_0}$ in a vacuum.

A.6 Resistance and reactance

We have seen that, for resistors, capacitors and inductors, the rms (and peak) voltages and currents are proportional. Such components are referred to as *linear*. We have also seen that, although the current is in phase with the voltage for a resistor, this is not the case for capacitors and inductors.

For these components, or their combination, the ratio $\dfrac{V_{\text{rms}}}{I_{\text{rms}}} \left(= \dfrac{V_0}{I_0}\right)$ is referred to as the

impedance, with the symbol Z. If the current and voltage are in phase, we call this ratio, the **resistance**, R. If they are $\frac{\pi}{2}$ out of phase, we call it the **reactance**, X. Like resistance, both impedance and reactance are expressed in ohms (Ω). Table A1 summarises the values of Z, R and X.

Component	R	X	Z	Notes
Resistor, R	R	–	R	I and V in phase
Capacitor, C	–	$X_{\text{C}} = \dfrac{1}{\omega C}$	$\dfrac{1}{\omega C}$	The current leads by $\dfrac{\pi}{2}$
Inductor, L	–	$X_{\text{L}} = \omega L$	ωL	The voltage leads by $\dfrac{\pi}{2}$

Table A1 Resistance, reactance and impedance

For individual components, we can use the values of reactance to relate I and V (rms or peak) just like resistances in a DC circuit.

Example

A capacitor, connected to a 6 V, 50 Hz supply has a reactance of 80 Ω.

(a) Calculate the current.

(b) If the frequency of the supply were changed to 100 Hz what would the current be?

Answer

(a) $I = \dfrac{V}{X_{\text{C}}} = \dfrac{6}{80} = 0.075$ A

(b) The reactance of a capacitor is inversely proportional to the frequency. The reactance at 100 Hz is thus half that at 50 Hz [40 Ω] so the current is doubled to 0.15 A.

We can now handle *individual* components in a circuit. The next problem is how to relate the current and voltage in a circuit with both resistive and reactive components. For example, consider the arrangement in Fig. A14.

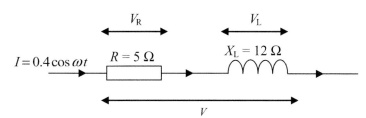

Fig. A14 Series RL combination

From the above information, you should be able to show that V_{R} and V_{L} are given by:

$$V_{\text{R}} = 2.0 \cos \omega t \quad \text{and} \quad V_{\text{L}} = -4.8 \sin \omega t \text{ [or } 4.8 \cos (\omega t + \tfrac{\pi}{2})].$$

But what about V, the pd across the combination? The answer is given on the right [in case you were worried]. There are several ways to solve this. We'll do so using *phasor* analysis.

▼ **Study point**

The reactance, X_{L}, of an inductor is **directly** proportional to the frequency;
X_{C} is **inversely** proportional to the frequency: ω (or f).

▼ **Study point**

In terms of the frequency, f:

$X_{\text{C}} = \dfrac{1}{2\pi f C}$ and $X_{\text{L}} = 2\pi f L$.

▼ **Study point**

Note that the question in the **Example** does not specify rms or peak values. It doesn't matter: 6 V rms will produce a current of 0.075 A rms; if $V_0 = 6$ V, $I_0 = 0.075$ A

Self-test A16

Calculate the capacitance of the capacitor in the **Example**.

Anxiety relief

In Fig. A14:

$V = 5.2 \cos (\omega t + 1.18)$

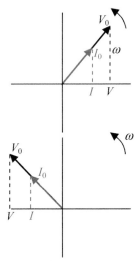

Fig. A16 I and V phasors for a resistor

A17 Self-test

In the same way as the Study point above, identify whether the values if I and V in Fig. A17(b) – (d) are positive or negative and increasing or decreasing.

A18 Self-test

In which diagram in Fig. A17(a) – (d) are the gradients of the $V(t)$ graph and the $I(t)$ graph both positive?

A.7 Solving RCL circuits

(a) Introduction to phasors

This is the technique most commonly used to solve AC circuits in A level Physics. Consider a vector of amplitude I_0 rotating about the origin, **O** as in Fig. A15. The x-projection of the vector is given by:

$$x = I_0 \cos \theta$$

If the vector is rotating with angular velocity ω, then

$$\theta = \omega t + \varepsilon,$$

where ε is the value of θ when $t = 0$. So $x = I_0 \cos (\omega t + \varepsilon)$.

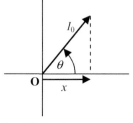

Fig. A15 Rotating vector

This means that x is varying just like the current in a circuit: one which has a peak value of I_0 and angular frequency (pulsatance) of ω. This suggests that we treat the vector I_0 as *representing* the current in magnitude and phase – and that the instantaneous value of I that we measure **is** the projection of this vector – a.k.a. *phasor*.

The top diagram in Fig. A16 shows the voltage and current phasors for a resistor at a particular instant (the current phasor is in red). The directions of the two phasors are the same because the voltage and current are in phase for a resistor. The values of V and I at the instant of the diagram are the projections on the x-axis.[4] In the first diagram the values of V and I are positive and decreasing. One quarter of a cycle later the phasors have moved round $\frac{\pi}{2}$ anticlockwise. The values of V and I are now both negative and becoming more negative.[5]

The I-V phasor diagram for a capacitor is more involved. The phasor diagrams sequence in Fig. A17 are at quarter-cycle intervals.

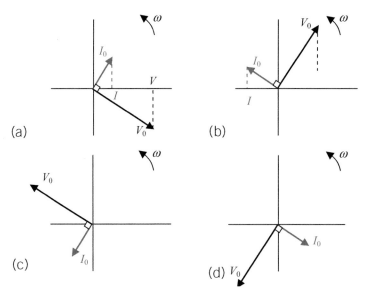

(a) (b) (c) (d)

Fig. A17 Capacitor phasor diagrams

---

4 Note that the apparent lengths of the current and voltage phasors are not significant – the lengths depend upon different scales.

5 It is not easy to express this clearly. V and I are both negative and becoming increasingly less than 0, i.e. they are decreasing but their magnitudes are increasing.

(b) Using phasors in RC circuits

Consider an RC combination connected across in an AC circuit, as shown in Fig. A18. The current and voltages shown are **not** the instantaneous values: for the moment we'll take them to be the peak values.

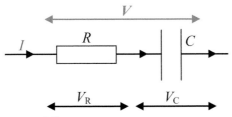

Fig. A18 RC combination

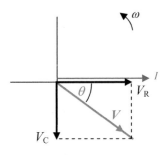

Fig. A19 V_R and V_C phasors

Because they are in series, the same current, I, is 'in' the two components, C and R.

Fig. A19 shows the V_R and V_C phasors, together with the I phasor at the instant when the I phasor is along the x-axis. The phasors are vectors, so we can add the two voltage phasors according to the parallelogram law of addition to give the total voltage, V, as shown (V in green).

We can write V_R and V_C in terms of I and the impedances. From above:

$$V_R = IR \quad \text{and} \quad V_C = IX_C = I\frac{1}{\omega C}$$

Using Pythagoras' Theorem and these expressions:

$$V = \sqrt{V_R^2 + V_C^2} = \sqrt{I^2R^2 + I^2\frac{1}{\omega^2 C^2}}$$

Simplifying, this leaves: $V = I\sqrt{R^2 + \frac{1}{\omega^2 C^2}}$. We can also write $V = I\sqrt{R^2 + X_C^2}$

We can use the phasor diagram, Fig. A19, to calculate the phase difference, θ, between the current and the voltage as follows:

$$\tan\theta = \frac{V_C}{V_R} = \frac{IX_C}{IR} = \frac{X_C}{R}$$

$$\therefore \theta = \tan^{-1}\left(\frac{X_C}{R}\right) \text{ or, because } X_C = \frac{1}{\omega C}, \theta = \tan^{-1}\frac{1}{R\omega C}$$

Question

When we talk about current and pd, we could mean instantaneous values, peak values or rms values. For which of these is the equation $V = I\sqrt{R^2 + \frac{1}{\omega^2 C^2}}$ valid?

Answer

We have derived this equation by considering peak values. But $V_0 = \sqrt{2}V_{rms}$ and $I_0 = \sqrt{2}I_{rms}$, so the equation is also valid for rms values of current and pd. It is **not** valid for instantaneous values. To see this, look back at Fig. A13: the instantaneous current and pd are zero at different times, so no equation of the form $V = kI$, where k is a constant, can be valid.

Summary

For a series RC combination, $V = IZ_{RC}$, where Z_{RC} is the impedance of the combination, $Z_{RC} = \sqrt{R^2 + X_C^2}$ and $X_C = \frac{1}{\omega C}$.

Self-test A19

Redraw Fig. A19 a quarter of a cycle later.

Self-test A20

State the voltage across the capacitor at the instant of Fig. A19.

Self-test A21

In terms of V_R and V_C, state the value of V:

(a) at the instant of Fig. A19;

(b) ¼ of a cycle after Fig. A19.

Self-test A22

Calculate the impedance of an RC combination with

$R = 100 \ \Omega$ and $X_C = 100 \ \Omega$.

Self-test A23

Calculate the phase angle between the current and voltage for the combination in Self-test A22.

Self-test A24

What value capacitor would have $X_C = 100 \ \Omega$ at a frequency (f) of 1 kHz?

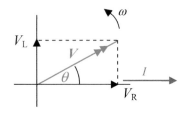

Fig. A21 V_R and V_L phasors

(c) Using phasors in *RL* circuits

We now consider the *RL* combination shown in Fig. A20. The only difference between this and the *RC* circuit is that V_L is $\frac{\pi}{2}$ **ahead** of *I*, rather than behind.

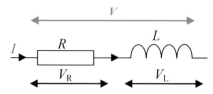

Fig. A20 *RL* combination

Fig. A21 shows the voltage phasor diagram. Note that, again, the diagram is drawn for the instant when the current phasor is along the *x*-axis. This is common practice but not essential. Also the current phasor is now shown slightly separate to avoid confusion.

Using the same working as in the previous section, it can be shown that:

$$V = I\sqrt{R^2 + X_L^2} = I\sqrt{R^2 + \omega^2 L^2}$$

and

$$\theta = \tan^{-1}\left(\frac{X_L}{R}\right) = \tan^{-1}\left(\frac{\omega L}{R}\right) \text{ [see Self-test A25].}$$

Again, *V* and *I* can be either rms or peak values – but not a mixture of the two and **not** instantaneous values.

A25 Self-test

In clear steps, derive the equations for *V* and θ for a series *RL* combination.

A26 Self-test

Show that, $V_0 = 5.2\,\text{V}$ and $\theta = 1.18\,\text{rad}$ for the *RL* combination in Fig. A14.

A27 Self-test

Use the result of Self-test A26 to write *V* as a function of time (no calculation needed).

Example

A variable frequency AC power supply with negligible internal impedance is connected across: (a) a series *RL* combination and (b) a series *RC* combination. Sketch graphs of *I* against ω.

Answer

(a) *RL*. $I = \dfrac{V}{\sqrt{R^2 + \omega^2 L^2}}$.

When $\omega = 0$, $I = \dfrac{V}{R}$. As ω increases *I* decreases and tends to 0 as $\omega \to \infty$.

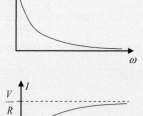

(b) *RC*. $I = \dfrac{V}{\sqrt{R^2 + \dfrac{1}{\omega^2 C^2}}}$

As $\omega \to 0$, $\dfrac{1}{\omega} \to \infty$, $\therefore I \to 0$.

As $\omega \to \infty$, $\dfrac{1}{\omega} \to 0$, $\therefore I \to \dfrac{V}{R}$

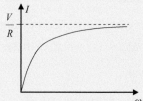

▼ **Study point**

The wire of a real inductor must possess resistance (unless it is a superconductor). We can consider a real inductor to be composed of a resistor and an inductor in series. The *I-ω* graph for a real inductor is therefore as shown in the **Example** with *R* being the resistance of the inductor wire.

(d) Using phasors in *RCL* circuits

We are now in a position to use phasors to solve the problem of series circuits containing all three components – resistors, capacitors and inductors.

Consider Fig. A22

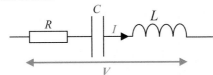

As usual, we'll draw the phasor diagram for the moment when the current phasor is along the *x*-axis. There are now 3 voltage phasors:

V_R in phase with I ∴ along the *x*-axis

V_C $\frac{\pi}{2}$ behind I, ∴ along the negative *y*-axis

V_L $\frac{\pi}{2}$ before I, ∴ along the positive *y*-axis.

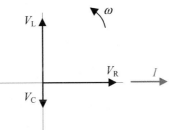

Fig. A22 Series *RCL* combination

These phasors are shown in Fig. A23. For the purposes of the diagram, we have assumed, for the moment, that $V_L > V_C$. Actually, as we shall see, the relative length of the phasors depends upon the frequency.

Fig. A23 *RCL* voltage phasors

To add these phasors we first combine V_L and V_C, i.e. we find $V_L - V_C$ and then use Pythagoras' Theorem to add this phasor to V_R. This is shown in Fig. A24.

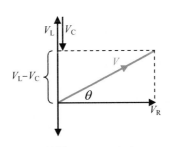

From the diagram:
$$V = \sqrt{V_R^2 + (V_L - V_C)^2}$$

From which, as before, $V = I\sqrt{R^2 + \left(\omega L - \frac{1}{\omega C}\right)^2}$.

So the impedance, Z, of this circuit is given by:

Fig. A24 *RCL* phasor analysis

$$Z = \sqrt{R^2 + \left(\omega L - \frac{1}{\omega C}\right)^2}$$ which is again of the form $Z = \sqrt{R^2 + X^2}$ but note that $X_{LC} = X_L - X_C$.

The reason for the '−' sign should be obvious, the voltage across the inductor and capacitor are π out of phase, i.e. in *antiphase*.

We can also calculate the phase angle, θ, as follows:

$$\theta = \tan^{-1}\frac{V_L - V_C}{V_R}, \text{ which leads to } \theta = \tan^{-1}\frac{X_L - X_C}{R} \text{ and so to } \theta = \tan^{-1}\frac{\omega L - \frac{1}{\omega C}}{R}.$$

(e) Solving *RCL* circuits using complex numbers

Physicists and engineers generally analyse AC circuits using complex numbers rather than phasors. The two treatments are entirely equivalent but complex numbers allow more complicated circuits to be solved. The mathematics of this analysis is to be found in Chapter 12 of the updated *Maths for A Level Physics* (2016) by Kelly and Wood.

Self-test A28

A 0.127 H resistor and a 30 Ω resistor are connected in series across a 12 V, 50 Hz supply. Calculate:

(a) the reactance of the inductor

(b) the impedance of the combination

(c) the current

(d) the phase angle between the current and voltage.

Self-test A29

Sketch a phasor diagram for the circuit in Self-test A28.

▼ **Study point**

It is useful to remember the formula for the impedance of an *RCL* circuit in the form:

$Z = \sqrt{R^2 + (X_L - X_C)^2}$. Also note that it doesn't matter if you write

$Z = \sqrt{R^2 + (X_C - X_L)^2}$ (can you see why this is the case?)

Self-test A30

A resistor of resistance 10 Ω is connected across a power supply in series with an inductor and a capacitor of reactances 50 Ω and 30 Ω respectively.

Calculate:

(a) the total impedance

(b) the phase angle between the voltage of the supply and the current.

Exercise **4A.2**

Use phasor analysis to answer the following questions.

1 A resistor, capacitor and inductor are connected in series across a power supply with a frequency of 50 Hz. The rms pd's across the components are 5.0 V, 20.0 V and 10.0 V respectively and current through the components is 2.0 A. Calculate:

(a) the supply voltage

(b) the impedance of the combination

(c) the resistance and reactance of each component

(d) the value of each component

(e) the phase difference between the current and voltage

(f) the mean power dissipated.

2 The frequency of the supply in question 1 is changed to 100 Hz but its voltage is unchanged. With no further calculation state the reactances of the capacitor and inductor, the total impedance, the current, power dissipated and the phase difference between the current and voltage.

3 The current though a 20 Ω resistor and an inductor of reactance 12 Ω in series is given, in amps, by $I = 0.1 \cos 300t$. Calculate:

(a) the rms current

(b) the inductance of the inductor

(c) the total impedance

(d) the peak and rms values of the supply voltage

(e) the instantaneous value of the supply voltage (i) when $t = 0$ and (ii) when $t = 2$ ms.

4 For the power supply in Q3, express the instantaneous pd, V, of the supply as a function of time in the form $V = V_0 \cos(\omega t + \varepsilon)$.

5 The pd of a supply, in volts, is given by $V = 24 \cos 500t$. The supply is connected across a series combination of 48 Ω resistor and a capacitor of reactance 36 Ω. Calculate:

(a) the rms voltage

(b) the capacitance of the capacitor

(c) the total impedance

(d) the peak current

(e) the rms current

(f) the instantaneous value of the current at $t = 0$

(g) the instantaneous value of the pd across each component at $t = 0$.

A.8 Resonance in *RCL* circuits

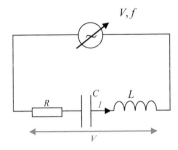

Fig. A25 Series resonance circuit

(a) Variation of current with frequency

We have seen that the relationship between V and I [either rms or peak values] for the circuit in Fig. A25 is:

$$I = \frac{V}{\sqrt{R^2 + \left(2\pi fL - \frac{1}{2\pi fC}\right)^2}}$$

The graph of I against f is plotted in Fig. A26 for the component values 12 V, 100 Ω, 0.10 H and 1.0 µC.

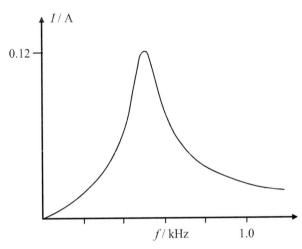

Fig. A26 Resonance curve for a series *RCL* circuit

The curve – called a **resonance curve** – has some obvious properties:

1 The current $I \rightarrow 0$ as $f \rightarrow 0$ or $f \rightarrow \infty$.

2 The current rises to a peak value, for $f \sim 500$ Hz in this case. This value of f is called the resonance frequency (or resonant frequency) with the symbol f_0.

3 The current has a peak value of (in this case) 0.12 A.

Why does I vary in this way? For the following reasons:

1 If we try and insert either $f = 0$ or let $f \rightarrow \infty$ in the equation for I, the denominator becomes infinite and hence I is zero.

2 The expression $\left(2\pi fL - \frac{1}{2\pi fC}\right)^2$ on the bottom line has a minimum value of zero.

   If $\left(2\pi fL - \frac{1}{2\pi fC}\right)^2 = 0$, then $f_0 = \frac{1}{2\pi\sqrt{LC}}$, which in this case is 503 Hz.

3 When $\left(2\pi fL - \frac{1}{2\pi fC}\right)^2 = 0$ the equation simplifies to $I = \frac{V}{R}$, which in this case is $I = \frac{12\text{ V}}{100\text{ Ω}}$, i.e. 0.12 A.

(b) The *Q* factor

The sharpness of the resonance curves in Fig. A27 is described by the *Q* factor: the greater the *Q* factor the sharper the curve. There are various definitions including:

$$Q = \frac{\omega_{\text{res}}}{\Delta\omega}, \text{ where } \Delta\omega \text{ is the } half\ power\ bandwidth, \text{ and}$$

$$Q = 2\pi \times \frac{\text{energy stored}}{\text{energy dissipated per cycle}}$$

Stretch & Challenge

Reproduce Fig. A26 using a spreadsheet program and then investigate the effects of varying R, C and L.

▼ **Study point**

The lower the value of R, the sharper the resonance curve. Unlike for mechanical resonance, the frequency, f_0, for the peak current doesn't change with resistance.

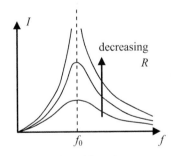

Fig. A27 Effect of R on the resonance curve

▼ **Study point**

When sketching resonance curves for *RCL* circuits remember that, for different values of R:

1 The resonance frequencies are all the same.

2 The graphs never cross [they meet at (0,0)].

3 The graphs all tend to the same gradient as $f \rightarrow 0$.

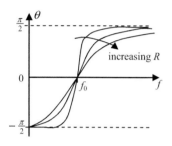

Fig. A28 Variation of phase angle with frequency

These are not further explored here but left for the interested reader to research. For our purposes we shall use the equations

$$Q = \frac{V_C}{V_R} \qquad \text{and} \qquad Q = \frac{V_L}{V_R}$$

where V_C, V_L and V_R are the rms pd <u>at resonance</u> across the capacitor, inductor and resistor respectively. The two equations are equivalent because $V_L = V_C$ at resonance. The Q factor is an important quantity for electrical engineers when designing resonance circuits, e.g. in radio receivers which need to respond strongly to one radio station but ignore others which have different frequencies.

(c) Variation of phase with frequency

In Section A.7(d) we saw that, in a resonance circuit, V leads I by θ, where

$$\theta = \tan^{-1} \frac{2\pi fL - \dfrac{1}{2\pi fC}}{R}.$$

The curves in Fig. A28 illustrate the variation of θ with f for different values of circuit resistance. As $f \to 0$, the numerator in the fraction tends to $-\infty$, so $\theta \to -\dfrac{\pi}{2}$. When $f = f_0$, the numerator is zero so $\theta = 0$. As $f \to \infty$, the numerator in the fraction tends to $+\infty$, so $\theta \to +\dfrac{\pi}{2}$.

Not that the variation in θ with f around f_0 becomes much more rapid as R becomes very small, i.e. when we have a large Q factor. In the theoretical case when $R = 0$, the graph becomes a step function between $\theta = -\dfrac{\pi}{2}$ and $\theta = +\dfrac{\pi}{2}$.

(d) Understanding the resonance circuit

To understand the circuit more completely it may help to stand back from the details of the maths and appreciate its physics qualitatively. As the frequency varies, ask what the controlling influence on the circuit is in various frequency ranges.

1   For **very low frequencies** we can forget about the inductor – its reactance approaches zero as we lower f to zero. The capacitor's reactance becomes larger, the smaller the frequency so at very low frequencies it dominates the circuit. For a capacitor the voltage is $\dfrac{\pi}{2}$ behind the current and at low frequencies the current is very small.

2   For **very high frequencies** the opposite argument applies: the inductor dominates, so V is $\dfrac{\pi}{2}$ ahead of the current and again I is very small.

3   When $f \sim f_0$, the effects of the capacitor and inductor cancel out because the voltages across these components are equal in magnitude but opposite in phase. This leaves a circuit which is dominated by the resistor. Hence $I \sim V/R$ and $\theta \sim 0$.

A.9 Using oscilloscopes

(a) Introducing oscilloscope controls and traces

An oscilloscope is a device for drawing a voltage–time graph of an electrical signal. Traditionally it is based upon a cathode-ray tube, like old-fashioned televisions, but computers can also detect and display electrical signals via a signal acquisition board which feeds in to the computer sound card via a USB port.

The pictures in Fig. A29 show (a) a simple CRO and (b) a typical computer-based oscilloscope screenshot.

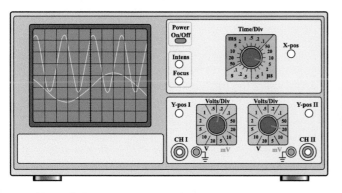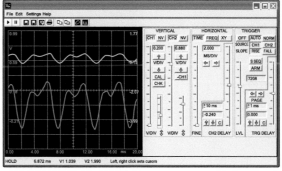

Fig. A29(a) A CRO and (b) A computer CRO program in action

An oscilloscope, whether hardware or software based has one or more input channels, which act in the same way as the terminals of a voltmeter. Both the devices in Fig. A29 have two input channels labelled CH1 and CH2; each input channel is produces one of the wave traces. The graph-grid on the screen has a vertical scale and a horizontal scale.

- The vertical scale is called Y-gain: for CH1, this is set at 0.2 V per division on both machines. The two input channels' Y-gains can be separately adjusted.

- The horizontal scale is controlled by the **time base**. The setting for this on the TeXtronic device is marked Time/Div and can be adjusted between 0.5 s/div to 0.5 μs / div. The time base setting for the two traces on an oscilloscope must be the same – this allows comparisons of the two signals to be made.

The vertical and horizontal positions of each trace can be adjusted [Y-Pos and X-Pos]. This helps to measure the amplitude and period against the fine scale divisions. The oscilloscope also has a **trigger** which acts to begin each sweep of the trace at the same point in the cycle of one of the inputs [usually CH1] to provide a steady display.

Example

Assuming that the lower-frequency signal on the CRO in Fig A29 is on CH1, give the frequency and peak voltage of this signal.

Answer

Period ~ 8.4 divisions (by eye); time base = 5 μs / div, from Fig. A29 (a)

$\therefore$ Period $= 8.4 \times 5 = 42$ μs

$\therefore$ frequency $= \dfrac{1}{42\mu s} = 0.024$ MHz $= 24$ kHz

Vertical height of wave [peak to peak] = 4.6 divisions (by eye).
Y-gain = 0.2 V/div

$\therefore$ Peak voltage $= \frac{1}{2} \times 4.6 \times 0.2$ V $= 0.24$ V $[V_{rms} = \dfrac{0.24}{\sqrt{2}} = 0.17$ V$]$

Fig. A30 shows how the Y-gain, X-Pos and Y-Pos controls can be used to facilitate measuring the voltage and period of an oscillation [see Study point].

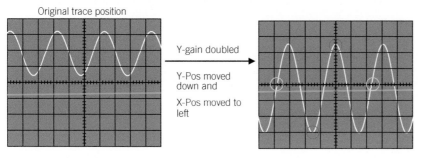

Original trace position

Y-gain doubled

Y-Pos moved down and

X-Pos moved to left

Fig. A30 Adjusting the trace to help taking readings

Self-test A35

State the vertical scale of CH2 on the CRO in Fig. A29.

Self-test A36

State the horizontal scale on the CRO.

Self-test A37

Find the frequency and peak voltage of the high frequency (CH2) signal on the CRO in Fig A29.

Self-test A38

The signals on the CRO program scope in Fig. A29(b) are not sinusoidal. Find the frequency of the signals.

▼ **Study point**

X-shift and Y-shift are alternative names for X-Pos and Y-Pos.

Self-test A39

Use the positions marked with the red circles to measure the peak voltage.

Use the positions marked with yellow circles to measure the period of the oscillations.

Y-gain = 0.5 mV / div

Time base = 1 mV / div

▼ **Study point**

Circuit symbol for an oscilloscope.

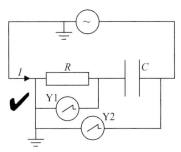

Fig. A31 Incorrect connections

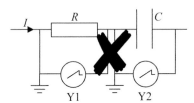

Fig. A32 Correct earth connections

▼ **Study point**

The pd across the resistor leads the pd across the capacitor, so the top trace is the pd across the resistor (Y1).

• Changing the Y-gain gives a greater number of divisions so the fractional uncertainty in the vertical distance is reduced.

• Moving the Y-Pos down positions the bottom of the traces on a grid line.

• Moving the Y-Pos to the left puts the top of the trace on the vertical scale. The period can be measured from the points where the trace crosses the central scale (again the distance should be as large as possible – so measure multiple cycles or adjust the time base).

(b) Using oscilloscopes to investigate AC circuits

A dual-beam oscilloscope is very useful for investigating AC circuits but it does have two restrictions:

1. It cannot measure current directly; however, it can measure the pd across a known resistor, which enables the current to be calculated. Remember that I and V for an ohmic resistor are in phase and related by $V = IR$.

2. Both input channels are earthed. This means that the earthed sides must be connected to the same point in the circuit. If the circuit in Fig. A31 were set up, the resistor R would be short-circuited by the two earths.

There is an additional problem that the output of signal generators is also often earthed on one side – all three earths must be connected to the same point in the circuit. Fig. A32 shows how a circuit should be set up to investigate a series RC combination.

The use of the oscilloscope in such a circuit is most easily illustrated by working through an example.

Example

We'll take the RC combination with $R = 10\ \Omega$ and show how it can be used to determine an accurate value for the capacitance of a nominally $33\ \text{nF}$ capacitor. The oscilloscope traces are shown in Fig. A33 together with the instrument settings. Determine the capacitance.

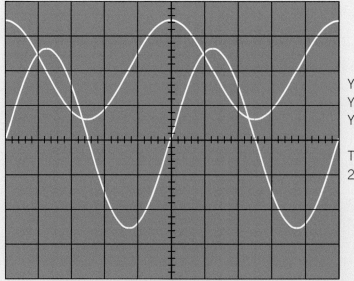

Y-gain:
Y1: 2 mV / div
Y2: 5 V / div

Time–base
2 ms / div

Fig. A33 RC circuit traces

1 Frequency: 2 cycles = 10 divisions = 20 ms. $\therefore$ Period = 10 ms
  $\therefore$ Frequency = 100 Hz.

2 V_R = peak resistor voltage = $\frac{1}{2} \times 2.9$ div $\times$ 2 mV/ div = 2.9 mV

3 I = peak current = $\frac{V_R}{R} = \frac{2.9 \text{ mV}}{10 \text{ }\Omega} = 0.3$ mA

4 V_{RC} = peak voltage across RC combination

   = $\frac{1}{2} \times 5.2$ div $\times$ 5 V/div = 13.0 V

5 Because $V_R << V_{RC}$ [3mV against 13 V]. We can write V_C = 13.0 V.

6 X_C = capacitor reactance = $\frac{V_C}{I} = \frac{13.0 \text{ V}}{0.29 \text{ mA}} = 44.8$ kΩ

7 So the capacitance, $C = \frac{1}{2\pi f X_C} = \frac{1}{200\pi \times 44.8 \times 10^3} = 36$ nF

▼ Study point

Notice that the resistor value in the example was chosen so that $R << X_L$. This enabled us to write $V_C = V_{RC}$ to a very good approximation.

The last example gives a method of calculating the reactance (and hence the capacitance or inductance) using the phase difference between the V_R and the V_{total} signals. It is useful when V_X and V_R are comparable in size. Alternatively, see the Study point opposite.

▼ Study point

In an RC circuit if V_{RC} and V_R are comparable in size, then we can use $V_{RC}^2 = V_R^2 + V_C^2$ to calculate V_C.

Example

CRO traces for V_R and V were examined for the combination shown in Fig. A34. X was known to be a capacitor, a resistor or an inductor.

The period of the AC voltage was found to be 10 ms and the V trace was 1 ms ahead of the V_R trace. Determine the nature and value of component X.

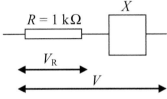

Fig. A34 RX combination

Answer

1 ms = $\frac{1}{10}$ = 0.1 cycle. $\therefore$ Phase difference, $\theta = 0.1 \times 2\pi = 0.628$ rad.

$\tan\theta = \frac{X}{R}$. $\therefore X = R\tan\theta = 1 \times 10^3 \tan 0.628 = 726$ Ω.

Because V is ahead of I [which is in phase with V_R] then X must be inductive.

$\therefore 2\pi f L = 726$. But $f = \frac{1}{T} = \frac{1}{0.01 \text{ s}} = 100$ Hz $\therefore L = \frac{726}{2\pi \times 100} = 1.2$ H

A42

Self-test

Let the value of V_R in the example be 1.5 V.

Calculate V_L and V.

Exercise **4A.3**

1 The oscilloscope screen shows two traces, A and B. The Y-gain of A is 2 V/div and that of B is 0.1 mV / div. The time base is 20 µs/div. The traces are both centred upon the horizontal midline.

Determine:

(a) the periods and the frequencies of the voltages V_A and V_B

(b) the time delay between traces A and B

(c) the phase difference between the two traces

(d) the rms voltages V_A and V_B

(e) the rms current and the mean power dissipated if V_A were applied across
(i) a 20 Ω resistor, (ii) a 0.1 H inductor, (iii) a 100 nF capacitor.

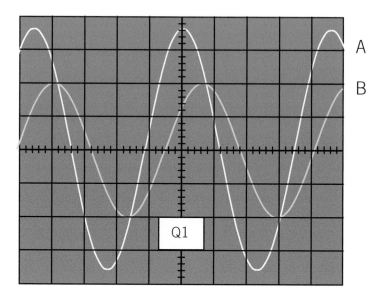

A

B

Q1

2 Suggest the Y-gain and time base settings for making the following voltage measurements. Justify your answers with reference to uncertainty.

(a) A pd of approximately 2 V rms with a frequency of approximately 50 Hz.

(b) A pd across a 500 Ω resistor with a 1 mA (rms), 200 Hz current.

3 State, with reasons, the changes would you make to the oscilloscope settings to make accurate voltage and frequency measurements of the trace shown.

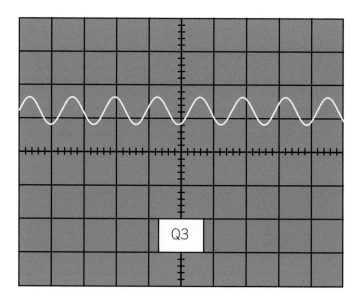

Q3

4 A 1.5 kΩ resistor is connected in series with an inductor of unknown value, L, across an AC power supply of unknown frequency, f. The CRO traces for V_R and V_{RL} are shown.

The Y-gain settings are the same; the time base is 0.5 ms / div.

(a) Determine L by measuring f and the phase difference between the voltage traces.

(b) Determine L by measuring f, V_R and V_{RL}.

Comment on the two methods.

Hint: Start by identifying the traces from the information given.

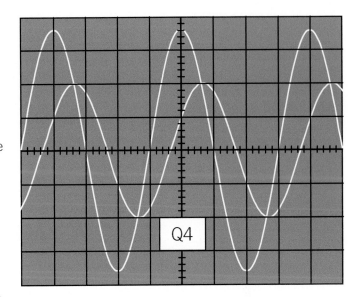

5 The Y-gain settings in Q4 were both 5 V / div. Determine the rms current and the power dissipated.

6 By calculation, determine how the traces in Q4 would change if the frequency of the supply were doubled. What changes (if any) would you make to the CRO settings to ensure the greatest possible accuracy? Assume that the voltage of the supply is independent of frequency.

7 In Q4, how would the traces be different if the inductor were replaced by a capacitor of the same reactance?

Option B: Medical physics

Medical diagnosis and treatment increasingly relies on physics and technology enabled by physics knowledge. Even though the technology may be operated by technicians, it needs to be set up and calibrated by medical physicists, e.g. the dose delivered by radioactive sources needs careful checking.

This topic covers four distinct areas of medical physics:

- X-rays
- Ultrasound
- Magnetic resonance imaging (MRI)
- Nuclear medicine.

Each of these is covered in its own section.

B.1 X-ray medicine

This is the largest of the topics and covers the production of X-rays as well as their use in diagnosis and treatment.

(a) X-ray nature and properties

With the possible exception of visible light, the division of the electromagnetic spectrum into different regions (infrared, visible, X-ray, etc.) is not, for the most part, an objective one. The black body emission spectrum (see AS textbook Section 1.6.1) of a **2000 K** object is mainly in the IR with a small amount of visible. As its temperature is raised to **6000 K** (roughly the temperature of the Sun's photosphere) the proportion of IR decreases, there is more visible and significant quantities of UV. There is also no scientific consensus on how to distinguish between X-rays and γ-rays. For the purposes of this book we shall adopt the definition that X-rays are produced by electrons and γ-rays by nuclear transitions, and that their ranges overlap.

The wavelength of the hard X-ray photons used in medicine is so small (see Self-test B1) that their particle nature is much more significant than their wave properties. The ionisation energy of atoms and molecules is in the eV range so medical X-rays can easily ionise biological molecules, with potentially serious consequences, e.g. causing cell damage through DNA mutations.

(b) X-ray production

X-rays in hospitals are produced by firing high energy (up to ~ 100 keV) electron beams at heavy metal targets, usually tungsten or a tungsten molybdenum alloy.

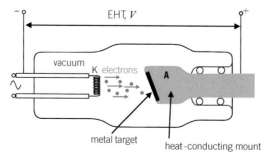

Fig. B1 Schematic diagram of an X-ray tube vacuum

The principle is that the electrons, emitted by the hot cathode, are accelerated by the potential difference between the anode (**A**) and cathode (**K**). They are rapidly decelerated when they hit the target, by colliding with its atomic electrons, and emit radiation in the process (see Study point). Because this is a random process, with some electrons losing their kinetic energy over many collisions and some in only a few, a continuous spectrum of radiation is produced.

A second process occurs when an energetic electron collides with an inner electron in the metal target atoms, knocking it out of the atom and leaving a vacant inner energy level. An electron from a higher energy level falls down into the vacant one, releasing the energy difference as a photon. This gives rise to a line spectrum that is characteristic of the target element(s).

Fig. B2 Gives the spectra for two different accelerating voltages, $V_1 > V_2$, with the same target element.

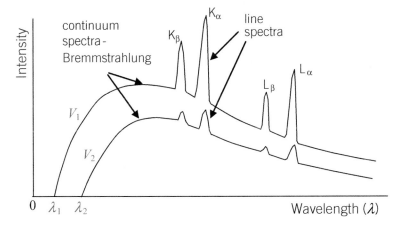

Fig. B2 X-ray spectra for two different voltages

The simplified atomic energy level diagram of tungsten in Fig. B4 can be used to illustrate the process of the production of the characteristic spectrum. If an incident electron ionises a tungsten atom by ejecting one of its K shell electrons, one of the L electrons can replace it by emitting a 59.3 eV photon (the difference in the levels).

Example

What other photons could result from the ejection of a K electron from a tungsten atom?

Answer

After the L electron has replaced the K electron, its place could be taken by an M electron with the emission of a $10.2 - 2.5 = 7.7$ keV photon.

Also an M electron could directly replace the K electron with the emission of a $69.5 - 2.5 = 67.0$ keV photon.

(c) Efficiency of X-ray production

Considered in purely energy terms, the X-ray tube is very inefficient. Typically only a very small fraction of the electron beam energy is converted to X-ray energy. This is because most of the electrons lose energy gradually, producing large numbers of IR photons. The efficiency increases with accelerating voltage: at 50 keV the efficiency is ~0.5%; at 100 keV it is ~1%.

▼ **Study point**

When charged particles are accelerated (or decelerated) they emit electromagnetic radiation in a process called *Bremmstrahlung* (German: braking radiation). Synchrotron radiation is a form of this; it is emitted when electrons follow curved paths in magnetic fields. The Diamond Light Source in Oxfordshire uses this to produce very bright X-ray beams for research.

▼ **Study point**

The X-ray intensity is sometimes plotted against photon energy rather than wavelength. In this case the graph is as in Fig. B3. Note the difference in the appearance of the continuum spectrum.

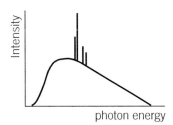

Fig. B3 Intensity v photon energy

▼ **Study point**

The lines in the X-ray line spectrum are labelled K, L, ... with subscripts α and β. The capital letters refer to the electron shell which the electrons drop down into and the α and β to energy levels within the shell.

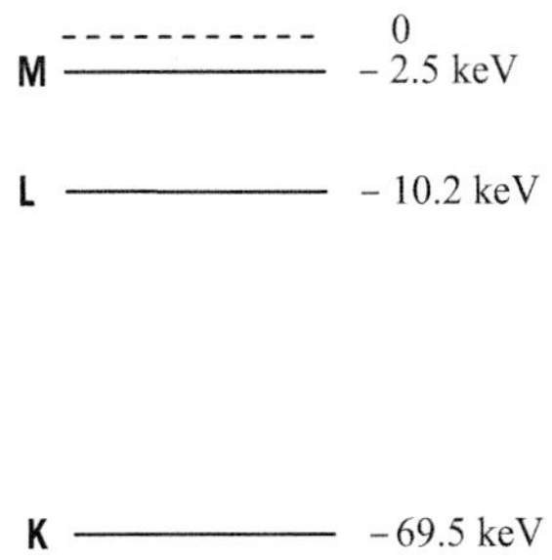

M ·········· 0
   ———————— – 2.5 keV

L ———————— – 10.2 keV

K ———————— – 69.5 keV

Fig. B4 Tungsten atomic energy levels

Example

Estimate the number of X-ray photons produced per second by a 100 keV tube operating with an anode current of 10 mA. State your assumptions.

Answer

Assuming an efficiency of 1% and a mean photon energy of 50 keV:

The number of electrons per second $= \dfrac{I}{e} = \dfrac{10\text{mA}}{1.6 \times 10^{-19}\text{C}} = 6.25 \times 10^{16}\ \text{s}^{-1}$.

$\therefore$ Incident power $= 100 \times 6.25\ 10^{16}\ \text{keV s}^{-1} = 6.25 \times 10^{18}\ \text{keV s}^{-1}$

$\therefore$ X-ray power $= 1\% \times 6.25 \times 10^{18} = 6.25 \times 10^{16}\ \text{keV s}^{-1}$

$\therefore$ X-ray photons $= \dfrac{\text{power}}{\text{mean photon energy}} = \dfrac{6.25 \times 10^{16}\text{keV s}^{-1}}{50\ \text{keV}} \sim 1 \times 10^{12}\ \text{s}^{-1}$.

This low efficiency means that a lot of heat needs to be conducted away from the anode – hence the conducting mount in Fig. B1.

B2 Self-test

A 100 keV X-ray tube is operating with an anode current of 10 mA. Estimate the rate at which heat needs to be conducted away from the anode.

B3 Self-test

(a) Calculate λ_{min} for an X-ray tube operating at 40 keV.

(b) What can you say about the pd needed to produce L photons from a tube with a tungsten target?

(d) Photon energy and beam intensity

Notice that the continuum spectrum is cut off at a minimum wavelength. This is because no radiation can be emitted with a photon energy greater than that of the incident electrons. For an EHT voltage, V, the kinetic energy of the electrons is eV, so the minimum wavelength, λ_{min}, is given by

$$eV = \frac{hc}{\lambda_{\text{min}}},$$

from which we can see that λ_{min} is inversely proportional to the accelerating voltage.

For a particular accelerating pd, the beam intensity is proportional to the number of electrons per second which hit the metal target. This is a function of the heater temperature: the greater the temperature, the more electrons are emitted per second and hence the greater the beam intensity. To obtain high definition X-ray images a combination of a high intensity beam and a small source spot on the anode is needed. The rotating anode tube has been developed to allow this (Fig. B5). The X-rays are produced from a small spot on the anode. The temperature of a stationary anode would be too great so the anode is made to rotate, allowing time for the heat to be conducted away.

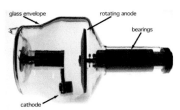

glass envelope   rotating anode

bearings

cathode

Fig. B5 Rotating anode X-ray tube

▼ **Study point**

If an X-ray fraction f_1 gets through object 1 and a fraction f_2 gets through object 2, the fraction which penetrates the two is $f_1 f_2$.

B4 Self-test

In an investigation into X-ray attenuation, 60% of a beam of monoenergetic photons penetrates a 2.0 cm thickness of a material. Calculate:

(a) the percentage getting through 4.0 cm of the material

(b) the attenuation coefficient of the material.

(e) X-ray attenuation

Section 3.5.4 dealt with the absorption of nuclear radiation. X-rays are absorbed in a similar way to that of γ-rays. If a mono-energetic beam of X-rays passes through a material, the probability of a photon being observed in any distance is constant, so the intensity, I, of the X-rays decreases according to an exponential decay law: $I = I_0 e^{-\mu x}$, where μ is called the **attenuation** (or absorption) **coefficient**. Different materials have different attenuation coefficients and these also depend upon the photon energy.

E_{phot} / keV	Attenuation coefficient / cm^{-1}					
	bone	muscle	adipose	blood	soft tissue	brain
10	18	5	3	6	7	5
30	1.6	0.40	0.38	0.50	0.30	0.40
60	0.45	0.20	0.19	0.21	0.21	0.18

Table B1 Attenuation coefficients of different materials

Example

The thickness of the human skull averages 6.3 mm. The width of the brain is 140 mm. Use the data in Table B1 to calculate the fraction of a beam of 60 keV X-ray photons which passes through the human head.

Answer

Total bone thickness = 12.6 mm = 1.26 cm

$\therefore$ Fraction penetrating the bone = $\dfrac{I}{I_0} = e^{-\mu x} = e^{-0.45 \times 1.26} = 0.57$

And: fraction penetrating brain = $e^{-0.18 \times 14} = 0.080$

$\therefore$ Total penetration fraction = $0.57 \times 0.080 = 0.046$.

The differences in attenuation coefficient are important in medical X-ray examinations.

(f) Low energy X-rays in radiography

Bones attenuate low energy X-rays much more than soft tissue because of their greater density; in particular their concentrations of mineralised calcium compounds. Fig. B6 shows a typical X-ray image of a broken collar bone, the sort of image which allows medical practitioners to decide on treatment – in this case, perhaps an intramedullary pin, to hold the two halves of the bone together to allow them to knit.

Soft tissues can also be revealed by X-rays as the chest X-ray (Fig. B7) shows. Although the muscles of the heart have a lower attenuation coefficient than the ribs, the heart has absorbed more X-rays because it is thicker. Even blood vessels are visible on the image.

In mammography, low energy X-rays are used to exploit the differential absorption of various tissues. The mammogram is examined for abnormal areas of density, which can indicate the presence of a cancer.

If real-time images are needed, e.g. of the blood vessels of a beating heart (cardiogram), **fluoroscopy** is used with a **contrast medium** and an *image intensifier*. The advantages of a digital image include rapid image acquisition (no chemical processes), greater sensitivity, image enhancement possibilities and the ability to transmit and store the image. Barium sulfate[1] is commonly used in alimentary canal imaging; iodine compounds in angiograms. There are many images available on the web showing fluoroscopic images of the alimentary canal and heart.

(g) X-ray computed tomography (CT scan)

In this imaging technique, the patient has many X-ray images taken from different angles when passing through the circular machine, as shown in Fig. B9. The X-ray tube and camera rotate around the patient to build up the images.

The cross-sectional images obtained are virtual slices of the body and the use of digital techniques allows structures with very little contrast to be distinguished. In the upper abdomen section in Fig. B10 you can probably recognise some ribs, a vertebra, the liver and the two kidneys – other structures in the image are the stomach and duodenum.

1 Sorry – the RSC has adopted US spelling.

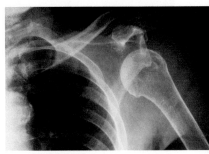

Fig. B6 Broken left clavicle

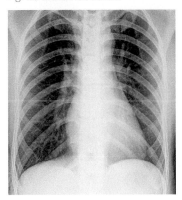

Fig. B7 Thoracic cavity X ray

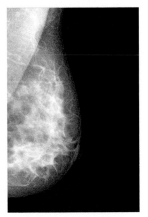

Fig. B8 Normal breast tissue

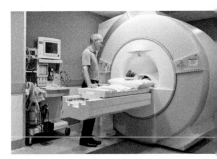

Fig. B9 CT scanner in action

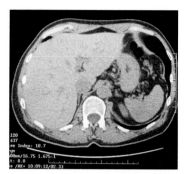

Fig. B10 – Upper abdomen CT section

B5

Self-test

Suggest why radiotherapy is not used for dispersed cancers such as leukaemia and for cancers which have metastasised.

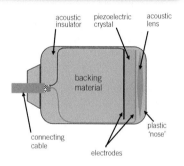

Fig. B11 Ultrasound transducer

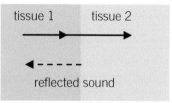

Fig. B12 Ultrasound reflection

── Terms & definitions ──

The **acoustic impedance**, Z, of a material is defined by
$$Z = c\rho$$
where ρ is its density and c the speed of sound in the material.

(h) X-ray radiotherapy

X-rays are also used to treat cancers. The DNA of dividing cells is more sensitive to ionising radiation than that of quiescent cells. Hence, even though X-rays are themselves carcinogenic, they can be used to kill cancer cells. In order to concentrate the dose on the tumour and to mitigate the damage to healthy tissue, several beams of X-rays are used. These are aimed at the tumour from several directions and the X-ray apparatus rotated around the tumour (or the patient rotated).

High energy X-rays are used for this treatment: for shallow cancers (e.g. in the skin), so-called *superficial* X-rays (50–200 keV) are used, with *megavoltage* X-rays (1–20 MeV) being deployed for deep cancers. These last are usually produced using particle accelerators rather than standard X-ray tubes. Radiation from cobalt-60 is often used (in which case it should be called γ-ray radiotherapy).

B.2 Ultrasound medicine

Sound with frequencies above 20 kHz, the upper limit for human hearing, is called ultrasound. Ultrasonic diagnostic and treatment procedures generally use ultrasound in the range 500 kHz – 5 MHz. As with X-rays, we'll start by considering how ultrasound is generated and detected.

(a) Generating and detecting ultrasound

Some crystals, e.g. lead zirconate titanate, generate electrical charge when stressed. This phenomenon is called *piezoelectricity*. Conversely, they deform when an electric field is applied. These piezoelectric crystals are used to generate and detect ultrasound. The principle of an ultrasound generator/detector is quite simple. Fig. B11 gives the structure of an ultrasound transducer.

A high frequency oscillating pd from the connecting cable is fed to the crystal, which generates the ultrasound sound waves (of the same frequency). These are emitted to the right: the backing material absorbs sound waves which propagate to the left and thus avoids reflected waves which could interfere with them. Incoming ultrasound waves from the right are absorbed by the crystal, distorting it and causing it to generate an oscillating pd which is fed out through the same cable to be analysed.

(b) The reflection of ultrasound

Ultrascans work in a similar way to echo location. The generator sends out a pulse of waves and receives the echo from an object – in this case the boundary between two different tissues. The time delay between transmission and echo reception gives us the distance to the boundary.

The important property of the tissues, which determines the fraction of the incident sound that is reflected, is the **acoustic impedance**, Z, of the two tissues. The bigger the difference in Z for the two materials, the greater the fraction reflected. The equation is:

$$\frac{I_r}{I_i} = \frac{(Z_2 - Z_1)^2}{(Z_2 + Z_1)^2}$$

where I_i and I_r are the incident and reflected intensities respectively.

Example

Use the values of Z in Table B2 to calculate the fraction of ultrasound reflected at the boundary between back muscle and the kidney.

Answer

$$\text{Fraction reflected} = \frac{(1700 - 1620)^2}{(1700 + 1620)^2} = 0.00058 \ (0.06\%)$$

Notice that it doesn't matter which way the sound wave is travelling because $(1700 - 1620)^2$ is the same as $(1620 - 1700)^2$.

Notice the low acoustic impedance of air. You should be able to show (Self-test B6) that over 99% of the sound will be reflected at an air/tissue boundary. If an ultrasound transducer is placed in contact with dry skin, there will always be a (very thin) layer of air between the two, i.e. there will be **two** air boundaries to cross and virtually no signal will be passed into the body. Hence examiners use **coupling agent** – a jelly with a similar acoustic impedance to that of the body to achieve maximum signal transmission.

(c) Frequency, pulses and repetition frequency

The reflections of continuous sound waves cannot be analysed, so the wave is broken up into a series of pulses.

Fig. B13 Pulsed ultrasound wave train

The resolution of the system is limited by the wavelength of the sound wave. For a resolution of 1 mm, the wavelength cannot be greater than this, so frequencies in the MHz range are employed (Self-test B7). Each pulse needs to travel through the body and back to the detector before the next is sent out. This takes approximately 0.1 ms (Self-test B8) so the gap between pulses must be at least this – so the repetition frequency (or pulse frequency) must be lower than 10 kHz.

(d) Ultrasound A scans

These are the simplest kind of scans. They are commonly used by optometrists to make measurements on the eye, especially the length of the eyeball. They can also be used to measure the position of a known object inside the body, e.g. a cancer, but they do not produce an image.

The output of the scan is connected to a CRO (see Option A) which displays the time of arrival of the reflected pulses. In Fig. B14 we see return pulses from the cornea, the front and back surfaces of the crystalline lens and from the retina at the back of the eye. The distances to these structures can be calculated from the time delays of the pulses.

Tissue	c	ρ	Z
Air	340	1.3	0.44
Lung	650	400	260
Fat	1470	920	1350
Water	1520	1000	1520
Brain	1560*	1030	1600*
Kidney	1558	1040	1620
Liver	1570	1060	1660
Muscle	1600*	1070	1700*
Bone	3400*	1600*	5400*

Table B2 Acoustic impedances

Units in the table:
c m s^{-1}; ρ kg m^{-3}; Z 10^3 kg m^{-2} s^{-1}
* These values vary a lot within and between tissues

Self-test **B6**

A medical dictionary gives the acoustic impedance of human skin as 1.6×10^6 kg m^{-2} s^{-1}. Calculate the fraction of an ultrasound wave which would be reflected between skin and air.

Self-test **B7**

Calculate the frequency of a sound wave with a wavelength of 1 mm in water.

Self-test **B8**

Calculate the time taken for a sound wave to travel 15 cm in water.

▼ **Study point**

Ultrasound waves also suffer attenuation. This limits the depth of penetration especially at high frequencies. Penetration of 10–15 MHz waves is limited to 2–3 cm.

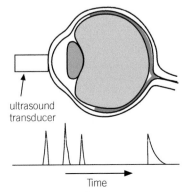

ultrasound transducer

Time

Fig. B14 A scan of the eye

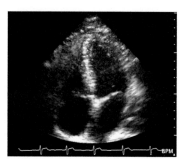

Fig. B16 Heart scan

(e) Ultrasound B scans

These scans give us the familiar images of foetuses in the uterus. They are produced by moving the scanner transducer over the area. The return signals from the different directions are combined electronically to produce the image on the display (Fig. B15).

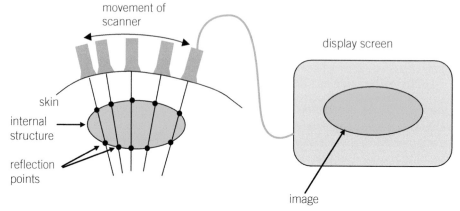

Fig. B15 Ultrasound B scan of an internal structure

Heart scans, such as in Fig. B16, need contrast agents because the blood/tissue boundary doesn't reflect well. A common agent is a liquid containing air microbubbles, which is injected into a vein.

(f) Doppler scans

Ultrasound is used in conjunction with the Doppler effect, which we met in Section 4.3.4, to study the speed of flow of blood cells. The basic principle is illustrated in Fig. B17.

For this application, the basic Doppler equation

$$\frac{\Delta f}{f} = \frac{v}{c}$$

needs to be amended in two ways.

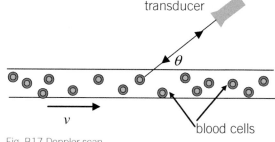

Fig. B17 Doppler scan

1. There are **two** Doppler shifts involved. The red blood cells 'see' a Doppler-shifted sound wave from the transducer, which they see as coming towards them. The reflected wave is further Doppler shifted because the source of the waves (the cells) is moving towards the transducer (in receive mode).

2. Only the radial component of the velocity of the cells is significant.

    So the relevant equation is: $\frac{\Delta f}{f} = \frac{2v \cos \theta}{c}$,

    where c is the speed of sound in the tissue.

    For small θ, the equation becomes $\frac{\Delta f}{f} = \frac{2v}{c}$

(g) Therapeutic uses of ultrasound

It has been mentioned that ultrasound is attenuated in tissues. This means that there is some energy transfer and that there are potential biological and therefore medical effects. These are insignificant at the intensities used for imaging but high power ultrasonic beams can be used in medical treatment. Here is a selection:

- For warming muscles and joints prior to manipulation by physiotherapists, or to promote healing by the warming effect.

- For descaling teeth by dental hygienists.

- Breaking up kidney stones and gall stones ('lithotripsy').

- Tumour ablation using high intensity focused ultrasound (HIFU).

- Assisting liposuction.

- Tooth and bone regeneration using low intensity pulsed ultrasound.

B.3 Magnetic resonance imaging (MRI)[2]

This technique, which produces the most exquisitely detailed images of the internal tissues of the body, employs advanced physics and some very expensive kit.

It involves lying still and being passed through an intense non-uniform magnetic field (Fig. B19) and having radio waves fired at you. However, as far as is known, the procedure carries very low risk (unless you have a heart pacemaker or forget to take your jewellery off!).

So what's going on?

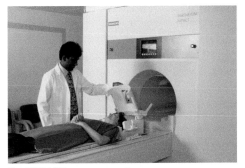

Fig. B19 MRI scan preparation

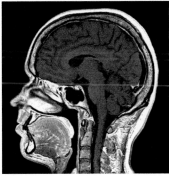

Fig. B18 MRI head scan

▼ **Study point**

Remember that protons and neutrons are composite particles, consisting of quarks that are charged. So a spinning nucleon consists of circulating charges and so it behaves like a small magnet.

(a) Magnetic nuclei

All nuclei spin. For nuclei such as $_1^1\text{H}$, with an odd number of protons or neutrons (see Study point), this results in the nucleus having a magnetic moment. In the absence of a magnetic field, the direction of the spin of hydrogen atoms in a material (such as water) is random. In a magnetic field, the nuclei line up approximately with the magnetic field lines. Actually, they can be pictured like a gyroscope, with the direction of the rotation precessing around the field lines (Fig. B21). There are two energy states available to the nuclei: parallel (low energy) and anti-parallel (high energy) to the magnetic field.

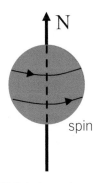

Fig. B20 Spinning nucleus

---

2 MRI scans used to be called NMR (nuclear magnetic resonance) scans. Rumour has it that the name was changed because 'an NMR' sounds like 'an enema,' which is quite a different medical procedure!

Self-test

Calculate ΔE for a 2.5 T magnetic field (typical MRI flux density). Express your answer in (a) J and (b) eV.

Self-test

Calculate the mean translational kinetic energy of a gas molecule at room temperature (300 K) and compare this with your answer to Self-test B12.

B14

Self-test

Use $E_{phot} = hf$ find the frequency of photons with an energy of $1.010\,168 \times 10^{-26}$ J.

---Terms & definitions---

The **Larmor frequency** is the photon frequency which resonates with the energy level difference between the magnetic moment states of a nucleus in a magnetic field.

The **characteristic time**, T, for the relaxation emission is the time for the intensity of the radiation to drop to $1/e$ of its original intensity.

Tissue	T/ ms
Fat	180
Liver	270
White matter	390
Spleen	480
Grey matter	520
Muscle	600
Blood	800
Cerebrospinal fluid	2000
Water	2500

Table B3 Relaxation time for body tissues in a 1 T field

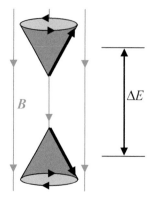

Fig. B21 Precessing nuclei

The energy difference, ΔE, between the two states is very small even by atomic standards: it is proportional to the magnetic field:

$$\frac{\Delta E}{B} = 1.010168 \times 10^{-26} \text{ J T}^{-1}$$

For a 1 T field, ΔE is orders of magnitude less than the Boltzmann energy, kT, at room temperature (see Self-test B13), so the numbers of nuclei in the two states are very nearly equal.

Very nearly – but not exactly. For a 1 T field there is a roughly 4×10^{-6} excess of the nuclei in the lower energy state, i.e. for every million nuclei, about 500 002 will be in the lower state and 499 998 in the upper state! So, in a magnetic field, water (because it contains hydrogen) will behave like a very weak magnet. This is true for lots of materials.[3]

(b) Nuclear magnetic resonance – Larmor frequency

Suppose we put a hydrogen-rich piece of material (e.g. living cells) into a magnetic field of flux density 1 T. Consider the effect of flooding this material with a beam of photons of energy 1.010168×10^{-26} J. This is the resonant frequency for the transition between the two energy levels. From our study of lasers (see AS book, Section 2.8.1) we expect to see absorption, spontaneous and stimulated emission of photons of this energy. These processes increase the population in the higher energy ('up') state. When we switch off the photon beam the hydrogen atoms will 'relax' into their previous state, with slightly more nuclei in the 'down' state than the 'up' state, i.e. by giving out a small burst of radiation. This emitted radiation shows the presence of hydrogen.

If you've done Self-test B14, you know that the frequency of photons needed for resonance between the two states is 42.6 MHz. This is called the **Larmor frequency** for hydrogen in a 1 T magnetic field. Generally for hydrogen in a magnetic field with flux density B (in tesla), the Larmor frequency is given by:

$$f_{Larmor} = 42.6 \times 10^6 \, B \text{ Hz}$$

Hence the Larmor frequency for hydrogen in a magnetic field with $B = 1.5$ T is $1.5 \text{ T} \times 42.6 \text{ MHz T}^{-1} = 63.9 \text{ MHz}$.

(c) Relaxation

The return of the nuclei to their pre-excited state is called *relaxation*. This occurs as an exponential burst of radiation with a **characteristic time**, T, of a few hundred milliseconds. The value of T depends upon the concentration of hydrogen nuclei in the material. This is because the magnetic fields of the hydrogen nuclei are weakly linked – the greater the concentration of hydrogen nuclei the longer the relaxation time. This is shown in Table B3: the greater the water concentration, the longer the relaxation time.

---

3 See the Institute of Physics 2011 schools lecture 6 for this demonstrated for a grape and a piece of Blu-tak.

(d) Application to imaging

There are two requirements for forming an image of the internal organs of the body:

1 The MRI scanner must 'know' where the signal is coming from. If all hydrogen nuclei radiate with the same frequency, we cannot tell where a particular photon has come from.

2 The MRI scanner must be able to distinguish different kinds of tissue.

The 'where' problem is solved by arranging for the magnetic field to be non-uniform. If the field increases in strength along the axis, the Larmor frequency will also increase. The subject is then flooded with a band of radio frequencies including all Larmor frequencies. Received photons must therefore come from the slice of the body with the appropriate field strength. In addition a small, non-uniform rotating field is applied across the subject. This enables the point of origin of the photons to be known across the slice.

The 'tissue' problem is solved by exploiting the different relaxation times of the tissues. If the signal is read and integrated over (say) the first 300 ms, the blood will not have given out much of its energy but the fat and liver will have given out most of theirs. Thus different tissues will have different image densities. In Fig. B22, the enlarged spleen is paler than the liver because it has a longer relaxation time.

▼ **Study point**

Although the detail is not great, the granulation of the liver showing cirrhosis is clearly visible in Fig. B22. The white mass to the lower right is an enlarged spleen due to portal hypertension.

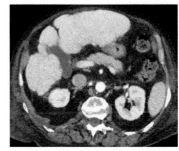

Fig. B22 Liver with cirrhosis

(e) Functional MRI (fMRI)

This is one of the most exciting recent developments in nuclear magnetic imaging. Using it, neurologists can identify the activity of different brain areas.

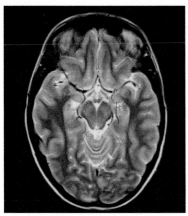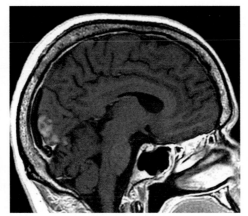

Fig. B23 FMRI showing the response to visual stimulation

The basis of the technique is that the active areas of the brain have an increased supply of oxygenated blood that has different MRI properties from de-oxygenated blood. Fig. B23 picks out the areas of the visual cortex that are active.

B.4 Comparison of Ultrasound, X-ray, CT and MR imaging

All four of these imaging techniques rely on sending waves (sound or electromagnetic) into a person and examining the reflected, transmitted or emitted radiation. Having looked at all four techniques, a brief comparison is called for.

Property	Ultrasound	Standard X-ray	CT scan	MRI
Radiation exposure	No ionising radiation	Exposure to ionising radiation	Significant exposure to ionising radiation (2–10 mSv) up to 5 years background	No ionising radiation
Application	Generally soft tissue including foetal; skeletal joints	Mainly bone breakages; with contrast agent can also be used for soft tissue	Bone injuries, lung and chest imaging, cancer detection, A&E investigations	Different kinds of soft tissue imaging, e.g. injuries, tumours
Biological effects	No known hazards in imaging	Carcinogenic effects and developmental defects in embryos	As X-ray	No known hazards; allergic reaction to contrast agents
Cost	Low cost	Low cost	About half the cost of MRI	High cost
Conditions	Short time; relatively painless (probe, e.g. rectal, may be required)	Very short time	Quite short time (5 min), ideally no movement but less of a problem than MRI	Long time; uncomfortable (no movement allowed); noisy; claustrophobia
3D imaging	Not possible	Not without moving the patient	Possible using a helical scan	Yes
Definition	Not high – depends on the skill of practitioner	High definition of bony structures	High definition of bony structures; moderate definition of soft structures (specially with contrast agent)	High definition (but requires stationary subject)
Contra-indications	None in imaging	Pregnancy	Weight limit of ~200 kg because of space and strength of moving table	Some metal implants; heart pacemaker; weight limit ~ 150 kg (space / strength of table)

B.5 Nuclear medicine

The topic of nuclear medicine includes the use of radioactive tracers for the investigation of organ function and body scanning as well as the treatment of medical conditions, mainly cancer. We shall start by discussing how nuclear radiation affects the human body.

(a) The effect of nuclear radiation on living material

As we discussed in Section 3.5, nuclear radiation (α, β and γ) is ionising. As it passes through living material, it ejects electrons from molecules, which can include structural and control molecules in cells as well as genetic material. This can damage or (in extreme cases) kill the cells. Cells that are undergoing mitosis (cell division) are particularly susceptible to damage. A large dose of radiation will therefore have consequences for the immune system, the reproductive apparatus and the epithelia (skin, lining of lungs, alimentary canal, vagina ...).

The **acute effects** (see Terms & definitions) are largely due to failure of the immune system to deal with the consequences of cell death due to the heating effect of a large radiation dose. The **short-term effects** are observed during high dose radiotherapy and include the frequently observed hair loss. The lung lining is very susceptible to radiation damage, which can result in inflammation and difficulty in gaseous exchange.

These effects are very strongly dependent on radiation exposure.

(b) Radiation exposure

There are several units of radiation exposure, each expressing a different aspect. Unfortunately there are also obsolete units which are also still in use. The fundamental quantity is the **absorbed dose**, D, which is the energy deposited per kilogram of tissue. The unit is the gray (**Gy**) which is equivalent to $J\ kg^{-1}$. (See Study point for the rad.)

We saw in Section 3.5 that α radiation is much more strongly ionising than β and γ, so its range is shorter. This results in greater damage to cells, i.e the damage in 1 kg of material due to the absorption of 1 mJ of α particles is much greater than that due to the same energy of β and γ. This is expressed in the *radiation weighting factor* – see Table B4 – and the **equivalent dose**, H.

$$\text{Equivalent dose, } H = DW_R,$$

The unit of equivalent dose is the sievert, Sv.

This is not quite the end of the story because different tissues (e.g. the lung lining, bone marrow) are more sensitive to radiation damage than others. In order to estimate the total **stochastic effect** on the body, the contribution from the radiation absorbed by the different tissues is expressed by the *tissue weighting factor*, W_T – see Table B5.

Tissue/organ	Tissue weighting factor, W_T
stomach, colon, lung, red bone marrow, breast, remaining tissues*	0.12
gonads	0.08
bladder, oesophagus, liver, thyroid	0.04
bone surface, brain, skin, salivary glands	0.01

Table B5 – Tissue weighting factors

* The 'remaining tissues' means all the tissues of the body not specified in the table, e.g. skeletal muscles, heart muscles.

▼ **Study point**

It is important to distinguish between the activity of a radioactive source, which is the number of emissions per second and dose of radiation which is a measure of the radiation received by a body.

Radiation	Weighting factor, W_R
X, γ	1
β	1
n	5–20
p	5
α	20
heavy nuclei	20

Table B4 Radiation weighting factors

▼ **Study point**

You may come across the rad as a unit of absorbed dose.

1 rad = 0.01 Gy

---Terms & definitions---

The **stochastic** effect of radiation is its random effect (i.e. the chance of its giving rise to cancer, etc.).

B15

Self-test

Show that the sum of the product of the tissues and W_T is 1.

▼ Study point

The rem is an obsolete unit of equivalent and effective dose. It is still in use in the US.

1 rem = 0.01 Sv.

B16 Self-test

In a treatment for a stomach cancer with γ rays, a patient's organs receive the following doses:

Stomach	3 mGy
Liver	2 mGy
Lung	2 mGy
Colon	0.5 mGy
Oesophagus	0.5 mGy

Calculate the total effective dose.

The procedure for calculating the **effective dose**, E, is to multiply the equivalent dose, H, received by each tissue by the relevant tissue weighting factor, W_T, and then add these together. The unit of effective dose is also the sievert, Sv. The result gives a way of comparing the likelihood of a cancer (or other stochastic condition) arising from the dose of radiation received.

Example

A patient is irradiated in the course of treatment. His lungs receive an equivalent dose of 4 mSv; the thyroid 3 mSv; the remaining tissues 2 mSv. Calculate the total effective dose and explain what it means.

Answer

$$E = 0.12 \times 4 \text{ mSv} + 0.04 \times 3 \text{ mSv} + 0.12 \times 2 \text{ mSv}$$
$$= 0.84 \text{ mSv}$$

This means that the probability of the patient getting cancer from the treatment is the same as if his whole body was uniformly irradiated with an equivalent dose of 0.84 mSv, i.e. an absorbed dose of 0.84 Gy (J kg^{-1}) assuming the radiation was β or γ.

(c) Gamma-emitting radio tracers

Radio tracers are chemical compounds in which one of the atoms has been replaced by a radioactive isotope. Such tracers have a wide industrial application, e.g. finding the location of cracks in buried gas pipelines. They are also used in medicine for locating medical problems and for imaging. One of the most common radiotracers used is the gamma emitting metastable isotope of technetium, Tc99m.

▼ Study point

Mo99 is prepared either by neutron bombardment of Mo98 or extracted from the fission products of a U235 nuclear reactor.

Technetium is not found in nature as it has no stable isotopes. Tc99m has a half-life of only 30 minutes, so it is not possible for hospitals to obtain it directly. Instead it is extracted in hospitals from a molybdenum generator (known colloquially as a moly cow) which contains the radioactive molybdenum isotope, Mo99, which has a half-life of 67 hours and decays by β$^-$ emission.

$$^{99}_{42}\text{Mo} \rightarrow {}^{99}_{43}\text{Tc}_\text{m} + {}^{0}_{-1}\beta$$

The m in nuclide Tc99m indicates that this nuclide is in an excited state. It decays to Tc99 with the emission of a 0.14 MeV photon (γ), which can be used for imaging.

B17 Self-test

(a) Identify the missing particle in the equation for the decay of Mo99.

(b) Write the equation for the decay of Tc99.

(c) Calculate the difference in mass of Tc99m and Tc. Express your answer in u.

B18 Self-test

I123 is prepared by bombarding Te124 with protons. The resulting excited nucleus immediately emits two neutrons. I123 decays by electron capture to Te123m, which decays almost immediately to Te123 with the emission of a 157 keV photon, which can be detected by a gamma camera. Write equations for these reactions.

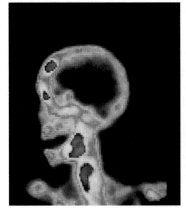

Fig. B24 Gamma camera image of a patient with a brain tumour.

The image in Fig. B24 is taken by a gamma camera (see below). The subject has been injected with a solution of sodium pertechnate (Na$^+$ TcO$_4^-$) which is absorbed more strongly by tumours because of their metabolic rates than normal tissue. Tc99m is also frequently used in myocardial perfusion scans which investigate heart function and blood flow.

Iodine-123 ($^{123}_{53}$I – half-life 13 hours) is also used as a medical tracer to investigate the functioning of the thyroid gland. Again it results in γ-emission (see Self-test B17) and, in addition to the short half-life, has the advantage over the previously used iodine-131 that there is no beta emission.

(d) Positron emission tomography (PET scan)

Self-test **B19**

Explain, using a relevant conservation law, why two photons are produced from the annihilation of an electron and a positron.

In PET scans, a biologically active molecule is tagged with a positron-emitting atom, such as fluorine-18. An example of such a molecule is fluorodeoxyglucose, which is a glucose molecule with a hydroxyl group being replaced by an **F18** atom. The two gamma photons produced when the positron and an electron mutually annihilate can be detected outside the body using a gamma camera.

Cancers contain actively dividing cells and so have a high requirement for energy and tend to concentrate glucose. Hence PET scans are useful for cancer imaging.

The patient in Fig. B25 has multiple myeloma. The dark areas in the images indicate concentrations of glucose.

The image on the left indicates widespread active cancers; the one on the right was taken after a single course of chemotherapy.

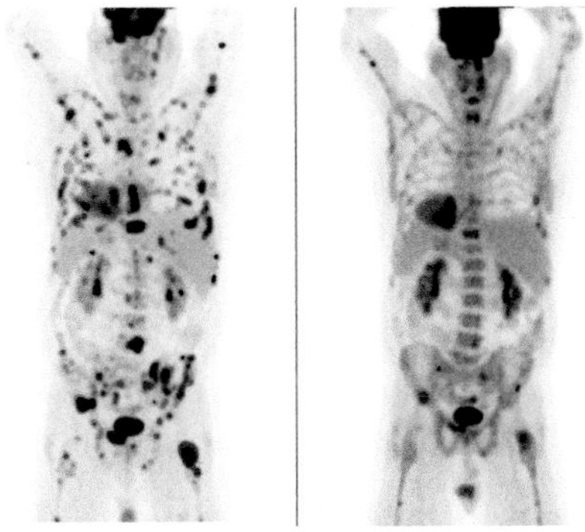

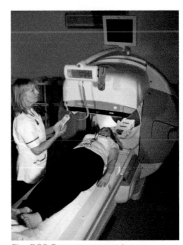

Fig. B25. PET scans during cancer treatment

(e) Gamma camera

Both the above two tracer procedures involve the detection of gamma rays and the production of images using a gamma camera. Fig. B26 shows a gamma camera in use and its principles are illustrated in Fig. B27.

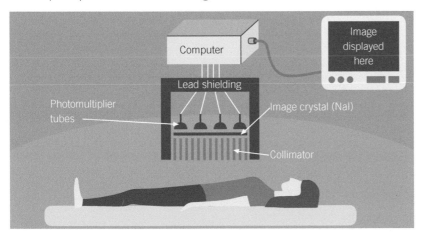

Fig. B27 Principle of operation of a gamma camera

There is no way of focusing γ-rays, so a lead collimator is used to ensure that only the γ rays travelling vertically upwards (in Fig. B27) hit the crystal.

The sodium iodide crystal flashes each time a photon hits it, the faint light is magnified by the photomultiplier tubes and the number of flashes in each pixel registered on a scintillation counter (not shown). The signal is fed to the computer which generates the image. In PET scans, the patient is surrounded by the detector (as in Fig. B26) and the computer only registers when it detects two photons on opposite sides at the same time. This allows it to reject stray background radiation photons.

Fig. B26 Gamma camera in use

▼ **Study point**

The compound eyes of insects work on the same principle as the collimator in a gamma camera. They have no lenses but only allow light rays to hit the retina which are travelling parallel to the axis of each tiny eye. The brain then constructs the image. The praying mantis even manages 3D images using these eyes.

Exercise 4B

1 The wavelengths of the K_α and K_β lines in the X-ray spectrum of copper are 154 pm and 139 pm respectively.

(a) Calculate the difference in the atomic energy levels which give rise to photons with these wavelengths. Express your answers in eV.

(b) Comment on the minimum tube pd which would be required to produce both K_α and K_β photons.

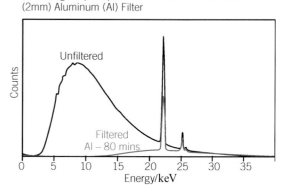

Mini-X-Ag Output Spectrum with and without 80 mil (2mm) Aluminum (Al) Filter

2 The spectrum is of a silver-target commercial X-ray tube without a filter (black curve) and with a thin aluminium filter (red curve). Note that the horizontal axis is of photon energy, not wavelength.

(a) Describe the effect of the filtering for different photon energies. Give a reason for this.

(b) Give an advantage in terms of the medical use of X-rays of filtering in this way.

(c) Estimate the operating pd of the tube.

(d) The spikes in the graph are the K_α and K_β lines. Calculate the wavelengths of these lines.

(e) Sketch filtered and unfiltered graphs of the output of this tube at this pd with X-ray wavelength on the horizontal axis.

3 A website gives the following percentage penetration of various X-ray photon energies through 10 cm of soft tissue:

Energy / keV	50	60	70	90	100	120	130	150
% penetration	3.6	5.3	6.7	9.5	10.2	11.5	12.0	13.0

Use the information in the table:

(a) to calculate the % of 100 keV photons which pass through (i) 20 cm, (ii) 5 cm and (iii) 15 cm of soft tissue

(b) to estimate the attenuation coefficient of 80 keV X rays in soft tissue.

4 A muscle with acoustic impedance 1.7×10^6 kg m^{-2} s^{-1} contains a cyst which contains fluid of density 1030 kg m^{-3} in which sound travels at 1530 m s^{-1}. Calculate the percentage of an ultrasound wave which (a) is reflected at the first boundary and (b) which penetrates the cyst. The attenuation within the tissues may be ignored.

5 In a Doppler scan, the frequency shift in 2 MHz waves is 346 Hz for blood which is flowing at 30° to the direction of the waves. Taking the speed of sound in blood to be 1550 m s^{-1}, calculate the speed of the blood flow.

6   The main field of an MRI scanner has a flux density of 1.60 T. A secondary field has a gradient of 40 mT m^{-1}.

(a) Calculate the Larmor frequency of hydrogen nuclei (protons) due to the main field.

(b) Calculate the difference in Larmor frequency between two positions which are 5 mm apart in the direction of the secondary (not-uniform) field.

(c) State the significance of the non-uniform field for locating a region in the body.

(d) The scanner can achieve a resolution of 1 mm. What frequencies of radio waves must it be able to distinguish?

7   In a course of treatment for a brain tumour, a beam of radiation is directed at the tumour from different directions. Explain briefly:

(a) How applying the radiation from different angles reduces the effect on the surrounding tissues.

(b) How charged particle therapy (proton beam) reduces the effect on surrounding tissues. [Hint: protons are absorbed in a similar way to electrons.]

8   The thyroid gland absorbs iodine. Hence this element can be used both as a tracer and for radiotherapy. There are several radioactive isotopes of iodine with different decay modes:

$^{123}_{53}$I decays by γ emission with a half-life of 13 hours

$^{131}_{53}$I decays by β^- emission to an excited state of xenon (Xe*) which in turn decays by γ emission with a half-life of 8.0 days.

(a) Compare these two isotopes for use as a tracer and for therapy.

(b) Write decay equations for the two-stage decay of $^{131}_{53}$I.

Option C: The physics of sports

This optional unit consists of basic physics concepts. The examiner can ask questions based upon these concepts in the context of any sport. For example, rotational dynamics can be assessed in the context of F1 cars or throwing the hammer. The selection of sports examples by the authors is entirely idiosyncratic and should not be taken as indicative of the actual contexts you will meet in a live examination.

Fig. C1 Stable or not?

C.1 The principle of moments

The moment of a force is a measure of its turning effect. The definition of the moment of a force about a point was covered in the Year 12 textbook – see Sections 1.1.6 and 1.1.7. Those sections are concerned with bodies in equilibrium, which is applicable to many sporting contexts. In many sports, rotating objects are far from being in equilibrium.

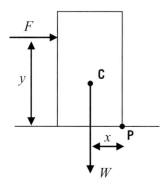
Fig. C2 Will F topple the box?

(a) Stability and toppling

In spite of her best efforts, Wendy's Laser dinghy (Fig. C1), is about to topple. This is a frequent occurrence in dinghy racing but very rare in F1 racing (or indeed in ocean yacht racing). We saw in Section 1.1.6(c) how to predict stability in situations where the only significant forces on an object are its own weight and the contact force with the ground. In sports this is rarely the case – toppling generally occurs when a force is applied from the side.

Will the box in Fig. C2 topple or not? We can use moments to find out.

If it topples, it will rotate clockwise about **P**. The resultant clockwise moment, M, is given by:

$$M = Fy - Wx$$

If $M > 0$, i.e. $F > \dfrac{Wx}{y}$

the box will topple. From this we can infer that the wider the box, the less likely it is to topple. So rugby players who are tackling or pushing will brace with one foot well in front of the other – to increase stability.

What about Wendy in Fig. C1 and the rather more sedate mariner in Fig. C3. What forces are involved?

The line drawing helps to clarify. The force of the wind acting on the sail exerts a clockwise moment about the centre of gravity, C. The buoyancy force and the weight of the dinghy pilot provide an anticlockwise moment.

But what if the wind blows a bit more strongly? The boat will heel over more, which will (a) reduce the wind area of the sail and (b) lower the height of the wind force above the centre of gravity, C – so stabilising the boat. Also the sailor might lean out a bit more: what effect will that have?

Fig. C3 A stable dinghy

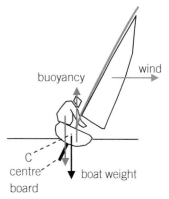
Fig. C4 Forces on the dinghy

Notice two features of the design of the dinghy:

1 The shape of the hull. As the boat heels, the buoyancy force (more or less in the centre of the submerged part) shifts to the right.

2 The centre board, which is a heavy steel plate. Some dinghies also have ballast in the bottom of the hull. These have the effect of lowering the centre of gravity.

The dinghy in Fig. C3 has an additional stability feature. The tubes on either side of the *gunwale* are floatation devices – called *dogs*. If the dinghy heels so that one is partly submerged (as shown), the buoyancy force on it provides an additional restoring moment.

Generally, the lower the centre of gravity of an object and the wider the base, the more stable it is. The box in Fig. C2 can rotate through $\tan^{-1}\frac{x}{h}$, where h is the height of the centre of gravity before it topples.

Rugby players who are going in for the tackle exploit this by keeping a low centre of gravity and bracing their feet apart in the front-rear direction.

The impact force, F, from the contact with the upright player is not far above the centre of gravity so has only a small moment. The player being tackled is much less stable – high centre of gravity – and can more easily be toppled.

Fig. C5 Ideal tackling position

(b) Moments in the human skeleton-muscular system

Skeletal muscles (as opposed to heart and gut muscles) apply tensile forces to bones via structures called tendons. Muscles never push – they only pull. They work in antagonistic (opposing) pairs, e.g. the biceps and triceps muscles of the upper arm.

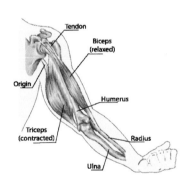
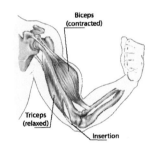

Fig. C6 Muscles of the upper arm working in opposition

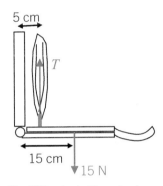
Fig. C7 Tension in biceps tendon

5 cm

T

15 cm

15 N

The arm is flexed by tightening the biceps, during which process the triceps is relaxed. Straightening the arm is achieved by the opposite process.

Example

Use Fig. C7 to estimate the tension in the distal biceps tendon when the forearm is held in the horizontal position.

Answer

Taking moments about the elbow joint: $T \times 5 \text{ cm} = 15 \text{ N} \times 15 \text{ cm}$

$$\therefore T = \frac{225 \text{ N cm}}{5 \text{ cm}} = 45 \text{ N}$$

Another antagonistic pair of muscles is the quadriceps / hamstring duo in the thigh. During running, these muscles contract alternately and overtraining can lead to severe tendonitis (inflammation of the tendon) especially of the quadriceps.

quadriceps

hamstring

Fig. C8 Antagonistic muscles in the thigh

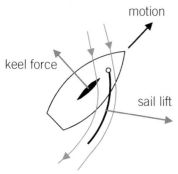

Fig. C9 Heading into the wind

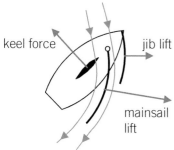

Fig. C10 Close hauled with jib

▼ **Study point**

Note that *e* is always positive. Taking the modulus of relative velocities means we do not have to be fussy about the order of the suffixes.

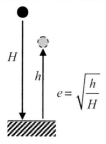

Fig. C12 Bouncing ball

C6 **Self-test**

A basketball umpire throws the ball up to 3 m. If both players miss the ball, how high will it bounce?

[*e* = 0.85]

Stretch & Challenge

A sphere of mass m_1 moving at speed *u* collides head on with a stationary sphere of mass m_2. Show that:

(a) the kinetic energy after the collision is given by

$$\tfrac{1}{2}m_1u^2\left[\frac{m_1 + e^2m_2}{m_1 + m_2}\right],$$ and

(b) that this tends to $\tfrac{1}{2}m_1u^2e^2$ as $m_2 \to \infty$.

(c) More moments when sailing

One of the most surprising things about sailing is that you can sail against the wind. The simple diagram of a sailing boat, Fig. C9, shows two of the horizontal forces on a boat with the wind coming from forward of the *beam* (sideways). The wind flowing over the sail produces a lift as shown. This is the same as the lift by an aeroplane wing (see Section C7). The keel of the boat produces a balancing sideways force which prevents the boat being dragged to the side (though all boats make some *leeward* motion) and the resultant is a forwards force.

If we think about the moments of the forces about the centre of gravity of the boat, that cannot be the whole story. The keel force passes roughly through the centre but, as drawn, the sail lift passes aft of the centre and therefore has an anticlockwise moment – so the head would turn into the wind. One way of counteracting this is to use a second sail before the mast, a jib: Fig. C10. This produces a clockwise moment, balancing the mainsail. It is also possible to steer away from the wind using the rudder, but this produces undesirable drag.

C.2 Collisions

(a) Coefficient of restitution

When two elastic objects collide, their separation velocity is less than their collision velocity. To a good approximation the ratio of their separation speed to the collision speed is a constant called the **coefficient of restitution**, *e*. This is clarified in Fig. C11.

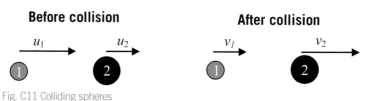

Fig. C11 Colliding spheres

The coefficient of restitution is defined by:

$$e = \frac{\text{relative velocity after collision}}{\text{relative velocity before collision}} = \left|\frac{v_2 - v_1}{u_2 - u_1}\right|$$

We can relate this to the relative bounce height of a ball: see Fig. C12.

From an initial height, *H*, the speed of impact = $\sqrt{2gH}$.

If it bounces to height *h* the speed just after impact = $\sqrt{2gh}$

Because the floor isn't moving we can write: $e = \dfrac{\sqrt{2gh}}{\sqrt{2gH}} = \sqrt{\dfrac{h}{H}}$

Note: From this it is tempting to conclude that e^2 is the energy efficiency of the bounce, i.e. the fraction of the original kinetic energy which remains after the collision. This is true for an object bouncing off a stationary very massive object (e.g. the Earth) but it is not true in general (see Stretch & Challenge).

Example

International Tennis Federation rules state that a tennis ball dropped from a height of 254 cm should bounce to between 135 cm and 147 cm when dropped on to a concrete floor.

Calculate the range of the coefficient of restitution specified.

Answer

$$e_{min} = \sqrt{\frac{135}{254}} = 0.73 \ (73\%) \qquad e_{max} = \sqrt{\frac{147}{254}} = 0.76 \ (76\%) \ [2 \text{ s.f.}]$$

The rules for basketballs are slightly more complicated (see Ex. 4C)

C7
Self-test
Calculate the impact and rebound speed for a tennis ball in the drop test if the coefficient of restitution is 0.75.

(b) Forces in collisions

We can estimate the mean force exerted in a collision using N2, which is conveniently written

$$Ft = mv - mu$$

The quantity $mv - mu$ or $m\Delta v$ is the change of momentum of an object which is also called the **impulse**. We can also call the product Ft the impulse. As the equation states, these two definitions give the same answer. If we make a reasonable assumption of the degree to which the tennis ball deforms on impact, we can estimate the mean force on the ball during the collision. This is our next example.

Study point

In calculating the times in the Example we have used

$$t = \frac{\text{distance}}{\text{mean speed}}$$

Essentially that means we have taken the accelerations to be constant, which is clearly not the case.

Example

Estimate the mean force exerted by a tennis ball of mass 58 g on the concrete floor during a drop test. Take e to be 0.75.

Answer

Using the answer to Self-test C7:

$$Ft = 0.058 \text{ kg} \left(5.295 - (-7.059)\right) \text{m s}^{-1} = 0.717 \text{ N s}$$

Note: u and v are vectors. We have taken upwards as positive, so the velocity changes from -7.059 m s^{-1} to 5.295 m s^{-1}.

Now we'll assume that the tennis ball deforms by 1 cm in this collision.

$$\therefore \text{ Mean time to decelerate to rest} = \frac{0.01 \text{ m}}{\frac{1}{2} 7.059 \text{ m s}^{-1}} = 2.83 \text{ ms.}$$

$$\therefore \text{ Mean time to accelerate from rest} = \frac{0.01 \text{ m}}{\frac{1}{2} 5.295 \text{ m s}^{-1}} = 3.78 \text{ ms}$$

(See Study point). So the total time, $t = 2.83 + 3.78$ ms $= 6.61$ ms

$$\therefore \text{ Mean force } F = \frac{mv - mu}{t} = \frac{0.717 \text{ N s}}{6.61 \text{ ms}} = 110 \text{ N}$$

This is the mean force exerted on the ball, so the mean force exerted **by** the ball (on the concrete floor) by N3 is -110 N, i.e. 110 N downwards.

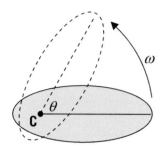

Fig. C13 Rotating object

C.3 Concepts and equations in rotational motion

Rotation is a feature of many sports. Before we analyse it we need to develop some concepts. To do this we recall that, in analysing linear motion we use the following quantities:

distance / displacement   speed / velocity   acceleration

force   momentum   kinetic energy   mass   time

With the exception of time, we shall look for rotational equivalents of all these quantities. We have met some of them in earlier parts of the A Level Physics course. For the moment we'll consider only rigid bodies.

Consider the object in Fig. C13 rotating about the point C. We can describe its orientation and angular motion by the angular position, θ, and angular velocity, ω. If ω is constant, it is defined by

$$\omega = \frac{\Delta\theta}{t} \quad \text{i.e.} \quad \omega = \frac{\theta_2 - \theta_1}{t}.$$

Similarly, we can define the angular acceleration, α, by analogy with linear acceleration by:

$$\alpha = \frac{\Delta\omega}{t} \quad \text{i.e.} \quad \alpha = \frac{\omega_2 - \omega_1}{t}$$

We can use linear equations to write equations of motion for rotation:

Linear equation	Condition	rotational equation
$x = vt$	Constant velocity	$\theta = \omega t$
$v = u + at$	Constant acceleration	$\omega_2 = \omega_1 + \alpha t$
$x = \frac{u+v}{2}t$		$\theta = \frac{\omega_1 + \omega_2}{2}t$
$x = ut + \frac{1}{2}at^2$		$\theta = \omega_1 t + \frac{1}{2}\alpha t^2$
$v^2 = u^2 + 2ax$		$\omega^2 = \omega_1^2 + 2\alpha\theta$

Table C1 Linear and rotational equations

Looking again at Fig. C13, we see that, even though the object has no linear motion it must possess kinetic energy because all its particles are moving.

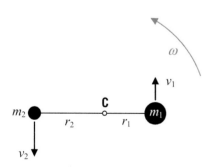

Fig. C14 Rotating spheres

C.4 Rotational kinetic energy

(a) Moment of inertia

What is the kinetic energy of the rotating system of connected spheres in Fig. C14? Taking the mass of the bar to be negligible:

$$E_k = \tfrac{1}{2}m_1 v_1^2 + \tfrac{1}{2}m_2 v_2^2$$

Expressing this in terms of the angular speed, ω, by using the relationship, $v = r\omega$,

$$E_k = \tfrac{1}{2}m_1(r_1\omega)^2 + \tfrac{1}{2}m_2(r_2\omega)^2$$

This can be rewritten as: $E_k = \tfrac{1}{2}(m_1 r_1^2 + m_2 r_2^2)\omega^2$

This is of the same form as the equation $E_k = \frac{1}{2}mv^2$ for linear motion, with ω replacing v and the quantity $m_1r_1^2 + m_2r_2^2$ replacing m. This is called the **moment of inertia** and has the symbol I. If there are more than two particles then we just add more terms to the sum (see Terms & definitions), and the rotational kinetic energy is given by:

$$E_{k\,rot} = \frac{1}{2}I\omega^2$$

Example

A speed regulator on a steam engine consists of two small metal spheres, each of mass 0.15 kg. Calculate the energy of rotation of the pair about the midpoint when they are rotating at 150 rpm and separated by a distance of 20 cm.

Answer

Distance of each sphere from midpoint = 10 cm = 0.10 m.

$\therefore$ Moment of inertia = $m_1r_1^2 + m_2r_2^2 = 0.15 \times 0.10^2 + 0.15 \times 0.10^2$
$= 0.003$ kg m²

Angular speed, $\omega = \dfrac{150 \times 2\pi}{60} = 15.7$ rad s⁻¹

$\therefore$ Kinetic energy = $\frac{1}{2}I\omega^2 = \frac{1}{2} \times 0.003 \times 15.7^2 = 0.37$ J

(b) Formulae for moment of inertia

Working out the moment of inertia of an extended solid object can be tricky, unless it has an easy symmetrical shape. Even then you probably need calculus. An exception is a thin uniform cylindrical ring (Fig. C15).

This is easy because all the particles are the same distance, r, from the axis of rotation, **C**. Hence $I = mr^2$. But even calculating I for a uniform disc needs integration. The formulae for this and a few other shapes are derived in the mathematics chapter – and some are left as exercises!

You won't need to remember formulae for moments of inertia – they will be given in the examination – but Table C2 has a few that you will come across. The formulae given are all for rotations about the centre of mass (centre of gravity). What if, say, a rod is rotating about its end (like a bowler's arm when delivering the ball)? Then the formula becomes $I = \frac{1}{3}ml^2$. Why is it bigger? This is left as a problem for you but, as a hint, consider the following diagram:

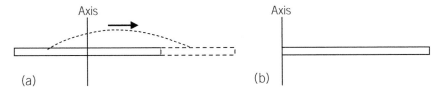

Fig. C16 Why is the moment of inertia in (b) greater than in (a)?

(c) Rolling objects

What if an object is spinning and moving along? What is its total kinetic energy?

Conveniently, the answer is the sum of the translational and rotational KEs, i.e.

$$E_k = \frac{1}{2}mv^2 + \frac{1}{2}I\omega^2,$$

where v is the speed of the centre of gravity and ω the angular speed (see Stretch & Challenge).

▼ Study point

The moment of inertia combines the mass of the object with the distance of its particles from the centre of rotation. The further out they are, the more they contribute to the moment of inertia.

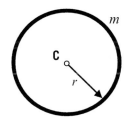

Fig. C15 Cylindrical ring

Object	I
Uniform ring	mr^2
Uniform disc	$\frac{1}{2}mr^2$
Solid sphere	$\frac{2}{5}mr^2$
Spherical shell	$\frac{2}{3}mr^2$
Uniform rod	$\frac{1}{12}mr^2$

Table C2 Some moments of inertia

▼ Study point

The moment of inertia of a cylindrical rod about its axis is the same as that of a disc.

Stretch & Challenge

The barbell below is rotating about its centre and moving along. Show that its kinetic energy is given by:

$$E_k = \frac{1}{2}mv^2 + \frac{1}{2}I\omega^2$$

Hint: Calculate the speed of each sphere at the instant of the diagram.

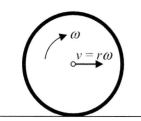

Fig. C17 Rolling wheel

For a rolling circular object, e.g. a bicycle wheel, the speed and angular speed are related by $v = r\omega$. This must be the case because the velocity of the rim is zero at the point of contact with the ground (Fig. C17).

Example

Estimate the kinetic energy of the road wheel of a car which is rolling at 30 m s⁻¹, using the following estimated quantities:

Mass of wheel, 15 kg; radius, 30 cm; moment of inertia, 1 kg m²

Answer

$$E_k = \tfrac{1}{2}mv^2 + \tfrac{1}{2}I\omega^2 = \tfrac{1}{2}16 \times 30^2 + \tfrac{1}{2}1 \times \left(\frac{30}{0.3}\right)^2 = 7200 + 5000 = 12\,200 \text{ J}$$

C8 Self-test

Show that the total kinetic energy of a rolling ring is mv^2.

C9 Self-test

Calculate the kinetic energy of a uniform cylinder of mass 5 kg which is rolling at a speed of 4 m s⁻¹.

Example

A *wood* (i.e. a lawn bowl) is launched by allowing it to roll down a slope of height 1.5 m. Approximating the wood as a uniform sphere, estimate its speed at launch.

Answer

Potential energy loss $= mg\Delta h = 1.5 \times 9.81m = 14.7m$

With a speed v, $E_k = \tfrac{1}{2}mv^2 + \tfrac{1}{2}I\omega^2$

But when the ball is rolling, $\omega = \frac{v}{r}$, and $I = \tfrac{2}{5}mr^2$

$\therefore E_k = \tfrac{1}{2}mv^2 + \tfrac{1}{2}\tfrac{2}{5}mr^2\left(\frac{v}{r}\right)^2 = \tfrac{7}{10}mv^2$

$\therefore \tfrac{7}{10}mv^2 = 14.7m,\ \therefore v = \sqrt{\dfrac{10 \times 14.7}{7}} = 4.58 \text{ m s}^{-1}$

Stretch & Challenge

An old-fashioned term for angular momentum is *moment of momentum*. Consider an object moving past a point, O, as shown in Fig. C18. Show that the moment of its momentum (i.e. px) is the same as $I\omega$. (You'll need a bit of trig.) Then go on to show that, if m is moving with a constant velocity, $I\omega$ is constant.

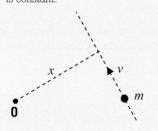

Fig. C18 Moment of momentum

C.5 Angular momentum

(a) Definition and torque

Linear momentum, p, is defined by the equation $\qquad p = mv$.

Similarly we define the angular momentum, J, by $\qquad J = I\omega$.

Note that, just as with kinetic energy, the angular equation has the same form as the linear equation with m replaced by I and v by ω. In order to change the angular momentum of a body we need a *couple*, which is two equal and opposite forces with different lines of action.

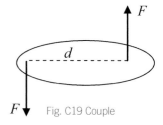

Fig. C19 Couple

The turning effect of a couple is its *moment* or *torque*, τ. A little thought should convince you that the moment of the couple about *any* point in Fig. C19 is given by:

$$\tau = Fd.$$

But can't we produce a rotation without having two forces, just by having a single force applied off centre, e.g. a snooker player applying side-spin? To do this the ball is cued slightly off centre – Fig. C20(a). The off-centre force, F, can be considered to be a force through the centre of mass (which produces linear acceleration) and a couple, which produces angular acceleration.

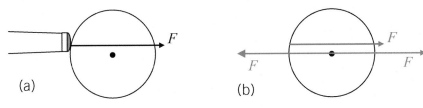

(a)

(b)

Fig. C20 A single force is equivalent to a force plus a couple.

The angular acceleration, α, is calculated from $\tau = I\alpha$

which is clearly the rotational equivalent of $\qquad F = ma$.

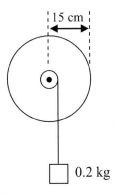

Fig. C21 Wheel and axle

Example

A smoothly mounted flywheel (a cylindrical disc) of mass 10 kg and radius 15 cm is accelerated by means of a 200 g mass attached to a cord which is wrapped around the 2 cm radius axle. Estimate the angular acceleration of the flywheel [see Study point].

Answer

Moment of inertia of the flywheel, $I = \frac{1}{2}mr^2 = 5 \times 0.15^2 = 0.1125$ kg m².

Torque produced by applied weight, $\tau = Fd = mgd = 0.2 \times 9.81 \times 0.02$
$$= 0.03924 \text{ N m}$$

$\therefore$ Angular acceleration $\alpha = \frac{\tau}{I} = \frac{0.03924 \text{ N m}}{0.1125 \text{ kg m}^2} = 0.35$ rad s⁻².

> ### ▼ Study point
>
> What are the approximations here (apart from the zero-friction mounting)?
>
> - We are ignoring the moment of inertia of the axle (a minor point).
>
> - The mass is accelerating, so there is a resultant force on it, so the tension in the cord is less than mg (but not by very much – see Self-test).

(b) Conservation of angular momentum

This is another similarity between linear and rotational motion. In the absence of externally applied torques, the angular momentum of a body system of particles is constant. This principle has universal application from orbiting planets[1] to pirouetting ballerinas and Olympic divers.

The ice-skater in Fig. C22 starts herself spinning (a) with her arms outstretched. She pulls in her arms (b), reducing her moment of inertia.

Momentum is conserved, so $I_1\omega_1 = I_2\omega_2$.

But $I_2 < I_1$, so $\omega_2 > \omega_1$, i.e. the skater speeds up. The same principle holds for a diver going into a tuck. A ballerina can use this trick to speed up to 400 rpm during a leap! Calculating the moment of inertia of the human body is quite tricky because its shape and density are both non-uniform, so we'll need to make some approximations:

Example

Estimate the moment of inertia of an upright 50 kg ice-skater (a) with her arms down (b) with her arms outstretched.

> ### Self-test C10
>
> Use the calculated value of α to find the acceleration of the mass and hence the tension in the cord. Comment on your answer.

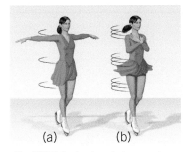

(a)   (b)

Fig. C22 Spinning ice-skater

> ### Self-test C11
>
> Give a unit for angular momentum.

> ### Self-test C12
>
> The ice skater in the example starts with her arms outstretched and is rotating at 1 revolution per second. Use the estimates in the example to calculate (a) her angular momentum and (b) her rotational speed when she pulls her arms in.

---

1 Kepler's second law of planetary motion is a consequence of this.

Answer

(a) We'll take the body to be a cuboid of width, w, 40 cm and thickness, t, 20 cm. The moment of inertia about the vertical axis is given by $I = \frac{1}{12}m(w^2 + t^2)$.

$$\therefore I = \tfrac{1}{12}50(0.4^2 + 0.2^2) = 0.83 \text{ kg m}^2$$

(b) We'll take the arms to be a rod of total length 1.5 m and mass 8 kg. The remaining body has mass 42 kg.

$$\therefore I = \tfrac{1}{12}42(0.4^2 + 0.2^2) + \tfrac{1}{12}8 \times 1.5^2 = 2.2 \text{ kg m}^2$$

C.6 Sports involving projectiles

Many sports involve projectile motion: javelin, rugby, golf, football (all codes). The year 12 course included the analysis of the motion of objects which move under gravity, with the assumption that there were no other significant forces. This is not a very good assumption in the case of most sports. Air resistance and aerodynamic lift play a part in the flight of a javelin; a tennis player can cause a ball to swerve by applying spin, in a similar way to a footballer taking a corner; arrows and darts have flights whose function is to control the motion of the projectile through the air.

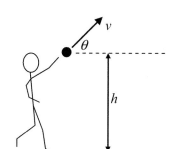

Fig. C23 Jessica Ennis preparing to put the shot

(a) Shot putting

Perhaps the only projectile sport in which the air plays an insignificant part (other than allowing the athlete to stay alive) is shot putting. The mass of the shot is 4 kg (women) and 7.26 kg (men) and the speed of projection is around 12 m s^{-1}. With these figures, air resistance is negligible.

If an object is projected from ground level with a speed v at an angle θ to the horizontal, its range, R is easily calculated, as follows:

Fig. C24 Projection from head height

Time in the air $= \dfrac{2v \sin \theta}{g}$, $\therefore R = v \cos \theta \times \dfrac{2v \sin \theta}{g} = \dfrac{v^2 \sin 2\theta}{g}$.

Here we have used the trig identity: $\sin 2\theta = 2 \sin \theta \cos \theta$. The maximum value of $\sin 2\theta$ is 1 (when $\theta = 45°$), so, with v at 12 m s^{-1}, the maximum range should be $12^2/9.81 = 14.7$ m.

But the world record (at the time of writing) is 23.12 m and most athletes seem to launch at about 35°, so what's going on? One clue is in Fig. C24. The launch is obviously not from ground level but from height, h, which for most athletes is about 2 m. The range formula now is a little more complicated but a few lines of algebra give:

$$R = \frac{v^2 \sin 2\theta}{2g} \left[1 + \sqrt{1 + \frac{2gh}{v^2 \sin^2 \theta}} \right].$$

The derivation is left as an exercise (see Stretch & Challenge).

Stretch & Challenge

Derive the formula for the range of a projectile released at height h, angle θ and velocity v.

Hint: Start by calculating the flight time.

Plotting R against θ for various release velocities is a nice spreadsheet exercise. The Brunel university biomechanics of athletics research group has produced Fig. C25 which shows this variation. Notice that the peak range occurs at about 42° now – less than 45° but still greater than the observed 35° at events. The answer lies in the fact that the speed of projection that athletes can achieve is not constant – it is a function of the angle of projection. The Brunel group has taken measurements on athletes and the following graph is typical:

Why should the launch speed decrease with θ? For one thing, the shot athlete has to accelerate the shot against gravity for upward angles. For another, the human body finds it easier to throw horizontally. The effect of this variation is to reduce the optimum launch angle to about 35° for most athletes as observed in international events.

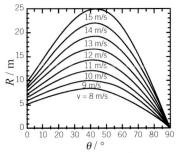

Fig. C25 Variation of R with θ and v

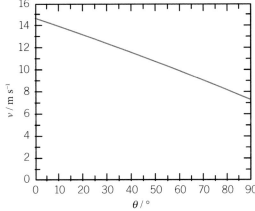

Fig. C26 Variation of v with θ.

C13

Self-test

Use the formula to calculate the range of a shot which is projected at 40° with a speed of 12 m s⁻¹ and compare it with the appropriate graph in Fig. C25.

C14

Self-test

Consider the athlete whose results are as Fig. C26. Calculate the ranges he should expect if launching the shot at (a) 45°, (b) 35°.

(b) Projectiles with non-gravitational forces

As we have already mentioned, most objects moving through the air experience other forces. In this section we'll focus on two of these, aerodynamic drag and lift (Fig. C27), in the context of javelin flight. Fig. C28 shows a typical flight path for a javelin.

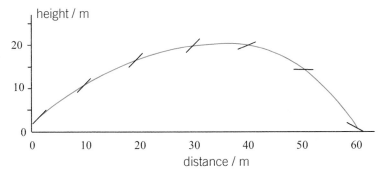

Fig. C28 Javelin flight path and attitude

The flight path clearly increasingly diverges from a parabolic curve as the javelin travels. This is because of the joint effects of drag and lift. Drag arises mainly because the javelin has to do work against the excess pressure in front of it; lift arises because some of the air molecules are deflected downwards by the javelin. Both these forces increase with the angle of attack (Fig. C27) and it happens that, for the javelin, they have roughly the same size (unlike the discus which has a lift-to-drag ratio of about 3).

As discussed in Section 1.3.5(c) of the year 12 textbook, drag is proportional to the square of the wind speed. It is given by

$$F_D = \tfrac{1}{2}\rho v^2 A C_D$$

where ρ is the air density, A is the effective area of the object (which increases with the angle of attack, α) and C_D is a constant called the drag coefficient, which is roughly unity (data sheets give a value of 1.2–1.4). During the first half of the flight of the javelin, both the air resistance and the lift are very small (see Self-test) but, because of the orientation of the javelin, these become increasingly important towards the end of the flight.

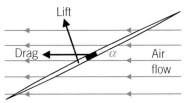

Fig. C27 Javelin lift and drag

▼ **Study point**

Fig. C27 shows the motion of the air relative to the javelin. Notice that the centre of lift is behind the centre of mass, which is in the hand grip.

┌─ **Terms & definitions** ─┐

The **angle of attack** is the angle between the air flow and the orientation of the javelin (or discus…).

C15

Self-test

Account for aerodynamic drag and lift in terms of Newton's first and second laws of motion.

C16

Self-test

Explain why the departure from a parabolic curve is only noticeable from the top of the flight path. (Hint: see Fig. C28.)

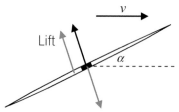

Fig. C29 Turning effect of lift

C17

Self-test

Suppose the distance of the centre of lift from the centre of mass in the javelin in the example is 0.11 m.

(a) Find the mean value of the lift.

(b) A new design of javelin with a lift equal to that in part (a) has the centre of mass 0.22 m behind the centre of mass. Suggest why it is unlikely to be a successful design.

Notice that the javelin changes its orientation through the flight.[2] In the absence of lift, this would not happen. Consider the javelin at the top of its flight. Because the flight has dipped down, the angle of attack is typically 30–40° at this point. Looking at Fig. C27 again, the drag is through the centre of mass, as is the weight (by definition) so these two forces do not cause a rotation. The lift, on the other hand, does: because the centre of lift is behind the centre of mass so, just as in Fig. C20, we can consider it to be a force through the CoM + a clockwise couple (Fig. C29), which gives the javelin a clockwise angular acceleration, which increases as the angle of attack increases on the downward leg.

Example

A javelin is launched at 25 m s^{-1} at an angle of 45°, with an initial angle of attack of 0. The throw distance is 60 m and it makes an angle of 45° on impact. Estimate (a) the flight time and (b) the mean torque applied by the lift.

Javelin moment of inertia = 0.42 kg m^2.

Answer

(a) Initial horizontal component of velocity = 25 cos 45° = 17.7 m s^{-1}

$$\text{Assuming constant velocity, time of flight} = \frac{60 \text{ m}}{17.7 \text{ m s}^{-1}} = 3.4 \text{ s}$$

(b) $\Delta\theta = 90° = 1.57$ rad. Using $\Delta\theta = \omega_1 t + \frac{1}{2}\alpha t^2$, with $\omega_1 = 0$

$$\alpha = \frac{2\Delta\theta}{t^2} = \frac{2 \times 1.57 \text{ rad}}{(3.4 \text{ s})^2} = 0.27 \text{ rad s}^{-2}.$$

Then $J = I\alpha = 0.42 \text{ kg m}^2 \times 0.27 \text{ rad s}^{-2} = 0.11$ N m.

C.7 Drag and lift in sports

These forces on objects which are moving through the air are important in ball games, athletics field sports, aviation and motor sports. The physics of this topic is far from easy and there are several misconceptions: this unit provides only a basic introduction.

Fig. C30 Aerofoil showing turbulent flow

(a) Aerodynamic drag (air resistance)

The air flow around the aerofoil in Fig. C30 is *turbulent*. The diagram below shows the same aerofoil with *lamina flow* and introduces a few terms.

▼ **Study point**

The aerofoil in Fig. C30 has a much greater attack angle than that in Fig. C31. Hence the air which separates at the front of the wing doesn't recombine smoothly and is turbulent. This effect happens at lower speeds with blunt ('non-aerodynamic') shapes and gives rise to a much greater pressure difference.

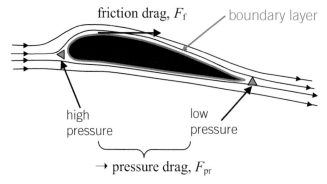

Fig. C31 Aerofoil with lamina flow

---

2 As a military weapon (in ancient Greece), it would have been pretty useless if the point didn't dip down towards the end of the flight.

There are two main drag components:

- The *friction drag* is due to the interaction of the moving air with the *boundary layer* which is a thin layer of stationary air in contact with the object.

- The *pressure drag* is due to the difference in pressure between the front of the object and the back.

Both these two components vary as v^2, the head-on area, A, of the object and the density, ρ, of the air, so the total drag force, F_D, can be expressed as $F_D = \frac{1}{2}\rho v^2 A C_D$. The constant C_D is called the drag coefficient, but unfortunately it isn't really constant. The variation of F_D with v for a football in Fig. C32 shows that the air resistance is roughly proportional to v^2 up to **20 m s⁻¹**, where the air flow is lamina. Then the flow becomes turbulent and the air resistance plummets with speed before starting to take off again. The lower speed region is dominated by the friction drag – at higher speeds the pressure drag dominates. This distinction controls the way that footballers curve the ball into the net (see later).

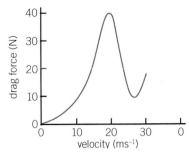

Fig. C32 Variation of drag force with velocity for a football

C18

Self-test

Does the low velocity portion of the drag graph in Fig. C32 really show proportionality to v^2.

Example

Assuming that C_D in the drag equation, $F_D = \frac{1}{2}\rho v^2 A C_D$, is constant, how would you expect the fuel consumption of an F1 car to vary with v?

Answer

Assuming that the variation of fuel consumption is dominated by the air resistance, the power is proportional to v^3, because $P = Fv$. This suggests that the fuel consumed per unit time is proportional to v^3. But the distance travelled is proportional to v, so the fuel consumed per unit distance should be proportional to v^2 (see Study point).

▼ **Study point**

It's a lot more complicated: rolling resistance is roughly constant; engines have an optimum running speed...... (so don't believe it!).

Question: How do we reduce the air resistance?

Answer: By making the object smoother.

Wrong! In the early twentieth century golfers found they could hit their old scuffed and cracked golf balls further than their nice new smooth ones. So what's going on?

It's all to do with the boundary layer. The dimples trap a turbulent layer of air next to the golf ball, which the air slides past smoothly and round the ball. This lamina flow separates much later from the ball leaving a smaller turbulent wake than the smooth golf ball. It is the pressure difference between the front and back which produces the drag. Because the low wake pressure acts over a smaller area in the dimpled ball, the drag is lower. This is summarised in Fig. C34.

Fig. C33 Golf ball with dimples

▼ **Study point**

Note that all the diagrams of objects moving through the air are drawn as if the object is stationary and the air is flowing past (e.g. Fig. C34). This is standard.

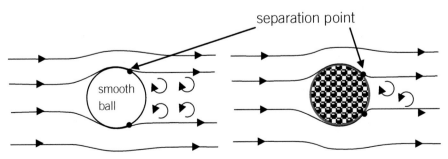

separation point

Fig. C34 Dimpled golf balls have a narrow wake, so a lower drag

Fig. C35 Aerodynamic spoiler

high speed,
low pressure

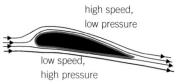

low speed,
high pressure

Fig. C36 Bernoulli explanation for lift

C19

Self-test

The air flows at 160 m s⁻¹ over the top of a wing of area 10 m². Underneath the speed is 140 m s⁻¹. Use Bernoulli's equation to calculate the lift.

$\rho_{\text{air}} = 1.3$ kg m⁻³.

Force exerted by air on wing ↑

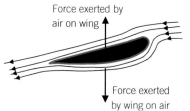

Force exerted by wing on air ↓

Fig. C38 Flying upside down

Force on spinning ball ←

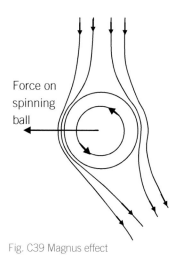

Fig. C39 Magnus effect

(b) Aerodynamic lift – explanation 1

The aerofoil in Fig. C31 experiences an upwards force that we call lift. The spoilers on the back of sports cars (Fig. C35) are there to create a downwards force to hold the back of the car on the road – this is also confusingly called lift. So how does it arise? Until recently the accepted explanation was the *Bernoulli effect* (pronounced burn-oo-i). If a stationary gas at a pressure p_0 accelerates to a velocity v, the pressure falls to p given by:

$$p = p_0 - \tfrac{1}{2}\rho v^2.$$

An alternative way of writing the relationship between pressure and velocity is

$$p + \tfrac{1}{2}\rho v^2 = \text{constant}.$$

This equation is an expression of the principle of the conservation of energy applied to moving gases. If we imagine multiplying the equation by the volume of a section of gas, the resulting pV is the internal energy and the $\tfrac{1}{2}\rho V v^2$ term is the kinetic energy, i.e. the total energy is constant. In any case we can interpret the equation as saying that if the velocity of a gas increases its pressure decreases.

Applying this to the aerofoil, the argument runs: the air flowing over the upper curved surface of the wing has a longer path than that under the wing, so its speed is greater and its pressure lower. This difference in pressure produces the lift.

So how does the plane fly upside down?

Fig. C37 Plane flying upside down

(c) Aerodynamic lift – explanation 2

The second explanation just uses Newton's 2nd and 3rd laws (N2 and N3). Look again at Fig. C38. The initial velocity of the air is horizontal to the right; it leaves the wing downwards to the right. The wing pushes the air downwards, i.e. it has exerted a downwards force on the air (N2); the air exerts an equal and opposite force on the wing (N3), i.e. an upwards force.

In order to fly upside down, the pilot in Fig. C37 has to angle the wings upwards against the design of the wing and plane, as shown in Fig. C38. This takes more energy than flying with the curved side of the wing on top.

(d) Bend it like Beckham – the Magnus effect[3]

Whether in tennis, football (see Robert Carlos's legendary 2009 free kick) or golf, it is well known that a spinning ball will swerve away from the forward spinning side, as illustrated in Fig. C39 – top spin will make a ball dip down; bottom spin will hold the ball up against gravity and a spin about a vertical axis will make it veer to one side. Why?

Both explanations rely on the experimental observation that the air which follows over the top of the ball (i.e. in the direction of the spin – Fig. C39) separates later than the air which flows underneath. This has the effect of bending the air flow downwards.

---

3 Named after the physicist Heinrich Gustav Magnus who described it in 1852.

Bernoulli explanation

Because of the direction change, the air which flows over the left of the ball (in Fig. C39) travels faster than the air on the right. So, by Bernoulli, the pressure on the left is lower than the pressure on the right. The pressure difference causes the force to the left.

Newton's explanation

Newton is said to have explained the effect in this way when watching tennis in Trinity College nearly two centuries before Magnus described it. The air is given a rightward momentum as it is deflected so the ball acquires an equal and opposite momentum.

Which explanation is right?

The boring answer is 'both.' But it is generally accepted that the N2/3 (or conservation of momentum) effect is much greater than the Bernoulli effect (by an order of magnitude).

But how do you make the ball spin?

Fig. C40 is an attempt to show the head of a tennis racket applying a slice to a tennis ball. The two forces arise because of the relative motion of the racket head and the ball. These forces will exist for a short time, Δt, and so each produce an impulse. What is their combined effect?

- The normal force pass through the centre of gravity and so just produces an impulse, $N\Delta t$, upwards and to the left and no rotation.

- As we saw in Section C5(a), we can think of the frictional force as being composed of a force, F, through the centre of mass and an anticlockwise couple, Fr, where r is the radius of the ball. So this produces an impulse $F\Delta t$ upwards and to the right and an anticlockwise impulsive torque (see Study point) $Fr\Delta t$.

The result of these is

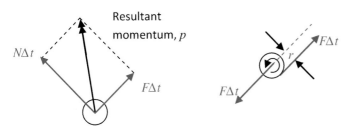

Fig. C41 Linear and angular momentum produced by the slice

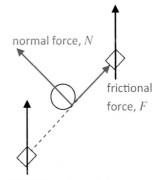

Fig. C40 Slicing a tennis ball

Exercise 4C

1. When sailing into the wind, the closer to upright a boat is, the faster it goes (for a particular angle to the wind). Explain this and also how it can be achieved by the sailor in Fig. C4.

2. The diagram shows various directions that a rudder, **R**, can take in sailing. Copy Fig. C9 and add the rudder in an appropriate direction to 'balance' the boat, i.e. stop it turning into the wind. Explain your answer and also explain why this will slow the boat.

3. Consider the dinghy in Fig. C4. A less well-designed dinghy has a higher centre of gravity. It is (on the diagram) just where the mast disappears behind the boat stern. Explain briefly, using moments, why this boat is less stable.

4. It is difficult to get an accurate measure of bounce height, so a tennis club has decided to test its match balls by dropping from the regulation height and timing the interval between the first and second bounce, using a microphone and a data logger.

   (a) Calculate the time intervals expected for the least and greatest legal bounce heights.

   (b) Suggest how accurately the timings need to be to detect a ball which bounces 1 cm lower or 1 cm higher than the allowed range.

5. The rules for basketball bounciness involve measuring to the top of the ball rather than the bottom. When dropped from a height of 6 feet (i.e. 72 inches) it should bounce to between 49 and 54 inches. Given that the circumference of a basketball is 30 inches, calculate the allowed range for the coefficient of restitution.

6. A sports aeroplane has two wings each of length 4.5 m. When flying horizontally they deflect the air within 1 m of the wing surface through an angle of $20°$ downwards. Calculate the lift achieved by the wings at an air speed of 60 m s^{-1}. Density of air $= 1.3$ kg m^{-3}.

7. A shot putter releases the shot from a height of 2.05 m, at a speed of 13.5 m s^{-1} and an angle of $35°$ to the horizontal.

   (a) Calculate the flight time of the shot and hence the distance achieved.

   (b) Investigate whether the athlete should consider changing the angle of release to $40°$ given that she can only release the shot at 13.0 m s^{-1} at this angle.

8. Use Fig. C32 and the answer to Self-test C17 to sketch a graph of the variation of the drag coefficient for a football with velocity.

9. Considering the Magnus force on a spinning object, theory suggests than the angle of deflection of the air is linked positively to the drag coefficient – the greater the drag, the more spinning ball deflects the air.

   How does this theory account for the trajectory of Roberto Carlos's free kick for Brazil against France in 1997?

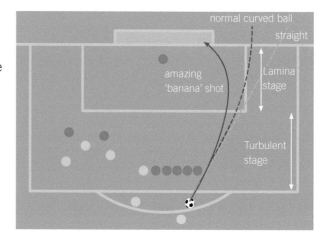

Option D: Energy and the environment

Unlike some of the other options, Option D contains little additional physics. The concepts which you have already covered, such as the radiation laws (Wien's, Stefan-Boltzmann, inverse square) are applied in the context of our need for energy and sustainability. Exceptions are Archimedes' principle and thermal conduction equations. Our starting point is the temperature of the Earth.

D.1 The Earth's rising temperature

The vast majority of atmospheric scientists are convinced by the evidence that anthropogenic[1] global warming is a fact. The published data include 'hockey stick' graphs of the rising greenhouse gas concentrations and the variations of the Earth's temperature (Figs D1 and D2).

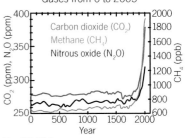

Concentrations of Greenhouse Gases from 0 to 2005

Carbon dioxide (CO_2)
Methane (CH_4)
Nitrous oxide (N_2O)

Fig. D1 Rising greenhouse gas concentrations

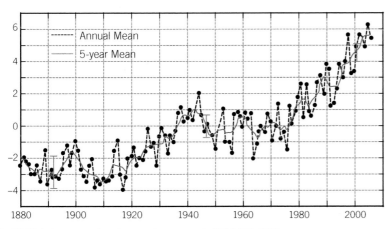

Fig. D2 Global-mean surface temperature anomaly (°C) 1880–2005

Before we examine the effect of greenhouse gases, we'll have a look at the radiative exchange between the Earth and space.

(a) The Earth's temperature with no atmosphere

We'll start by looking at how the temperature of the Earth is established in a stable period. We know the atmosphere helps keep us warm. What would its mean temperature be without an atmosphere? As Fig. D3 shows, there would be a balance between the power input from the Sun. We receive 1.36 kW m^{-2} of radiation (mainly in the near UV, visible and near IR). This is the value of the **solar constant**, G.

∴ The total power received is given by, $P = \pi R^2 G$

Because the Earth is rotating, it will radiate from all around. If the Earth has a mean temperature, T, it radiates a power given by the Stefan-Boltzmann law

$$P = 4\pi R^2 \sigma T^4$$

▼ **Study point**

Note the different time scales on the two graphs – comparison graphs over the last two millennia are available. The word *anomaly* means the difference between the actual temperature and a baseline value. 1960 is often used, as in Fig. D2.

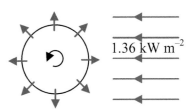

Fig. D3 Radiation balance

Terms & definitions

The **solar constant**, G, is the radiative power from the Sun per unit area incident on a surface at right angles to the solar radiation. Its value at the radius of the Earth is 1.36 kW m^{-2}.

▼ **Study point**

The Earth rotates so the temperature of the day and night hemispheres will be similar.

1 Anthropogenic = caused by humans.

If we assumed that the Earth absorbed all the incident radiation, these two values would be equal for equilibrium. But this is not the case: the Earth has an *albedo* of about 0.3, i.e. it reflects 30% (most of which is from clouds see Fig. D4) of the incoming radiation, so the power actually absorbed is $0.7 \times \pi R^2 G$.

We can equate the absorbed power to the radiated power:

$$\therefore \; 4\pi R^2 \sigma T^4 = 0.7 \times \pi R^2 G$$

Inserting the value for G and simplifying gives

$$T^4 = \frac{0.7 \times 1360 \text{ W m}^{-2}}{4 \times 5.67 \times 10^{-8} \text{ W m}^{-2} \text{ K}^{-4}},$$

which leads to a value of T of 255 K, i.e. −18 °C.

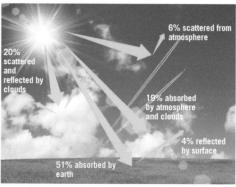

Fig. D4 Reflection of solar radiation

Conclusion: Without an atmosphere the Earth would be frozen. In fact the mean temperature is a relatively comfortable 288 K (+15 °C). You should do Self-tests D1 & D2 before proceeding.

Note. Kirchhoff's law of radiation says that if the Earth only absorbs 70% of the incident radiation, the emissivity of the Earth should also be 70% of the black body value. However, the 30% albedo is only the figure for the visible and near IR and UV parts of the em spectrum, which are the main components of the incoming solar radiation. In the thermal infrared the Earth behaves more like a perfect black body. This is similar to a white-painted domestic radiator – it behaves like a black body in the thermal infrared.

(b) The effect of the atmosphere

It appears that Mars (Self-test D1) has a temperature that is in agreement with the calculation above. Why the difference? The biggest factor is the density of the Martian atmosphere at the surface, 0.020 kg m^{-3}, which is less than one sixtieth that of the Earth. We need to look at some of the properties of the Earth's atmosphere.

Composition (dry atmosphere): Nitrogen (78.08%), Oxygen (20.95%), Argon (0.93%), Carbon dioxide (0.04%) + trace gases, e.g. methane. Air also contains a variable amount of water vapour – typically 1% at sea level.

Transparency: Fig. D5 has a wealth of information. Let's examine it in detail:

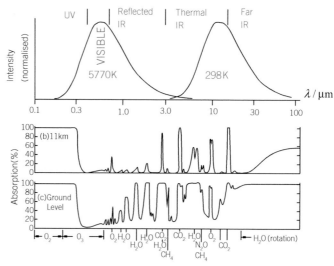

Fig. D5 Absorption of radiation by atmospheric molecules

- The air at ground level is opaque to radiation of wavelength less than $0.3\ \mu m$ (in the UV) and longer than $20\ \mu m$ (in the far IR).

- Radiation from the Sun, which is mainly $0.3–2\ \mu m$ penetrates the atmosphere (with absorption in bands between 1 and $2\ \mu m$).

- Much of the radiation from the ground, mainly $5–50\ \mu m$, is absorbed by the atmosphere.

- Solar radiation which is reflected from the ground (near UV, visible and near IR) escapes largely into space.

- Various bands of wavelengths are wholly or partially absorbed by H_2O, CO_2, CH_4, O_3 and N_2O (see Study point).

A $6.5\ \mu m$ photon, for example, emitted from the ground is quite likely to be absorbed by a water molecule, promoting it to a higher vibrational energy level. The molecule can de-excite in two ways:

1   Collision with another molecule – increasing the translational and/or rotational kinetic energy of the molecules.

2   Re-radiation. This can occur in any direction, so roughly 50% of these photons will be reabsorbed by the ground.

The effect of these two processes is to raise the temperature of atmosphere and ground – the **greenhouse effect**. The equilibrium temperature is higher than would be the case without an atmosphere – but the rate of energy loss to space (emitted + reflected) is still equal to the net incoming power from the Sun.

It should be noted that the atmosphere also gets directly heated by the Sun: by absorption of solar radiation in one of the molecular bands between 1 and $3\ \mu m$. The heated atmosphere will lose this energy as above – some of it resulting in raised ground temperature and some escaping to space, reducing the magnitude of the **greenhouse effect**.

(c) Anthropogenic global warming – with positive feedback

In spite of its low atmospheric concentration, carbon dioxide makes a major contribution to the **greenhouse effect**. Its natural concentration has varied widely over geological time, as shown in Fig. D7.

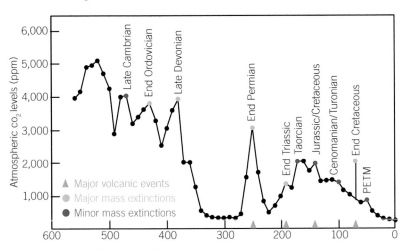

Fig. D7 Variation of CO_2 concentration in the phanerozoic aeon

▼ **Study point**

Diatomic molecules don't absorb photons in the near UV, visible and IR regions of the e-m spectrum because the jumps in the rotational, vibrational or electron energy levels don't match the photon energies. Molecules with more than two atoms have additional vibrational energy modes (Fig. D6) which absorb photons in the near and thermal infrared.

Fig. D6 Vibrating CO_2 molecules

D4

Self-test

N_2O absorbs radiation with a wavelength of $7.5\ \mu m$. Explain this in terms of the energy levels of an N_2O molecule.

Terms & definitions

The **greenhouse effect** is the name given to the process of the warming of a planet by radiation from its atmosphere to a temperature above what it would be without an atmosphere.

D5

Self-test

The concentration (vertical) scale in Fig. D7 is in ppm (parts per million). What would the labels be in %?

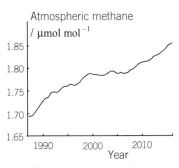

Fig. D8 Rising methane levels

237

▼ **Study point**

The 400 ppm level of CO_2 was recorded for the first time in Mauna Loa in May 2013.

The pre-industrial CO_2 concentration was 280 ppm. Burning of fossil fuels has increased this to 400 ppm, which has increased the absorption of thermal infrared radiation. A more potent greenhouse gas is methane. Increased intensive agriculture, especially animal rearing, has increased methane levels (Fig. D8).

There are feedback mechanisms which amplify the warming effect:

• Melting ice cover reduces the fraction of radiation reflected back into space.

• Melting permafrost in Canada and Siberia releases methane.

Such feedback is called *positive feedback* because its effect is in the same direction as its cause.

(d) Melting ice and rising sea levels

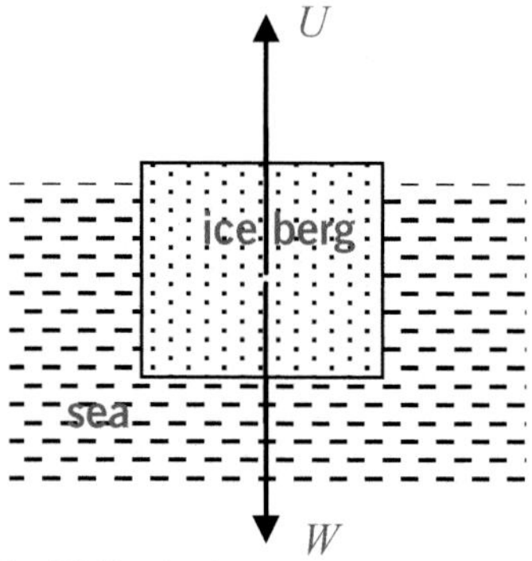

The current rate of sea level rise is estimated as 2.9 ± 0.4 mm year^{-1} and appears to be accelerating at 0.013 mm year^{-2} from the 20th-century average of 1.7 mm year^{-1}. This appears to be due mainly to melting ice. However, the melting of floating ice (mainly in the Arctic ocean) does not contribute directly to this. Why not?

Fig. D9 Floating ice

Archimedes' principle states, 'When a body is wholly or partially immersed in a fluid, it experiences an upthrust, U, which is equal to the weight of fluid displaced by the body.' Applying this to a floating object, e.g. the iceberg in Fig. D9, this means that the weight, W, of the berg is equal to the weight of the sea water which would occupy the volume of the iceberg below the water line. Because ice is less dense than North Atlantic sea water (930 kg m^{-3} v 1030 kg m^{-3}) it floats with about 10% above the water line. Its volume when melted is the same as that of the sea water and so it doesn't cause the sea water to rise.

D6
▼ Self-test

Calculate the volume of:

(a) an iceberg of mass 10^6 tonne.

(b) the volume of sea water of mass 10^6 tonne.

Example

An iceberg has a surface area of 1.0 km^2 and a height of 100 m. Calculate, (a) the mass of the iceberg and (b) the height of the iceberg above the water.

Answer

(a) Using the above values for density:
$$M_{ice} = \rho V = 930 \text{ kg m}^{-3} \times 1.0 \times 10^6 \text{ m}^2 \times 100 \text{ m}$$
$$= 9.30 \times 10^{10} \text{ kg}$$

(b) By Archimedes' principle the iceberg displaces its own mass of sea water, which has a volume given by
$$V = \frac{M_{ice}}{\rho_{sea water}} = \frac{9.30 \times 10^7 \text{ m}^3}{1030 \text{ kg m}^{-3}} = 9.03 \times 10^7 \text{ m}^3$$
$$\therefore \text{ Fraction under water} = \frac{9.03 \times 10^7}{1.0 \times 10^8} = 0.903.$$

$\therefore$ Fraction above water = 0.097.

$\therefore$ Height above water = 0.097×100 m = 9.7 m

▼ **Study point**

When the iceberg in the example melts it will occupy a volume of 9.03×10^7 m^3 once it has mixed with the seawater and attained the same density. There is a possible second order effect due to a drop in the salinity of the seawater.

D.2 Energy sources

This is a short review of **renewable** and **non-renewable** energy sources, which can be used for transport, electricity generation, heating and cooling. Most renewable energy sources are, directly or indirectly, solar energy. Apart from thermal and photovoltaic solar power, this includes wind, wave, hydroelectric and biomass. Geothermal energy relies on heat from within the Earth, the origin of which is the decay of radioisotopes within the core (supplemented by the release of latent heat from the growth of the inner core). Sections (a) – (c) are concerned directly with solar power.

(a) The origin of solar power

Without the input of radiation from the Sun, the Earth would be lifeless rock, with a surface temperature of a few tens of kelvin. The origin of this radiation is the energy released from the nuclear fusion of hydrogen-1 nuclei (protons) to helium-4 (alpha particles):

- The *proton-proton chain* produces 98.3% of the He4 nuclei. This is the dominant fusion route for **main sequence** stars of mass less than 1.3 times that of the Sun.

- More massive main sequence stars produce most of their energy by the CNO (carbon-nitrogen-oxygen) cycle. This accounts for only 1.7% of the energy output of the Sun.

The initial step of the proton-proton chain is the fusion of two protons to produce deuterium:

$$\ _1^1H + \ _1^1H \rightarrow \ _1^2H + \ _1^0e^+ + \nu_e$$

We can tell from the change in quark flavour (d→u) and the involvement of a neutrino that this is controlled by the weak interaction. As a result of this, even under the conditions of temperature and pressure, the lifetime of a proton at the centre of the Sun is of the order of 10^9 **years**. The next step is more rapid – the mean lifetime of the $_1^1H$ has been estimated to be 4 seconds!

$$\ _1^2H + \ _1^1H \rightarrow \ _2^3He + \gamma$$

There are now two main routes (branches) called ppI and ppII:

ppI: $\quad \ _2^3He + \ _2^3He \rightarrow \ _2^4He + 2\ _1^1H$

ppII: $\quad \ _2^3He + \ _2^4He \rightarrow \ _4^7Be + \gamma$

$\quad\quad\quad \ _4^7Be + \ _{-1}^0e^- \rightarrow \ _3^7Li + \nu_e$

$\quad\quad\quad \ _3^7Li + \ _1^1H \rightarrow 2\ _2^4He$

These reactions can be detected from laboratories on Earth by the characteristic neutrino energies (see Exercise 4.D). For energy calculation purposes we can summarise the pp chain as:

$$4\ _1^1H + 2\ _{-1}^0e^- \rightarrow \ _2^4He.$$

(b) The solar constant

The solar power received per unit area on a surface at right angles to its direction of travel (the **solar constant**) is 1360 W m^{-2} at a distance of 1 AU from the Sun. The energy released in the production of a He4 nucleus from four protons is 26.73 MeV (Self-test D8) of which 2.3% is carried away by neutrinos. The rest is energy of e-m radiation, initially γ in the core but emitted mainly as near UV, visible and IR radiation from the photosphere. The total power emitted as e-m radiation by the Sun is $3.83 \times 10^{26} \text{ W}$, which we can relate to the solar constant using the inverse square law.

— Terms & definitions —

An energy source is regarded as **renewable** if it is replenished naturally in a human lifetime (some organisations use 50 years as the time scale). **Non-renewable** sources are regarded as effectively not capable of replenishment.

A **main sequence** star is one which generates most of its energy from the fusion of hydrogen to helium in its core.

Solar trivia

The temperature, pressure and density of the centre of the core are 15 MK, 26.5 PPa and 150 g cm^{-3} respectively.

99% of the power generation comes from the central $0.24R$ (1.4% of the volume) which also contains 40% of the mass.

The power production density is only 280 W m^{-3}, which is less than the metabolic production of a human being – but the Sun is rather larger.

D7
Self-test

A website says that the second stage of the proton-proton chain is controlled by the strong interaction. Comment on this.

▼ Study point

There is also a minor branch called ppIII and a hypothesised branch, ppIV, which has not been observed.

D8
Self-test

Test the value of 26.73 MeV for the proton-proton chain using the following data:

$m_p = 1.007\ 276\ 47 \text{ u}$

$m_{He4} = 4.001\ 506\ 47 \text{ u}$

$m_e = 0.000\ 548\ 60 \text{ u}$

$1 \text{ u} \equiv 931.5 \text{ MeV}$

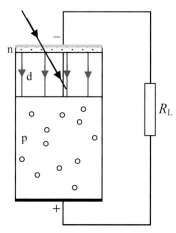

Fig. D10 PV cell schematic

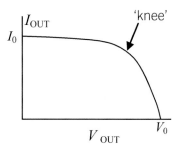

Fig. D11 PV cell characteristic

Example

Use solar luminosity to calculate the solar constant, G.

[1 AU = 1.50×10^{11} m]

Answer

Solar luminosity, $L_e = 3.83 \times 10^{26}$ W.

At the radius of the Earth's orbit, this radiation is spread over a sphere of surface area A, given by $4\pi r^2$.

$$\therefore G = \frac{L_e}{4\pi r^2} = \frac{3.83 \times 10^{26} \text{ W}}{4\pi \times (1.50 \times 10^{11} \text{ m})^2} = 1350 \text{ W m}^{-2}$$

Note: This shows that the given data are reasonably consistent!

(c) Photovoltaic (PV) cells[2]

A PV cell is a p-n junction diode (Fig. D10, but see also Fig. 3.5.9 and AS textbook Fig. 2.7.24). Diffusion of the doping carriers creates an electric field across the depletion zone (red arrows). If a sufficiently energetic (>0.7 eV) photon enters through the transparent window (grey) it can create an electron-hole pair. The electron is attracted upwards by the electric field and the hole downwards. Some of these mobile charges will be lost to recombination but some will create a current in the external circuit through the load (R_L). Hence power can be extracted.

The graph in Fig. D11 is a typical current-voltage characteristic for a PV cell. As with a cell, the pd falls with current but not linearly. At small voltages the current falls gradually with pd, then it starts to fall more sharply in a region which is called the *knee*. The maximum current, I_0, is when the PV cell is shorted; it depends on the light intensity (roughly proportional) and is in the mA range. The open-circuit pd, V_0, is roughly equal to the band gap for the p-n junction – it is about 0.6 V for a silicon PV cell.

To make a practical solar panel, the PV cells are typically arranged in a series / parallel combination. Fig. D12 gives the principle of this. In this case the maximum p.d. available would be about $3V_0$ and the maximum current $2I_0$ (from the usual rules for combination of pd and current in series and parallel.

The electronic controls of PV panels are set to maximise the power output. Where on the characteristic should the control box select to do this? The red graph in Fig. D13 gives the output characteristic of a PV module comprising a large number of PV cells.

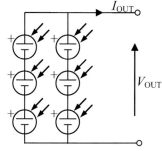

Fig. D12 PV cells

---

2 The electricity generated by photovoltaic panels provided 99% of the domestic electricity consumption in June 2015.

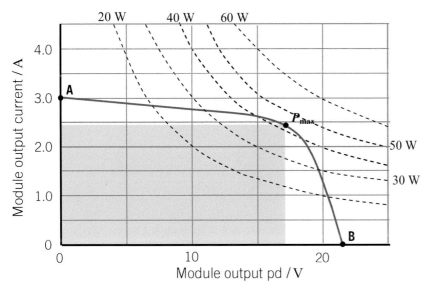

Fig. D13 Output of a PV module

▼ **Study point**

The pecked curves in Fig. D13 are of VI = 20 W, 30 W .. 60 W. For example, the output is 20 W at (7.0 V, 2.86 A) and approximately (20 V, 1.0 A).

Self-test **D11**

Under the same lighting conditions as those for which the red graph in Fig. D13 was obtained, each PV cell has I_0 and V_0 values of 30 mA and 0.60 V. Calculate the number of cells in the module and suggest how they are connected.

The power output is the product of pd and current: $P_{OUT} = V_{OUT} I_{OUT}$.

At **A** (short circuit), $V = 0$, so $P = 0$. Similarly $P = 0$ at **B**.

The grey rectangle is the author's estimate of the one under the graph with the maximum area, i.e. the maximum value of $V \times I$ (see Self-test D12).

Self-test **D12**

Estimate P_{max} for the PV module in Fig. D13.

(d) Power from wind

The red graph in Fig. D15 is a typical power curve for a wind turbine.

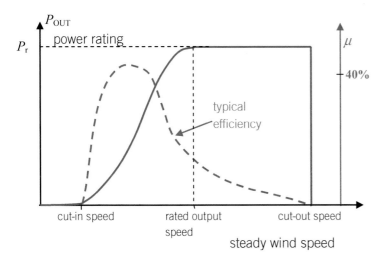

Fig. D15 Typical characteristic of a wind turbine

Fig. D14 Gwynt y Môr offshore windfarm

Below the **cut-in speed** (typically 3–4 m s⁻¹) the turbine blades produce insufficient torque to rotate against friction. Above the **rated output speed** (typically about 12 m s⁻¹) the turbine is designed to extract a constant power from the wind, often by decreasing the angle that the blades make to the wind (see Option C; section 7). Turbines are designed to stop rotating and turn their blades edge on to the wind at the **cut-out speed** which is often ~25 m s⁻¹.

— **Terms & definitions** —

The **power rating** of a wind turbine is the maximum power at which it is designed to operate.

The minimum wind speed at which the turbine generates is the **cut-in speed**.

The **cut-out speed** is the wind speed above which the turbine produces zero output.

The **rated output speed** is the minimum wind speed for which the turbine outputs at its power rating.

▼ Study point

The curve in Fig. D13 is for a *steady* wind speed. The inertia of the turbine makes it slow to respond to rapidly fluctuating winds, so reducing the output. The gradual tail-off of the power output below the rated output speed is due to the gradual introduction of the output power control measures.

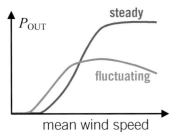

Fig. D16 Effect of wind-speed variation

Stretch & Challenge

The wind-speed probability function works like this:

If the value of the function is $p(v)$, the probability of the wind being within a small range Δv centred on v is $p(v)\Delta v$.

Suppose for $p(10) = 0.12$. State the probability of the wind being between 9.75 m s^{-1} and 10.25 m s^{-1}. If the power output at 10 m s^{-1} is 2.4 MW calculate the contribution of 10.0 ± 0.25 m s^{-1} winds to the annual energy output.

Fig. D18 Tidal stream generator in Strangford Lough

We can calculate the power available from a steady wind of speed v, to a turbine of area A, by identifying it with the kinetic energy of the incident air per unit time.

$$\text{Volume of air incident per unit time} = Av$$

∴ $\quad$ Mass of air incident per unit time $= \rho A v$, $\quad\quad$ where ρ = air density

∴ $\quad$ KE of air incident per unit time $= \frac{1}{2}mv^2 = \frac{1}{2}(\rho A v)v^2 = \frac{1}{2}\rho A v^3$

i.e. $\quad P_{IN} = \frac{1}{2}\rho A v^3$

The efficiency, μ, is given by: $\mu = \dfrac{P_{OUT}}{P_{IN}} = \dfrac{P_{OUT}}{\frac{1}{2}\rho A v^3}$. It is usually expressed as a percentage.

The efficiency is also known as the *power coefficient*, c_p. The maximum value of efficiency is always less than 50% (see Fig. D15). Reasons for this (apart from friction in the turbine) include: the kinetic energy of the air cannot be reduced to zero as it passes the blades; the turbulent wake of one turbine interferes with the others downwind.

In order to estimate the practical power output of a wind turbine, several factors need to be taken into account in addition to its characteristic:

1 The effect of fluctuating wind speed. As shown in Fig. D16, the general effect of fluctuations is to reduce the power output to below that of a steady wind. The output power for a range of mean speeds from below the cut-in speed upwards is actually greater than that of the equivalent steady speed; this is because of the disproportionate effects of the occasional high-speed gusts. For speeds slightly below the rated output speed, the power output reduction procedures give an output below that of the steady speed value.

2 The daily variation of wind speed.

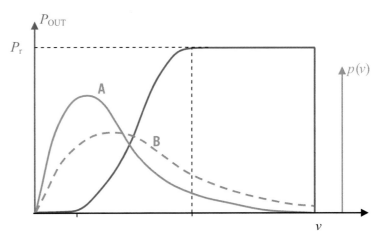

Fig. D17 Wind-speed probability functions

Windier places give a higher energy output. The green curves in Fig. D17 are for two different locations: B is windier on the average than A. The green curves are the *wind-speed probability functions* for A and B. The area under both graphs is 1 (one), i.e. there is a 100% probability of the wind-speed being between 0 and ∞. If we use them to plot a graph of $p(v) \times P_{OUT}(v)$ against v, the area underneath it gives the mean power output.

Tidal stream power stations operate on the same principle as wind trubine. Narrow tidal races, such as at St David's head in Pembrokeshire, provide ideal conditions. They have the advantages of tidal predictability and the much greater energy density of flowing water (because of its higher density).

(e) Hydroelectric power

The release of the gravitational potential energy in water as it flows from an upper reservoir to a lower level is used to generate electricity by driving turbines. This principle is used in hydroelectric, pumped-storage and tidal-barrage power stations.

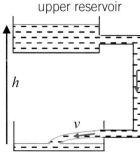

Fig. D20 $E_p \rightarrow E_k$ in water

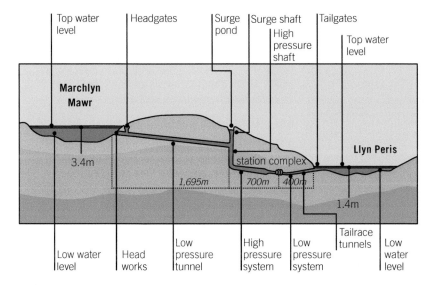

Fig. D19 Dinorwig pumped-storage power station

Consider the flow of water from the upper to the lower reservoir in Fig. D20. Unless the pipe is very long compared to its diameter, we can ignore *pipe losses*, so the rate of loss of potential energy per second is equal to the gain of kinetic energy. This means we can use the equation $\frac{1}{2}mv^2 = mgh$ to calculate the speed of flow, i.e. $v = \sqrt{2gh}$. The volume flow rate of the water is Av (where A is the cross-sectional area of the pipe) which allows us to calculate the rate of transfer of energy (see Self-test D15).

If we want to extract power, P_{OUT}, from the system, which is the point of a hydroelectric system, we need to build in a turbine, **T** (Fig. D21). We can now model the energy transfer as:

GPE loss from upper reservoir	=	Work done on turbine	+	KE gained by water

Now we see that the KE gained by the water must be less than it would have been without the turbine, i.e. we capture the GPE by allowing the water to flow more slowly and not gain as much KE. Using this model, then, if the volume flow rate is V then, in time Δt:

GPE loss $= \rho Vgh\Delta t$ and KE gain $= \frac{1}{2}(\rho V)\left(\frac{V}{A}\right)^2 \Delta t = \frac{1}{2}\frac{\rho V^3}{A}\Delta t$

Note that, here the symbol A represents the cross-sectional area of the water tunnel at the turbine itself. So, the maximum output power possible (i.e. ignoring energy losses in the turbine, pipe losses etc.) is given by

$$P_{OUT} = \rho Vgh - \frac{1}{2}\frac{\rho V^3}{A}$$

And the maximum possible efficiency is found by dividing the power out by the power in,

i.e. $\mu_{max} = 1 - \frac{1}{2}\frac{V^2}{Agh}$.

The graph in Fig. D22 shows this output function for a turbine with a 50 m head of water and a turbine cross-sectional area of 1 m².

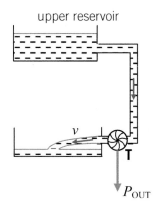

Fig. D21 Power from a head of water

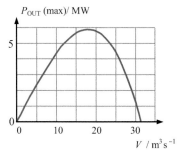

Fig. D22 Model output power for a hydroelectric turbine

D15

Self-test

A reservoir drains through a 20 cm diameter pipe to an outlet 20 m below. Calculate:

(a) the speed of the water through the pipe,

(b) the volume flow rate,

(c) the mass flow rate, and

(d) the rate of energy transfer.

[$\rho_w = 1000$ kg m⁻³]

D16

Self-test

Show that the equation for P_{OUT} is homogeneous.

Show that this model predicts that the maximum power output of a hydroelectric turbine occurs when

$$V = \sqrt{\frac{2gh}{3A}}.$$

Show than the efficiency of the energy transfer at this volume flow rate is independent of h, A (and g).

D17 Self-test

A tidal barrage has an area of 0.25 km^2. The mean height difference between the water levels inside and outside is 3 m.

(a) Estimate the energy available assuming a 50% conversion efficiency.

(b) Estimate the mean power output.

Fig. D19 is of a **pumped-storage** power station. Electricity itself cannot be stored and battery technology is not yet sufficiently developed to allow it to be used for grid backup,[3] so pumping water from the lower to the upper reservoir provides a way of using power input to the grid (e.g. from wind and nuclear power providers) at times of low demand.

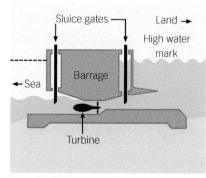

Fig. D23 Tidal barrage principle

Tidal barrages have recently been proposed as hydroelectric energy sources. A barrage is constructed with inbuilt turbines as shown in Fig. D23. As high tide approaches, the water level is higher outside the barrage than inside, so the sluice gates are opened allowing water to flow and its potential energy to be tapped.

When the water levels are equal the sluice gates are closed until the tide drops; they are then opened (as in Fig. D23) and electricity is again generated from the water flow out past the turbines. As with tidal stream power stations they have the advantage of predictability.

(f) Enriching uranium

Nearly all nuclear reactors in the world operate by the fission of U235. The principles of operation of nuclear fission power stations, including the need for moderator, control rods and the physics of the fission reactions themselves are not covered here. Natural uranium comprises 99.3% of U238 which absorbs neutrons but does not fission. Most reactor designs require about 3%–5% U235 so **enrichment** is required. Historically the separation was based on gaseous diffusion but the current technology uses gas centrifuges. This is a physical process based on the small mass difference between the two isotopes.

Fig. D24 US gas centrifuge facility

D18 Self-test

Calculate the percentage difference in the molecular masses of UF_6 containing the two different uranium isotopes.

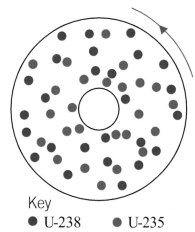

Key
● U-238   ● U-235

Fig. D25 Gas centrifuge principle

The chemically identical isotopes are reacted with fluorine to produce the gaseous compound uranium hexafluoride, UF_6. The mixture of gases is fed into a spinning tube, which is shown schematically in Fig. D25.

The heavier U238 molecules are preferentially spun to the outside and the slightly enriched mixture (i.e. with a higher percentage U235) extracted along the central tube and fed to the next tube in a cascade. The outside gas with less U235 is returned to the previous centrifuge. Each tube enriches the uranium by a factor of something like 1.3. The separation is this low because of the very small percentage difference in the masses of the molecules of the different isotopes.

D19 Self-test

The enrichment per stage in a gas centrifuge is a factor of 1.2.

(a) State the enrichment after 2 stages, 5 stages, n stages.

(b) Calculate the number of stages needed to achieve a U235 concentration of 5% from an initial feedstock concentration of 0.7%

Gas diffusion used the same starting point and was based upon rate of diffusion through a semi-permeable membrane being proportional to the square root of the molecular mass (*Graham's law of diffusion*).

3 There are suggestions that, as the number of battery-driven cars increases, these could eventually be used for grid backup purposes, with the advantage that a distributed energy store provides.

(g) Breeding nuclear fuel

Although U238 is not fissile, it does capture neutrons, giving rise to U239 which decays in two stages by β^- emission to give the fissile Pu239.

$$^{238}_{92}U + ^{1}_{0}n \rightarrow ^{239}_{92}U \xrightarrow[2.35 \text{ min}]{\beta^-} ^{239}_{93}Np \xrightarrow[2.3 \text{ day}]{\beta^-} ^{239}_{94}Pu$$

A smaller fraction of U238 undergoes multiple neutron capture before decaying twice to give rise to the non-fissile Pu240 and the fissile Pu241. Thus there is a gradual build-up of plutonium isotopes in fuel elements; it can be removed chemically at reprocessing. By the time the nuclear fuel element is removed for reprocessing, about half the fissile isotopes have undergone fission but half remain.

The fissile isotopes of plutonium can be incorporated into fuel rods together with U235 in so-called MOX (*mixed oxide*) fuel. Most European PWRs (*pressurised water reactors*) operate with some MOX fuel rods; typically 30%. It has been proposed to use plutonium from nuclear weapons decommissioning in the production of MOX fuel, thus removing the existence of this highly dangerous material.

(h) The nuclear fusion triple product

In order to obtain fusion using, for example, $^{2}_{1}H + ^{3}_{1}H \rightarrow ^{4}_{2}He + ^{1}_{0}n$ three conditions have to be satisfied:

1 A high enough temperature, T, to enable the reactant nuclei to approach close enough to overcome the Coulomb repulsion and allow fusion to occur; the strong nuclear interaction is short range so the nuclei have to approach within about 10^{-14} m; a typical temperature requirement is 100 MK (i.e. 10^8 K).

2 A high enough particle density, n, to allow a high enough collision rate between reactant nuclei; any individual collision of sufficient energy might not result in a fusion reaction, e.g. because the collision is slightly oblique.

3 A long enough **confinement time**, τ_E, which is a measure of how long the fuel maintains its internal energy; the hot fuel radiates away its energy by X-rays.

Each of these conditions needs to be achieved separately. The aim of fusion engineers is to maximise the product, $n T \tau_E$, which is called the *triple product*: the higher this product, the more likely it is for a sustainable nuclear fusion reaction to occur.

Example

It has been estimated that the minimum value of the triple product for the deuterium-tritium reaction is 3×10^{21} keV s m^{-3}, where the temperature is expressed in keV. If a fusion reactor can attain a value of $n\tau_E$ of 3×10^{20} m^{-3} s, calculate the kelvin temperature required.

Answer

$n T \tau_E = 3 \times 10^{20}$ m^{-3} s $\times T = 3 \times 10^{21}$ keV m^{-3} s

$\therefore T = \dfrac{3 \times 10^{21}}{3 \times 10^{20}}$ keV $= 10$ keV $= \dfrac{10 \text{ keV} \times 1.6 \times 10^{-16} \text{ J keV}^{-1}}{1.38 \times 10^{-23} \text{ J K}^{-1}} = 96$ MK

Terms & definitions

A **fissile** nuclide is one that undergoes fission when it absorbs **thermal neutrons**, with the emission of enough neutrons to maintain a chain reaction.

Thermal neutrons are neutrons with a kinetic energy corresponding to temperatures of the order of 300 K.

Study point

The presence of Pu239 and Pu241 in spent nuclear fuel rods is of significance for nuclear non-proliferation because it is the fuel of choice in nuclear fission bombs and also in the primary stage of nuclear fusion bombs (*hydrogen bombs*).

Terms & definitions

In nuclear fusion, the **confinement time**, τ_E, is defined by:

$$\tau_E = \frac{W}{P_{loss}}$$

where W is the energy density and P the power loss per unit volume (e.g. by conduction and radiation) from the fusion fuel.

Self-test D20

Scientists working on nuclear fusion often express temperature in the energy units of keV, by using the Boltzmann constant, e.g. in these units room temperature (~300 K) is $\dfrac{300 \text{ K} \times 1.38 \times 10^{-23} \text{ J K}^{-1}}{1.6 \times 10^{-16} \text{ J keV}^{-1}}$.

(a) Evaluate 300 K in keV.

(b) The temperature of the Sun's core is 15 MK. Express this in keV.

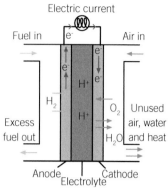

Fig. D26 Principle of a proton-exchange fuel cell

D21 Self-test

A fuel cell produces 5 kW of electrical output. Assuming a 40% efficiency, calculate the mass of hydrogen it consumes per hour.

Take the heat of combustion of hydrogen to be 290 kJ mol⁻¹.

D.3 Fuel cells

The fuel cell is a battery with continuously replenished reactants. In the fuel cells which are proposed for transport use (cars, buses, goods vehicles) the fuel is hydrogen with the oxidant being oxygen. The principle of operation is:

- Hydrogen is drawn in to one side of the fuel cell, the anode.

- The hydrogen is oxidised by a catalyst (the *anode catalyst*, usually powdered platinum).

- The hydrogen ions (i.e. protons, H^+) diffuse into the electrolyte through a non-conducting barrier, leaving the negatively charged electrons.

- The electrons travel along by electric wires into an external circuit, where they do work before entering the cathode.

- At the cathode, oxygen combines, under the influence of a catalyst (the *cathode catalyst*, often nickel or a nano-material) with the electrons and diffused protons from the anode to produce water, which is the waste product.

The half-reactions at the electrodes are:

Anode: $\quad H_2 \rightarrow 2H^+ + 2e^-$

Cathode: $\quad \frac{1}{2}O_2 + 2H^+ + 2e^- \rightarrow H_2O$

So the overall reaction is the familiar $2H_2 + O_2 \rightarrow 2H_2O$, but because the energy extraction is not via heat, the efficiency of the fuel cell is not subject to the same restriction. Theoretical maximum efficiency is of the order of 80% and practical fuel cells regularly achieve 40% transfer to electrical energy.

The major advantage of the fuel cell is its lack of greenhouse gas emissions but this is only possible if the hydrogen can be produced without CO_2 emission, which requires electricity generated by non-fossil-fuel-based technologies (PV cells, wind, hydro…). Research cars using fuel cells currently obtain their hydrogen from hydrocarbons and so lead to carbon dioxide waste.

D.4 The conduction of heat

According to a 2008 DEC[4] report, the annual UK energy use is 6.5 EJ. The domestic use accounts for 30% of this figure of which half is for space heating. We need to heat our homes because the indoor temperature is higher than the outdoor temperature for most of the year. It is this temperature difference that drives heat loss, mainly by gas movement (convection) and conduction through the walls / windows. Indoor temperatures have risen markedly in the last 40 years, as Table D1 shows. This tends to increase the rate of heat loss and therefore requirement for heating, with the consequent greater use of fossil fuels and increased greenhouse emissions. However, insulation standards and the general leakiness of houses have improved markedly over the same period, with the consequence that the domestic energy demand has only increased marginally.

D22 Self-test

Express the annual domestic space heating energy use in standard form.

Location	Winter temperature / °C	
	1979	2008
Living room	18.3	21.3
Hall	15.8	21.1
Bedroom	15.2	21.0

Table D1 Mean British winter room temperatures

4 Department of Energy and Climate Change.

(a) The conduction equation

The conduction of heat through good conductors can be investigated using the set-up which is shown schematically in Fig. D27 and in a practical version in Fig. D28.

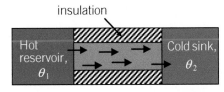

Fig. D27 Measuring conductivity

Fig. D28 Searle's apparatus

A constant temperature gradient is established between the ends of a specimen of material and the flow of heat measured. One way of maintaining the temperatures θ_1 and θ_2 constant is to use a jet of steam in the hot reservoir and a stream of cold water in the sink. The heat flow is measured by the small temperature rise in the water flow in the sink.

The results of such experiments are the conduction equation:

$$\frac{\Delta Q}{\Delta t} = -AK\frac{\Delta \theta}{\Delta x} \text{ (see Study point)}$$

where A and x are the cross-sectional area and distance in the direction of heat flow. The constant K is the coefficient of thermal conductivity and is characteristic of the material and has the unit $\text{W m}^{-1}\text{ K}^{-1}$. The values of K in this unit range from 430 (silver) and 400 (copper) through glass (1–2) and brick (0.1–0.7) to rock wool (0.05) and air (0.025). The range is not nearly as great as the range of electrical conductivities.

In order to measure the thermal conductivities of the insulators, a different experimental arrangement is needed. The specimen is usually a thin disc (large value of A and small value of Δx) rather than a long bar.

(b) Insulation and heat loss

Consider the house wall in Fig. D29. It was built in the days before cavity wall insulation. Using the symbols on Fig. D29, the rate of heat <u>loss</u> through an area, A, of wall is given by:

$$\frac{\Delta Q}{\Delta t} = -AK_b\frac{\theta_1 - \theta_2}{d_b}$$

where K_b is the thermal conductivity of the brickwork.

What is the effect of putting on a layer of insulation, such as expanded polystyrene on the interior wall (Fig. D30)? The temperature between the insulation and the bricks must now be in between θ_1 and θ_2 so $\Delta\theta$ for the wall is now less, so the rate of heat loss is less.

We can calculate the temperature at the join by noticing that the heat flow through the bricks and insulation must be the same.

Hence: $$AK_b\frac{\theta_1 - \theta_2}{d_b} = AK_i\frac{\theta_1 - \theta}{d_b}$$

We can solve this equation for θ and then use the heat flow equation for either the bricks or the insulation (it doesn't matter which).

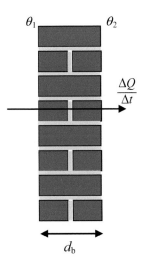

Fig. D29 Uninsulated wall

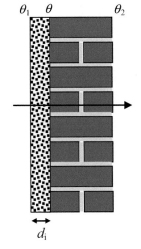

Fig. D30 Insulated wall

> **▼ Study point**
>
> Note that the use of ΔQ rather than Q for heat flow is not consistent with the first law of thermodynamics equation. However, in this context it is conventional.

> **Self-test** **D23**
>
> Show that $\text{W m}^{-1}\text{ K}^{-1}$ is the basic SI unit of the coefficient of thermal conductivity.

> **Self-test** **D24**
>
> Use the conduction equation to explain the use of a thin disc when determining K for poor conductors.

> **▼ Study point**
>
> Don't learn the final equation for multi-layer calculations. Just apply the basic conduction equation to the two layers as in the main text.

▼ Study point

Notice that in finding θ, the unit of thickness was left as mm. This does not matter as this unit cancels on the two sides. In calculating the power loss, m² had to be used to fit in with the unit of K.

Example

A 110 mm thick brick wall ($K_b = 0.5$ W m⁻¹ K⁻¹) has a 50 mm layer of expanded polystyrene ($K_b = 0.033$ W m⁻¹ K⁻¹) added as insulation. Calculate the heat loss per unit area for the insulated wall when the temperatures of the inside and outside surfaces are 24 °C and 0 °C respectively.

Answer

First find the temperature, θ, of the insulation-wall boundary:

Equating heat loss through each layer: $A \times 0.5\dfrac{\theta - 5}{110} = A \times 0.033\dfrac{24 - \theta}{50}$

Solve this to give $\theta = 7.41$ °C – this is left as an exercise.

∴ heat loss through brick $= 0.5$ W m⁻¹ K⁻¹ $\dfrac{(7.41 - 5.00)\ \text{K}}{110 \times 10^{-3}\ \text{m}} = 11$ W m⁻²

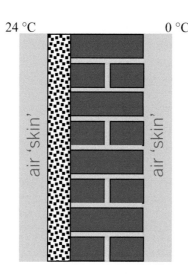

24 °C   0 °C

Fig. D31 The insulating effect of air

The real heat loss is likely to be much lower than the results of the calculation suggest. The actual temperature difference across the wall will be much lower than the 24 °C in the example because there will be a 'skin' of stationary air in contact with both the internal and external surfaces, providing two extra layers, which are good insulators. This is illustrated in Fig. D31. Most of the total temperature drop occurs across these air skins. The inside surface of the wall will be several degrees cooler than the room temperature; the outside surface will be warmer than the air temperature outside. On windy days, this effect is reduced. This is taken into account in the determination of U values (see below).

(c) U values

Building regulations for insulation purposes are framed in terms of the **U values** of building elements, such as walls, doors and windows, rather than the thermal conductivities of the materials themselves. Because of the problems of weather conditions, U values are specified under standardised conditions: usually 24 °C (297 K) temperature difference, under wind-free conditions at 50% humidity.

The U value of a structure is defined by: $\dfrac{\Delta Q}{\Delta t} = UA\Delta\theta$

Notice that we have dropped the minus sign (the direction of the heat transfer is by inspection).

A related quantity is *thermal resistance*, R, which is defined as U^{-1}. The useful aspect of thermal resistance is that the thermal resistance of a layered building element (such as the wall with insulation in the previous example) is the sum of the individual thermal resistances (as with electrical resistance).

── Terms & definitions ──

The U **value** of a building element (e.g. a wall) is the rate of heat transfer per unit area per unit temperature difference.

UNIT: W m⁻² K⁻¹

Working with **U values**: $\frac{1}{U} = \frac{1}{U_1} + \frac{1}{U_2} + \dots$ (as with conductances)

Example

In the brick wall and insulation example, the U values are

$U_b = 4.5$ W m^{-2} K^{-1}; $U_i = 0.66$ W m^{-2} K^{-1}. Calculate the rate of heat transfer.

Answer

$\frac{1}{U} = \frac{1}{U_1} + \frac{1}{U_2} = \frac{1}{4.5} + \frac{1}{0.66} = 1.74.$ $\therefore U = 0.58$ W m^{-2} K^{-1}.

$\therefore$ Per unit area $\frac{\Delta Q}{\Delta t} = U\Delta\theta = 0.58 \times (24 - 5) = 11$ W m^{-2}.

Exercise 4D

1 A student measures the wavelength of the peak of the solar spectrum to be 493 nm. Use Wien's constant [2.90×10^{-3} m K] to calculate the temperature of the Sun's photosphere.

2 A website gives the diameter and temperature of the photosphere of the Sun as 1.392 million km and 5778 K respectively. The mean Sun–Earth distance is 1.496×10^8 million km. Use these data to calculate: (a) the power output of the Sun, and (b) the Solar constant.

3 A student drops a piece of ice into a measuring cylinder containing 200 cm³ of water. The water level rises to a reading of 240 cm³.

   (a) State the volume of ice which is submerged.

   (b) Calculate the total volume of the ice.

   (c) State and explain what happens to the water level in the measuring cylinder when the ice melts.

   (d) Briefly state the relevance of this experiment to our understanding of the effect of melting ice on rising sea levels.

$$[\rho_{ice} = 917 \text{ kg m}^{-3}; \rho_{water} = 1000 \text{ kg m}^{-3}]$$

4 The dwarf planet, Ceres, has a mean orbital radius of 2.77 AU.

   (a) Calculate the ratio of the intensity of sunlight at Ceres to that at the orbit of the Earth.

   (b) Estimate the mean temperature of the surface of Ceres given that it possesses no atmosphere.

5 A tidal flow turbine of diameter 5 m is placed in Ramsey Sound where the peak tidal current in 3 m s⁻¹. Estimate the mean power energy available given a power coefficient of 0.4.

6 The low power turbine at Cwm Rheidol power station uses a water head of 7.5 m. The flow speed is approximately 6 m s⁻¹ through a turbine with a cross-sectional area of 4 m². Estimate:

   (a) the mass flow rate

   (b) the loss in gravitational potential energy per second

   (c) the gain in kinetic energy per second of the water

   (d) the power generated, assuming a turbine and generator combined efficiency of 80%.

   (e) the overall efficiency achieved.

7 Thorium-232 has been successfully used in a breeder reactor[5]. It is present in the Earth's crust in much larger amounts than any uranium isotope. Its half-life is 1.41×10^{10} years. Not itself fissile, it absorbs thermal neutrons, with an energy of around 0.025 eV, giving an excited nuclide which decays by γ emission followed by two β^- decays to a fissile isotope of uranium.

   (a) Use the Boltzmann constant to justify the figure of 0.025 eV for thermal neutrons.

   (b) Write reactions for the three decays mentioned above.

   (c) Explain why a thorium-232 reactor requires the presence of a fissile nuclide such as U235.

   (d) Why is Th232 more prevalent that U235 or U238 in the Earth's crust?

Exercise 4D

8   The nuclear material in a nuclear reactor consists initially of 5% of U235 and 95% of the non-fissile U238. Up to 30% of the energy produced in a reactor comes from reactions other than the fission of U235. Briefly explain this.

9   A 2 m × 1 m, singled-glazed window is made of glass. The coefficient of thermal conductivity of the glass out of which a window is made in 1.0 W m^{-1} K^{-1}. The thickness of the glass is 0.6 mm.

(a) Calculate the rate of heat flow through the window if a temperature difference of 20 °C is maintained between its faces.

(b) The U-value of the window is 4.8 W K^{-1}. Use this value to calculate the rate of heat loss on a calm day from a room with this window when the inside and outside temperatures are 25 °C and 5 °C respectively.

(c) Explain the (huge) difference in the rates of heat loss in the answers to parts (a) and (b).

(d) Calculate the actual temperature difference between the inside and outside faces of the glass window pane in part (b).

(e) A double-glazed unit is made from two 4 mm sheets of this glass separated by a (12 mm) layer of air. The U-value of the unit is 2.8 W m^{-2} K^{-1}. Assuming the U-value of the glass panes is inversely proportional to their thickness, estimate the U-value of the air layer.

Mathematical and data-handling skills

The content of Units 3 and 4 of the A level specification requires more advanced analytical skills than that of Units 1 and 2. There are two main areas of increased mathematical demand, which are developed in this chapter to be applied in the relevant sections of Units 3 and 4:

- The study of rotations and oscillations in Unit 3 and the optional topic of AC electricity requires a deeper understanding of trigonometrical functions.
- Exponential and logarithm functions are essential to the description of growth and decay, such as in the study of damped oscillations, radioactive emissions and the absorption of γ radiation. These two functions are also used in analysing data which involves power law and exponential decay.

Content

M1 **Trig functions**

M2 **Ellipses**

M3 **Rate of change**

M4 **The exponential function**

M5 **Logarithms**

M6 **Using logs to linearise functions**

Exam tip

As with the Year 1 book, this chapter is not a substitute for a maths textbook. For practice in maths techniques you should use a maths textbook, e.g. *Maths for A Level Physics* (Kelly & Wood).

Most of the mathematical skills and data-handling needed for A Level physics are the same as those for AS and are covered in the Year 1 & AS textbook. That book was written with A Level as well as AS students in mind, so the demands of the exercises and examination-style questions are at an appropriate level. However, there are some additional skills specified for A Level and some of the contexts in the non-AS section of the course require a deeper understanding of the skills that were covered at AS.

Most of the analytical work on orbits in Sections 4.2 and 4.3 is in terms of circles, but some understanding of ellipses is required. Section M2 gives a reference for a few terms.

M1 Trig functions

The trigonometric sine and cosine are used to model the behaviour of oscillating systems. It is important to know how to sketch and interpret them as well as to understand the relationships between them.

M1.1 The form of sine and cosine functions

Simple harmonic oscillating motions and alternating currents (and voltages) can be expressed in the form:

$$f(t) = A \cos(\omega t + \varepsilon) \quad \text{and} \quad f(t) = A \sin(\omega t + \phi).$$

Here $f(t)$ is the displacement, velocity, acceleration, current or pd expressed as a function of time, A is the amplitude of the oscillating quantity, ω the **angular frequency** and ε (and ϕ) are constant phase angles.

We recall from Section 4.4.4 of the Year 1 & AS textbook, that the extreme values of the sine and cosine functions are ±1 and the function repeats every 2π radians. Fig. M.1 shows these functions with ε and ϕ both zero.

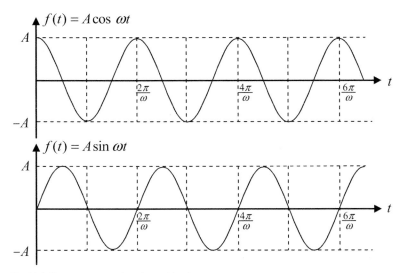

Fig. M.1 Sine and cosine functions with phase constant zero

What if the phase constant, ε in $f(t) = A \cos(\omega t + \varepsilon)$ is not zero? If $\varepsilon > 0$, the function $f(t) = A \cos(\omega t + \varepsilon)$ is the same as $f(t) = A \cos \omega t$ but shifted earlier in time by $\Delta t = \varepsilon/\omega$ (see Study point). A negative phase constant shifts the graph to the right. Alternative expressions for Δt are:

$$\Delta t = \frac{\varepsilon}{\omega} = \frac{\varepsilon}{2\pi f} = \frac{\varepsilon T}{2\pi}$$

Similarly, $f(t) = A \sin(\omega t + \phi)$ is the same as $f(t) = A \sin \omega t$, just shifted earlier in time by $\Delta t = \phi/\omega$.

CALCULATOR TIP

For oscillations calculations, your calculator should be set in radian mode.

---Terms & definitions---

The **angular frequency** or **pulsatance**, ω, of an oscillation is $2\pi f$ where f is the frequency.

UNIT: s^{-1} or rad s^{-1}.

Self-test M1

Use Fig. M.1 to give three values of t for which:

(a) $A \cos \omega t = A$

(b) $A \cos \omega t = 0$

(c) $A \sin \omega t = A$

Self-test M2

The displacement, x, of an oscillating object is given by $x = A \cos \omega t$, where A is 12 cm and ω is 31.4 rad s^{-1}.

(a) State the amplitude of the oscillation.

(b) Calculate the frequency of the oscillation.

(c) Calculate the first two times after $t = 0$, for which $x = 0$.

▼ Study point

$A \cos(\omega t + \varepsilon)$ is shifted **earlier** than $A \cos \omega t$ (if ε is positive) because a smaller value of t is now needed to give the same value of the argument of the cosine.

Self-test M3

An alternating current, I, with I in mA, and t in s, varies as $I = 0.1 \sin(1000t + 0.5)$

(a) State the pulsatance.

(b) Calculate the frequency, f.

(c) Calculate the period, T.

(d) Calculate the times between $t = -T$ and $t = T$ for which $I = 0$.

M1.2 The relationship between the sine and cosine functions

(a) Phase

The shape of the cosine function is the same as that of the sine function; it is just shifted earlier in phase by $\frac{\pi}{2}$ (i.e. ¼ of a cycle), that is $T/4$ earlier in time. So we can write:

$$\cos \omega t = \sin\left(\omega t + \tfrac{\pi}{2}\right) \qquad \text{or, equivalently,} \qquad \sin \omega t = \cos\left(\omega t - \tfrac{\pi}{2}\right)$$

This means that we can always write any sine function as a cosine function instead, with the appropriate phase angle.

We can go further than this. If we take a sine or cosine function and shift it along in either direction by $T/2$ in time, i.e. by π radians, we get the negative of the same function. That means

$$\cos\left(\omega t \pm \pi\right) = -\cos \omega t \qquad \text{and} \qquad \sin\left(\omega t \pm \pi\right) = -\sin \omega t$$

This gives us further flexibility in writing sines or cosines.

(b) Gradients

By applying calculus techniques, which are beyond the scope of this book, we can show that the gradient of a sine curve is a cosine curve and vice versa. More precisely:

$$\text{gradient of } \sin t = \cos t \qquad \text{and} \qquad \text{gradient of } \cos t = -\sin t$$

Multiplying t by ω in $\sin \omega t$ and $\cos \omega t$ squashes the oscillations horizontally by the factor ω so increases the gradient by the same factor. Multiplying by A, as in $A \sin \omega t$ and $A\cos \omega t$, stretches the oscillations vertically by the factor A, so also increases the gradient by the same factor. Hence:

$$\text{gradient of } A \sin \omega t = A\omega \cos \omega t$$

and

$$\text{gradient of } A \cos \omega t = -A\omega \sin \omega t$$

In studying simple harmonic motion we usually write the displacement, x, as:

$$x = A \cos \omega t, \qquad \text{or, with phase constant} \qquad x = A \cos\left(\omega t + \varepsilon\right)$$

The velocity is the gradient of the displacement–time graph, so

$$v = -A\omega \sin \omega t \quad \text{or} \qquad\qquad v = -A\omega \sin\left(\omega t + \varepsilon\right)$$

And acceleration is the gradient of the velocity–time graph, so

$$a = -A\omega^2 \cos \omega t \text{ or} \qquad\qquad a = -A\omega^2 \cos\left(\omega t + \varepsilon\right)$$

Example

The displacement (in **cm**) of a body of mass 0.002 kg is given by

$$x = 2.5 \cos\left(600t + 0.8\right),$$

Find the maximum magnitude of the resultant force on the body and the first time (> 0) when this occurs.

M4 Self-test

Write the function for the alternating current in Self-test M.3 as a cosine function.

M5 Self-test

Two students come up with the following answers for an oscillation:

$x = 10 \cos\left(100t - 1.0\right)$

$x = 10 \sin\left(100t + 0.57\right)$

Can they both be right?

Stretch & Challenge

Estimate the gradient of $\sin t$, at $t = 0$ by considering the interval -0.1 rad to $+0.1$ rad. Repeat this for smaller and smaller intervals, e.g.

-0.01 rad to $+0.01$ rad;

-0.001 rad to $+0.001$ rad…

How does this support the suggestion that the gradient of $\sin \theta$ is $\cos \theta$. Try repeating for other central values, e.g. $\theta = 0.5$ rad.

▼ **Study point**

If $\quad x = A\cos \omega t \quad$ and
$\quad a = -A\omega^2\cos \omega t$
then $\;a = -\omega^2 x$.

So the acceleration is directly proportional to the displacement and in the opposite direction. This kind of motion (shm) arises when an object is displaced slightly from an equilibrium position.

Answer

From above, the acceleration $a = -2.5 \times 600^2 \cos(600t + 0.8)$ cm s^{-2}

$\therefore a_{max} = 2.5 \times 600^2 = 900\,000$ cm s^{-2} = 9000 m s^{-2}

$\therefore$ By N2: $F_{max} = m\,a_{max} = 0.002$ kg $\times 9000$ m s^{-2} = 18 N

This occurs when $\cos(600t + 0.8) = \pm1$, i.e. when $600t + 0.8 = n\pi$, where $n = 0, \pm1$ (see Study point).

$\therefore t = \dfrac{n\pi - 0.8}{600}$, so the smallest $t > 0$ is when $n = 1$,

i.e. $t = \dfrac{\pi - 0.8}{600}$ s = 0.39 ms.

> ▼ **Study point**
>
> It helps to be familiar with the values of θ for which trig functions are 0 and ±1. For example,
>
> $\cos\theta = 1$ when θ is ... $-2\pi, 0, 2\pi, 4\pi$...
>
> $\cos\theta = -1$ when θ is ... $-3\pi, \pi, 3\pi, 5\pi$...
>
> $\cos\theta = 0$ when θ is ... $-\frac{\pi}{2}, \frac{\pi}{2}, \frac{3\pi}{2}, \frac{5\pi}{2}$...
>
> So $\cos\theta = \pm1$ when $\theta = n\pi$

(c) Pythagoras' theorem in terms of trig functions

Consider the angle θ in the right-angled triangle in Fig.M.2

Pythagoras' theorem: $\qquad\qquad a^2 + b^2 = c^2$

Divide by c^2: $\qquad\qquad\qquad \dfrac{a^2}{c^2} + \dfrac{b^2}{c^2} = 1$

But $\dfrac{a}{c} = \sin\theta$ and $\dfrac{b}{c} = \cos\theta$, so $\quad \sin^2\theta + \cos^2\theta = 1$

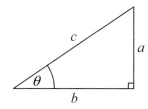

Fig. M.2 Right-angled triangle

A common application for this equation in simple harmonic motion is the relationship between velocity and displacement:

Ignoring the phase constant: $x = A\cos\omega t$ and $v = -A\omega\sin\omega t$

$\sin\omega t = \pm\sqrt{1 - \cos^2\omega t} \quad \therefore v = \pm A\omega\sqrt{1 - \dfrac{x^2}{A^2}}$

which can be simplified to $v = \pm\omega\sqrt{A^2 - x^2}$

M1.3 Graphs of $\sin^2\theta$ and $\cos^2\theta$

A body undergoing simple harmonic motion has a velocity which varies with time as a sinusoid (i.e. a sine or cosine function). Its kinetic energy therefore varies as the square of this. The potential energy is also the square of a sinusoid (it is the area under the F–x graph which is $\frac{1}{2}kx^2$). The graphs of $\sin^2\theta$ and $\cos^2\theta$ are just sinusoids themselves. The graphs in Fig. M.3 show how $\cos^2\theta$ is related to $\cos\theta$.

> ▼ **Study point**
>
> Notice that the period of the $\cos^2\theta$ graph is half that of the $\cos\theta$ graph; $\cos^2\theta$ is never less than 0 (why not?). The mean value of $\cos^2\theta$ is 0.5.

> **Self-test**   **M6**
>
> Sketch a graph of $y = 2\sin\pi t$ for t between 0 and 4 seconds. Sketch $y = 4\sin^2\pi t$.
>
> Arrange the graphs like Fig. M.3 to show the connection between them.

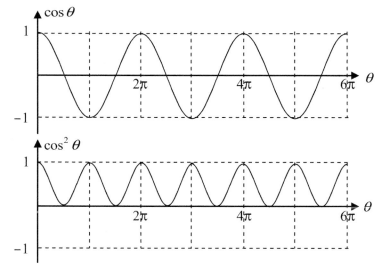

Fig. M.3 The relationship between $\cos\theta$ and $\cos^2\theta$

M2 Ellipses

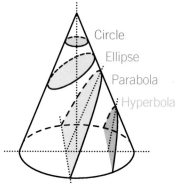

Fig. M.4 Conic sections

Ellipses belong to a family of curves called *conic sections* because they can all be produced by cutting circular cones in different ways – see Fig. M.4. They are important because orbits around a single object are always conic sections:

- Kepler established by observation that planetary orbits are elliptical with the Sun at one of the **foci**. The orbits of comets and planetary moons are also ellipses.

- Newton derived elliptical orbits mathematically using his laws of motion and showed that higher energy paths can be hyperbolic ($E > 0$) or parabolic ($E = 0$).

One way of defining an ellipse is given in **Terms & definitions**: this is clarified in Fig. M.5.

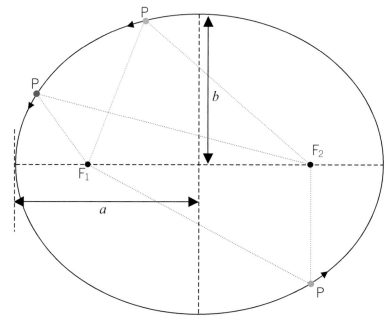

Fig. M.5 Constructing an ellipse

┌─ **Terms & definitions** ─┐

An **ellipse** is the path of a point which moves so that the sum of its distances from two fixed points, called the **foci**, is constant.

Note: 'Foci' is the plural of focus.

The **semi-major axis**, a, of an ellipse is half the longest diameter.

The **semi-minor axis**, b, of an ellipse is half the shortest diameter.

The **eccentricity**, e, of an ellipse can be calculated using the equation:

$$e = \frac{\sqrt{a^2 - b^2}}{a}$$

The point **P** moves in such a way that the sum of its distances from the two **foci**, (PF_1 + PF_2) and F_2 is a constant. Thus **P** traces out an **ellipse**: the total lengths of the green, red and blue dotted lines are all the same.

The **eccentricity** of an ellipse is a measure of how 'squashed' it appears. A circle has $e = 0$. By taking measurements on Fig. M.5, you should be able to use the definition to show that the eccentricity of the ellipse is 0.6. The maximum possible eccentricity for an ellipse is 0.99..... As far as this A level physics specification is concerned, all you need to know about elliptical orbits is that the period and energy are independent of the eccentricity: they depend only upon the masses of the two objects and the semi-major axis. Proving this takes a few lines of algebra [see Stretch & Challenge in Section 4.3.2].

M3 The exponential function

This function is often referred to as the *growth function*. There are many physical situations in which a variable gets larger or smaller (usually smaller) at a rate which is proportional to the value of that variable.

Examples

- Radioactive decay: the activity of a sample of a r/a nuclide (i.e. the number of decays per second) is proportional to the number, N, of undecayed nuclei in the sample. The activity is the rate of *decrease* of N, so

$$\frac{dN}{dt} = -\lambda N, \text{ where } \lambda \text{ is the } \textit{decay} \text{ constant.}$$

- Capacitor discharge: the rate of *decrease* of charge, Q, on a capacitor is proportional to the charge on the capacitor,

$$\frac{dQ}{dt} = -\frac{Q}{RC} \text{ where } C \text{ is the capacitance.}$$

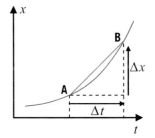

Fig. M.6 Rate of change

M3.1 Rate of change

In Section M1, we wrote 'the gradient of' several times and we now consider how to write 'the gradient of' mathematically.

The mean rate of change of the variable x between points **A** and **B** in Fig. M.4 is given by the gradient of the **chord AB**.

Hence: Mean rate of change between **A** and **B** is $\frac{\Delta x}{\Delta t}$.

The *instantaneous* rate of change of x at any point, e.g. **A**, is the gradient of the tangent at that point and is written $\frac{dx}{dt}$. Usually (as in Fig. M.6) this rate of change varies with time, i.e. it is itself a function of time. Note that x can be any variable. It could be displacement, current, velocity, activity of a radioactive material..... So in these cases the rate of change would be written $\frac{dx}{dt}$, $\frac{dI}{dt}$, $\frac{dv}{dt}$ and $\frac{dA}{dt}$ respectively.

so, if $x = A \cos \omega t$, $v = \frac{dx}{dt} = -A\omega \sin \omega t$, etc.

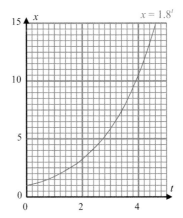

Fig. M.7 A power law

M3.2 Investigating a power law

The graph in Fig. M.7 is of a power law: $x = 1.8^t$. The value 1.8 is quite arbitrary – we could choose any positive number greater than 1.

The actual results are given at the end of the section but you should have found this:

$$\frac{x(4)}{x(2)} = 3.24 \quad \text{and} \quad \frac{\frac{dx}{dt}(4)}{\frac{dx}{dt}(2)} = 3.24$$

In other words the rate of increase of x is proportional to x. You can try this with any values of x. You should find from your results that $\frac{dx}{dt} \approx 0.6x$.

It also works for any base value, e.g. $x = 5^t$ will give $\frac{dx}{dt} = 1.6x$ (2 s.f.)

M8 Self-test

(a) Use your calculator to find e^{-1}, $e^{-1.5}$, e^{-2}, e^{-3}, $e^{-4.605}$, e^{-5}, e^{-10}, e^{-100}

(b) What is the relationship between e^5 and e^{-5}?

(c) What is the relationship between e^{-10} and e^{-100}?

▼ **Study point**

The graph in Fig. M.6 is of $x = Ae^{+\lambda t}$. It passes through the point $(0, A)$, **the blue ring**, and tends to 0 as $t \to -\infty$.

When $t = \frac{1}{\lambda}$, $x = Ae \sim 3A$

M9 Self-test

The number of bacteria in a colony increases from 100 to 10,000 in 4 hours. Assuming exponential growth, calculate:

(a) the number of bacteria in 8 hours

(b) [slightly tricky] the number of bacteria in 10 hours

(c) how long it would take for the number to reach 10^{10}.

M10 Self-test

(a) Use the value of x after time $3/\lambda$ in Fig. M.8 to estimate the value of x after a time of $6/\lambda$, $12/\lambda$ and $60/\lambda$

(b) Check your answers by using the formula $x = Ae^{+\lambda t}$.

M3.3 Euler's constant, e

There is one base value, which we call e for which the constant of proportionality is exactly 1. The value of e is 2.718 281 828 46 to 12 significant figures – like π it is an irrational number.

So if $\quad x = e^t, \quad \dfrac{dx}{dt} = e^t, \quad$ i.e. $\quad \dfrac{dx}{dt} = x$

The rate of growth is equal to the value of the function.

And if $\quad x = Ae^{kt} \quad \dfrac{dx}{dt} = Ake^{kt} = kx$

And, because most of our needs for this function involve decays,

if $\quad x = Ae^{-kt} \quad \dfrac{dx}{dt} = -Ake^{-kt} = -kx$

So if we have a physical situation in which we know the rate of change of a quantity is proportional to its value, we can find the function which the variable obeys.

Example

The discharge current, I, from a capacitor is given by $I = \dfrac{Q}{RC}$. Express Q as a function of time.

Answer

The discharge current is the rate of decrease of charge, i.e. $I = -\dfrac{dQ}{dT}$

$\therefore \dfrac{dQ}{dT} = -\dfrac{1}{RC}Q$, so from above $Q(t) = Ae^{-\frac{1}{RC}t}$.

M3.4 Graphs of exponential functions

(a) Exponential growth

Functions of the form $x = Ae^{+\lambda t}$ with λ a positive constant, all have the same shape as the graph in Fig. M.8.

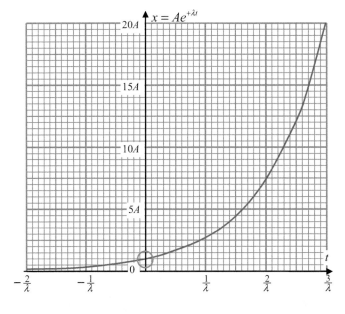

Fig. M.8 Exponential growth function $x = Ae^{\lambda t}$, for $\lambda > 0$

If you look carefully at the graph of $x = Ae^{\lambda t}$ you will notice that the time taken for the function to increase by a given factor is constant. Looking at the graph, the time to double from A to $2A$, from $5A$ to $10A$ and $10A$ to $20A$ are all a shade under $\frac{0.7}{\lambda}$ (actually 0.693 to 3 s.f.). Use this idea to answer Self-test M9 and M10.

(b) Exponential decay

This is much more common in A level physics problems than exponential growth. Functions of the form $x = Ae^{-\lambda t}$ all have the same form as the graph in Fig. M.9.

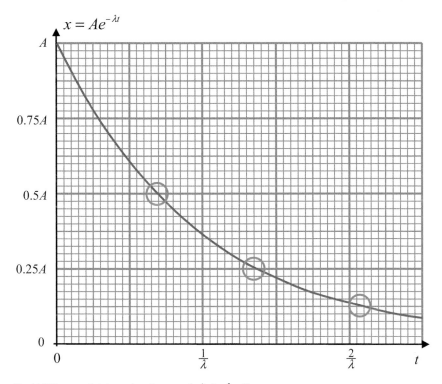

Fig. M.9 Exponential decay function $x = Ae^{-\lambda t}$, for $\lambda > 0$

The characteristic feature of an exponential decay is that the time taken for x to fall by any given factor is always the same. For the decay of radioactive nuclides, *half-life* is given. From Fig. M.9, this is approximately equal to $\frac{0.7}{\lambda}$ (as the doubling time for exponential growth). Again from the graph, the activity drops to 0.37 of its original value in a time of $1/\lambda$. To perform such calculations without the aid of a graph, we need the help of logarithms.

Self-test M11

Use the points in the blue circles in Fig. M.9 to show that the time for the exponential decay function to halve is constant. Confirm this by taking further readings at, e.g. $x = 0.6A$, $0.3A$ and $0.15A$.

Self-test M12

The activity of a radioactive nuclide drops to 20% of its original value in 32 days. Use Fig. M.7 to determine:

(a) the half life (in days) and

(b) the decay constant (in day^{-1}).

Self-test M13

A capacitor of capacitance C, discharges through a resistor of value $10\,\text{k}\Omega$. It takes 25 s for the pd across the capacitor to drop from 10 V to 3 V.

The equation for the discharge is

$$V = V_0 e^{-\frac{1}{RC}t}.$$

Use Fig. M.7 to determine:

(a) the decay constant, λ and

(b) C.

Answer to Stretch & Challenge in Section M3.2

$x = 1.8^t$ can be written $x = e^{t\ln 1.8}$, so $\dfrac{dx}{dt} = (\ln 1.8)e^{t\ln 1.8} = x\ln 1.8$

$\therefore$ Ratio of $\dfrac{dx}{dt}$ = ratio of x because the ln 1.8 values cancel.

Note that $\ln 1.8 = 0.59$ (2 s.f.)

M4 Logarithms

▼ **Study point**

We normally write ln (x) without the brackets, i.e. ln x, unless there would be confusion.

We write the log of $a+b$ as ln $(a+b)$ but ln $2a$ would be taken to mean ln $(2a)$ just as sin $2x$ means sin $(2x)$.

M4.1 Definition of the 'natural' logarithm function

Just as $\sin^{-1}$ is the inverse function to the sine function, meaning that $\sin^{-1}(\sin\theta) = \theta$ or $\sin(\sin^{-1}\theta) = \theta$, the *natural logarithm* function, ln, (or *logarithm to the base e*), is the inverse function to the exponential function, e^x : i.e.

$$\ln(e^x) = x \qquad \text{and} \qquad e^{\ln x} = x.$$

Remember these defining equations. Because it is the inverse function of e^x it is often on the same calculator key, with either of the functions needing the inverse key. Finding the logarithm of a number is often referred to as, 'taking logs'.

M14

Self-test

Evaluate:

(a) ln 2,

(b) ln e^2, i.e. ln (e^2),

(c) $e^{\ln 5}$,

(d) 2 ln 10,

(e) $3e^{2\ln 12}$.

You are (for the moment) allowed to use your calculator for (a) and (e). If you must, use your calculators for the others then try and explain the answer!

Example

From Self-test M8 we know that $e^{4.605} = 100$. Write this in terms of the logarithm function.

Answer

'Taking logs' of both sides of the equation:

$$\ln(e^{4.605}) = \ln 100$$

But from the definition of the logarithm function, $\ln(e^{4.605}) = 4.605$

$$\therefore \ln 100 = 4.605$$

▼ **Study point**

Check all these examples. They do work. Make sure you know how the author was sure they would, e.g. $125 = 5^3$

M4.2 Properties of logs

You need to know the following properties of logs:

1 $\ln ab = \ln a + \ln b$ e.g. $2 \times 5 = 10$, so $\ln 10 = \ln 2 + \ln 5$

2 $\ln\dfrac{a}{b} = \ln a - \ln b$ e.g. $\ln 6 = \ln 30 - \ln 5$

3 $\ln\dfrac{1}{x} = -\ln x$ e.g. $\ln 5 = -\ln 0.2$

4 $\ln x^n = n \ln x$ e.g $\ln 125 = 3 \ln 5$

5 $\ln(\sqrt[n]{x}) = \dfrac{1}{n}\ln x$ e.g. $\ln 100 = \dfrac{\ln 1000000}{3}$

These rules can be derived from the definition of the log function[1].

Example

Work out the answer to Self-test M14 (e) without using a calculator.

Answer

$2 \ln 12 = \ln 12^2 = \ln 144$ Rule 1

$e^{\ln 144} = 144$ Definition of ln

$\therefore 3e^{2\ln 12} = 3 \times 144 = 432$

M15

Self-test

Given that ln $x = 5$,

(a) Write down ln x^2.

(b) Use your calculator to find x.

---

1 See Chapter 4 of Mathematics for A Level Physics by Kelly and Wood.

M4.3 Logs to the base 10

The logarithm function can be, and often is, defined using 10 as the base. The log to the base 10 of x is written $\log_{10} x$ or just $\log x$. The definitions are

$$\log_{10}(10^x) = x \quad \text{and} \quad 10^{\log x} = x.$$

Notice that we have left the subscript $_{10}$ off the second one. It is optional. The properties 1–5 from Section M.4.2 apply equally to the log function as to the ln, e.g.

$$\log ab = \log a + \log b.$$

M4.4 Using logs to find unknown exponents

If we want to solve an equation such as $a = kx^2$, or $b = m\sqrt{\dfrac{x}{n}}$ where x is the unknown quantity, we just re-arrange them and take the square root or square them as appropriate, i.e.

If $\qquad a = kx^2 \qquad$ then $x^2 = \dfrac{a}{k} \qquad$ so $x = \pm\sqrt{\dfrac{a}{k}}$

If $\qquad b = m\sqrt{\dfrac{x}{n}} \qquad$ then $b^2 = m^2\dfrac{x}{n} \qquad$ so $x = \dfrac{b^2 n}{m^2}$

Similarly we can re-arrange the Pythagoras' relationship $a^2 + b^2 = c^2$ to give $a = \pm\sqrt{c^2 - b^2}$, as in Section M1.2(c).

If the unknown quantity is in the exponent we solve the equation by taking logs.

Example 1

Find x if $2^x = 35$

Answer

Taking logs gives: $\qquad \ln 2^x = \ln 35$

But $\ln 2^x = x \ln 2 \qquad \therefore x \ln 2 = \ln 35$

$\therefore x = \dfrac{\ln 35}{\ln 2} = 5.13$ (to 3 s.f.)

Example 2

The intensity, I, of a parallel beam of X-rays decreases with distance, x, into a material according to the equation $I = I_0 e^{-\lambda x}$, where λ is a constant called the attenuation coefficient. Find λ if the intensity after 8.0 cm is 10% of the initial value.

Answer

At $x = 8.0$ cm, $I = 0.1 I_0$. $\therefore 0.1 = e^{-8.0\lambda}$

Taking logs: $\ln 0.1 = \ln(e^{-\lambda \times 8.0 \text{ cm}})$

So from the definition of the ln function: $\ln 0.1 = -\lambda \times 8.0$ cm.

$\therefore \lambda = \dfrac{\ln 0.1}{-8.0 \text{ cm}} \dfrac{-2.30}{-8.0 \text{ cm}} = 0.29 \text{ cm}^{-1}$ (to 2 s.f.)

Self-test **M16**

The power, P, available from a wind turbine is given by

$P = \frac{1}{3}\rho\pi r^2 v^3$.

(a) Make r the subject of the equation.

(b) Make v the subject of the equation.

Self-test **M17**

You can also solve Example 1 by taking logs to the base 10. Do this and show that you get the same answer.

Self-test **M18**

The pd, V, across a capacitor falls with time t according to the equation $V = V_0 e^{-\lambda t}$, where V_0 is the initial pd. Show how we can use logs to make λ the subject of the equation.

M5 Using logs to linearise functions

M5.1 Graphs of exponential functions

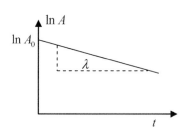

Fig. M.10 Semi-log graph

Logs can be used to transform a decay curve into a linear graph, which is: (a) easier to draw, (b) enables the putative exponential form of the decay to be verified and (c) allows the decay constant to be determined.

We start from the suggested decay equation, e.g. for the activity, A, of a radioisotope: $A = A_0 e^{-\lambda t}$.

We take (natural) logs of this equation: $\ln A = \ln(A_0 e^{-\lambda t})$

Applying Rule 1 from Section M4.2 $\qquad \ln A = \ln A_0 + \ln e^{-\lambda t}$

Applying the definition of natural logs $\qquad \ln A = \ln A_0 - \lambda t$

Comparing this with a linear equation $\qquad y = mx + c$

The **red arrows** indicate the variables and the **blue arrows** the constants in the equations. Hence, if the suggested equation is correct, a graph of $\ln A$ against t will be a straight line of gradient $-\lambda$ with an intercept of $\ln A_0$ on the $\ln A$ axis.

M5.2 Graphs of power functions

Sometimes we suspect that two variables, say x and y are connected by a power function,
$$y = Ax^n$$
but we don't know what value n should have, possibly because we have no working theory. Taking logs can also be used to find n.

Taking logs of the putative equation $\qquad \ln y = \ln A + n \ln x$

Comparing this with a linear equation $\qquad y = mx + c$

In the same way as for the semi-log graph, we see that, if the suggested equation is correct, a graph of $\ln y$ against $\ln x$ will be a straight line. Its gradient will be n. The intercept on the $\ln y$ axis (i.e. the value of $\ln y$ when $\ln x = 0$) is $\ln A$.

Note: Often, when plotting a log-log graph, the data points are a long way from the origin. This means that the equation of the line needs to be calculated using one of the techniques in Section 4.5.4 of the Year 1 / AS textbook. See also Question 10 in the following exercise.

Terms & definitions

A graph of the logarithm of a variable against a linear variable is called a **semi-log graph** or a **log-linear graph**.

A graph of the logarithm of one variable against the logarithm of another variable is called a **log-log graph**.

M19 — Self-test

Suppose two variables, x and y, are connected by the equation
$$y = 10x^3.$$
What will be the gradient of a graph of $\ln y$ against $\ln x$? What will be the intercept on the $\ln y$ axis?

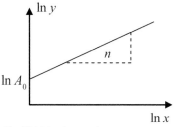

Fig. M.11 log-log graph

Exercise M

1 Express the following functions as functions of the form $\pm\sin \omega t$ or $\pm\cos \omega t$.

(a) $\sin (\omega t-\frac{\pi}{2})$ (b) $\cos (\omega t+\frac{3\pi}{2})$ (c) $\cos (\omega t+2\pi)$ (d) $\sin (\omega t-\frac{3\pi}{2})$ (e) $\sin (\omega t-\pi)$

2 Express the following functions in the form $\cos(\omega t \pm \varepsilon)$.

(a) $-\cos \omega t$ (b) $-\sin \omega t$ (c) $\sin \omega t$ (d) $-\cos (\omega t-\frac{\pi}{2})$ (e) $\sin (\omega t-\frac{\pi}{2})$

3 A particle oscillates with its displacement, x, given by $x = 5 \cos 314t$. Calculate the times, between -20 ms and $+20$ ms for which (a) $x = 0$, (b) $x = 5$, (c) $x = 2.5$ and (d) $x = -2.5$.

4 Give the times for the same displacements as in Q3 if the particle now oscillates with the function
(i) $x = 5 \cos (314t + \frac{\pi}{2})$, (ii) $x = 5 \cos (314t + \frac{3\pi}{4})$, (iii) $x = 5 \cos (314t - \frac{\pi}{3})$.

5 An alternating electric current varies with time as $I = 3.5\cos(800\pi t - \frac{\pi}{4})$.

(a) Calculate the frequency, f, and the period, T.
(b) Rewrite the equation for I without using π (i.e. put in the value 3.142).
(c) Calculate the value of I at (i) $t = 0$, (ii) $t = 0.625$ ms and (iii) $t = 1.25$ ms.
(d) Calculate the times between $t = 0$ and $t = 5$ ms when the current is instantaneously zero.
(e) Calculate the times between $t = 0$ and $t = 5$ ms when $I = +3.5$ A.

6 Use your calculator to find the values of (a) $e^{3.5}$, (b) $4\,e^{-2.5}$ (c) $\ln 25$, (d) $\frac{6}{7\ln 5}$, (e) $2.5\ln(e^{2.5} - 6)$.

7 Use the definition and rules of natural logs to simplify and evaluate the following expressions
(a) $\ln (e^{-1})$, (b) $5\ln (e^{3})$, (c) $e^{\ln 6}$, (d) $e^{3\ln 6}$, (e) $4e^{(\ln 2+\ln 3)}$.

8 Solve the following equations for x
(a) $25 = 10\ln x$, (b) $0.6 = 8\ln (2x - 3)$, (c) $6 = 4\ln\left(\frac{1}{1-x}\right)$, (d) $80 = 500e^{-0.2t}$, (e) $4\times10^{-6} = 25e^{-0.001t}$.

9 The pd, V, across a capacitor falls with time t according to the equation $V = V_0 e^{-\lambda t}$, where V_0 is 12.0 V and λ is 0.005 s^{-1}. Calculate (a) the pd after 5 minutes and (b) the time taken for the pd to fall to 1.0 V.

10 The illuminances (light-sensor readings), E_v, at a fixed distance from a filament lamp, for different supply voltages, V, are as follows:

V / V	6.0	8.1	9.4	10.1	11.1	12.1	12.9	13.4
E_v /lux	20	58	99	127	172	221	274	307

The background illuminance, i.e. with the light off, is 2 lux when the lamp is off. It is suspected that the E_v from the lamp depends on V by a relationship of the form $E_v = kV^n$ for some unknown values of k and n. By plotting a log-log graph, show that this is approximately correct and use your graph to determine values of n and k.

11 A capacitor discharges through a 4.7 kΩ resistor. In 10 s the current, I, decreases from 8.0 mA to 1.5 mA.

(a) Determine the gradient of a graph of $\ln I$ against time.
(b) Calculate the capacitance of the capacitor given that the decay equation is $I = I_0 e^{-\frac{t}{RC}}$.

Exam practice questions

Question 1: Vibrations and circular motion

(a) A group of students performed an experiment to determine the acceleration due to gravity using a simple pendulum.

The pendulum bob was a small heavy metal sphere with an inbuilt ring, to which the thread was attached. The students measured the length, l, of the thread between the base of the supporting cork to the top of the ring, as shown.

They measured the time for 10 complete oscillations. In an initial experiment, they estimated the uncertainty in the timing of measurements to be ± 0.2 s.

They obtained a series of results for values of l between 0.250 m and 0.800 m.

l / m	Time for 10 oscillations / s	Period²,$(T$ / s$)^2$
0.250	10.29 ± 0.20	1.06 ± 0.04
0.300	11.31 ± 0.20	1.30 ± 0.05
0.400	13.05 ± 0.20	1.70 ± 0.05
0.500	14.36 ± 0.20	2.06 ± 0.06
0.700	16.85 ± 0.20	
0.800	18.20 ± 0.20	

(i) Describe a technique which allows the timings to be made to this degree of accuracy. [2]

(ii) Complete the Period² column, add the results and error bars to the graph of T^2 against l. [2]

(iii) Theory predicts that the relationship between T and l can be written as

$$T^2 = \frac{4\pi^2}{g}(l + \varepsilon),$$

where ε is the distance of the centre of mass of the pendulum bob below the point of attachment.

Use the plotted results to determine the value of g together with its absolute uncertainty. [5]

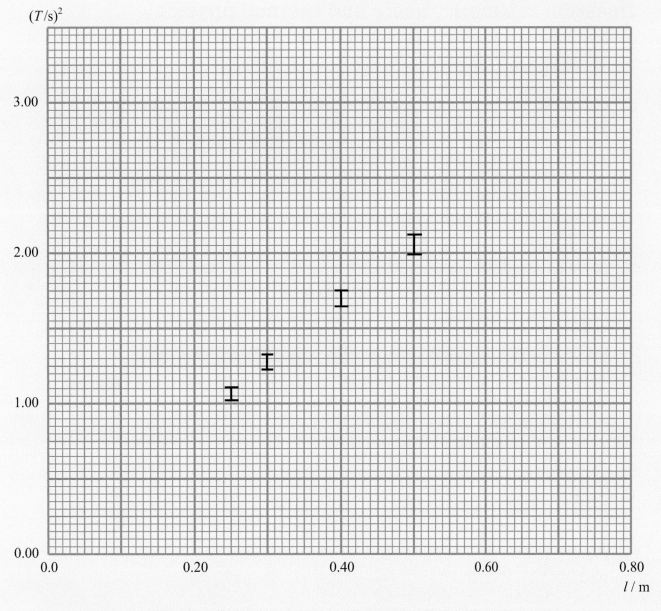

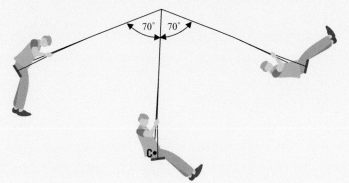

(b) An 80 kg man swings on a playground swing with an amplitude of 70°. The position of the centre of mass, **C**, of the man and swing seat is 3.0 m from the point of suspension.

    (i) Show that the man's speed at the lowest point in the swing is approximately 6 m s⁻¹.   [3]

    (ii) Calculate the tension in the supporting ropes of the swing seat at the lowest point.   [3]

    (iii) State the direction of the resultant force on the man at the extreme points of the swing. Explain your answer.   [2]

Question 2: Kinetic theory and thermal physics

(a) A helium cylinder of volume 9.00 m³ stores 75 kg of gas at a pressure of 5.0×10^6 Pa. Calculate the mean molecular kinetic energy of the helium.
$[M_{He} = 4.00 \times 10^{-3} \text{ kg mol}^{-1}]$ [3]

(b) The molecular mass of oxygen is 16 times that of hydrogen. Two identical cylinders, A and B, contain the same number of moles of gas at the same temperature. Cylinder A contains hydrogen; cylinder B contains oxygen. The molecules of hydrogen have an rms speed of 2 000 m s⁻¹.

    (i) Calculate the rms speed of the oxygen molecules.   [1]

    (ii) Compare the mean change in momentum of hydrogen molecules, with that of oxygen molecules when they hit the walls of the container.   [1]

    (iii) How does the kinetic theory of gases explain the fact that the pressure in the two containers is the same?   [3]

(c) The first law of thermodynamics can be written

$$\Delta U = Q - W$$

Complete the table, indicating for the given process, whether ΔU, Q and W are positive (> 0), negative (< 0) or zero ($= 0$). One cell is already filled in.   [3]

Process	ΔU	Q	W
A sample of ideal gas expands at constant temperature	$= 0$		
A perfectly insulated sample of ideal gas is compressed			
A sample of water at 100 °C is boiled into steam at 100 °C			
An iron wire at constant temperature has an electric current in it			

(d) In an experiment to determine the specific heat capacity of iron, a group of students poured 200 g of iron filings at 100 °C into a 150 g of water at 18.0 °C which was held in a glass beaker of mass 50 g. After stirring, the students determined the final temperature of the mixture to be 28.5 °C.

    (i) Calculate the specific heat capacity of iron.   [3]

        Data: specific heat capacity of glass, $c_g = 670 \text{ J kg}^{-1}\text{K}^{-1}$;

              specific heat capacity of water, $c_w = 4190 \text{ J kg}^{-1}\text{K}^{-1}$.

    (ii) In order to improve the accuracy of the result, one of the students suggested using a larger mass of iron filings. Another suggested using water with a lower initial temperature. Discuss these suggestions.   [3]

Question 3: Nuclear physics

(a) Radon is a naturally occurring radioactive gas that emits α particles.

   (i) Explain briefly which organ(s) of the body would be most at risk from radon gas. [3]

   (ii) The half-life of radon is 3.3×10^5 s. Calculate how long it will take a sample of radon gas to decay to:

      (I) 25% of its initial activity [1]

      (II) 20% of its initial activity. [3]

(b) A beam of radiation of initial intensity I_0 consists of one type of radiation (α, β or γ). The intensity, I, of the beam is measured to decrease exponentially with penetration distance in aluminium according to the equation

$$I = I_0 e^{-\mu x}$$

   (i) The absorption coefficient, μ, for the beam of radiation is 10 m^{-1}. Show that the intensity drops by approximately 3% in 3 mm of aluminium. [1]

   (ii) Explain briefly which radiation (α, β or γ) is present in the beam. [2]

(c) A possible reaction for future power stations on Earth is the fusion of deuterium, ^2_1H, and tritium, ^3_1H.

$$^2_1\text{H} + {}^3_1\text{H} \rightarrow {}^4_2\text{He} + \text{X}$$

   (i) Identify X and the interaction responsible for the reaction. Give your reasoning. [3]

   (ii) The masses of the particles involved in the reaction are

      $m_{^2_1\text{H}} = 2.0136$ u,

      $m_{^3_1\text{H}} = 3.0155$ u,

      $m_{^4_2\text{He}} = 4.0015$ u,

      $m_x = 1.0087$ u

      Calculate the energy released in the above reaction. (1 u $\equiv$ 931 MeV) [3]

   (iii) Explain briefly how you know that the nuclear binding energy of ^4_2He is greater than that of ^2_1H and ^3_1H combined. [3]

Question 4: Capacitors

A capacitor is charged to a potential difference of 10 V using a battery. At time $t = 0$, the capacitor is disconnected from the battery, and connected to a resistor. The charge, Q, on the capacitor plates then varies as shown.

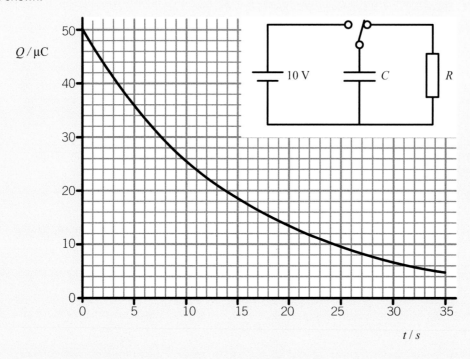

(a) Show that the capacitor's capacitance, C, is 5.0 μF. [1]

(b) Use the exponential nature of the decay to determine the time constant, RC. [2]

(c) **(i)** Draw a tangent to the graph at $t = 7.5$ s and **hence** determine the current at $t = 7.5$ s, showing your working. [3]

      (ii) Check your answer to (c)(i) by a different method, giving your reasoning and conclusion clearly. [4]

(d) Calculate the energy dissipated in the resistor in the first 7.5 s of discharge. [2]

Question 5: Electrostatic fields

Two small, positively charged spheres are placed in empty space as shown.

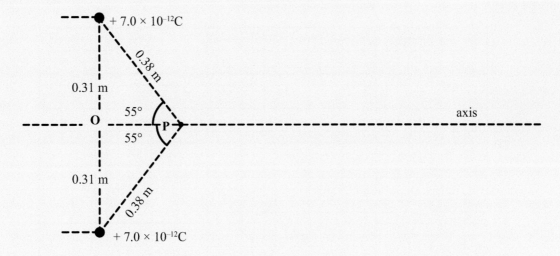

(a) **(i)** Put arrows at point **P** to show the *electric field strengths* at **P** due to each charge. [1]

(ii) Calculate the **resultant** electric field strength at **P**. [2]

(iii) The graph shows how the resultant electric field strength, E_{res}, varies with distance along the axis (see diagram) from point **0**.

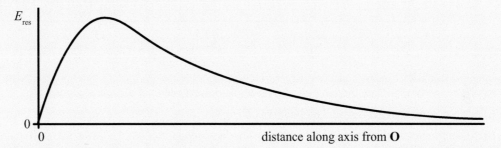

Explain briefly why:

(I) E_{res} is zero at **0**. [1]

(II) E_{res} decreases with distance along the axis, at large distances from **0**. [1]

(b) A positive ion with a charge of 4.8×10^{-19} C and mass 4.5×10^{-26} kg enters the region shown in the diagram, travelling along the axis. At point **0** the ion is moving to the right with a speed of 2000 m s^{-1}. Gravitational and resistive forces are negligible.

(i) Calculate the *acceleration* of the ion as it passes through point **P**.

[Make use of your answer to (a)(ii).] [2]

(ii) Describe how the speed of the ion changes as it travels along the axis from **0** until it is well beyond **P**. [2]

(iii) Calculate the total energy (kinetic energy + electrical potential energy) of the ion as it passes through point **0**. [See diagram.] [4]

(iv) Hence find the maximum speed eventually reached by the ion. [3]

Question 6: Orbits and the wider universe

(a) The plot of $\ln T$ against $\ln a$, where T is the orbital period and a the orbital radius, is for the five largest moons of Uranus.

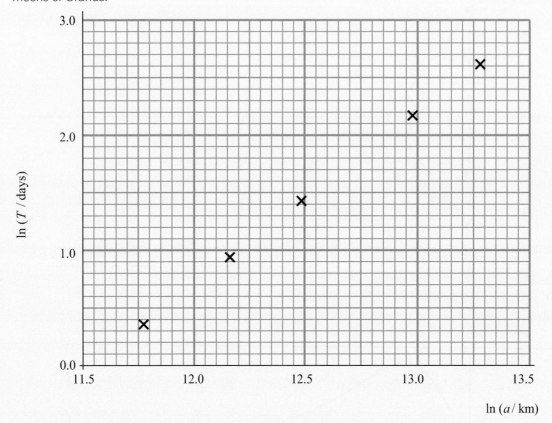

Kepler's laws of planetary motion can also be applied to satellites of the same planet.

(i) State Kepler's third law of planetary motion. [1]

(ii) Use the above plot to investigate whether the five largest moons of Uranus satisfy this law. [3]

(iii) Explain clearly how a graph of $\ln T$ against $\ln a$ for the moons of Jupiter would compare with this graph for Uranus. [3]

(b) From measurements of the mass of stars and clouds of interstellar gas within the orbit of the Sun about the centre of the galaxy, astronomers calculate that the orbital speed of the Sun should be 108 km s^{-1}. Its actual speed is measured at 208 km s^{-1}.

The distance to the centre of the galaxy is $2.51 \times 10^{20} \text{ m}$.

(i) State why astronomers only need to consider the mass within the orbit of the Sun. [1]

(ii) Explain how the data give evidence for the existence of Dark Matter. Use calculations to support your answer. [4]

(c) The most recent value of the Hubble constant, H_0, is $2.20 \times 10^{-18} \text{ s}^{-1}$. State the meaning of H_0, and explain how its value can be related to an estimate of the age of the universe.

A calculation should form part of your answer. [4]

Question 7: Magnetic fields

(a) The force on a current-carrying wire in a uniform magnetic field is investigated with the apparatus shown. For various angles, θ, between the wire XY (of length 0.025 m) and the magnetic field direction, the current, I, is adjusted so that the force, F, on the wire is always 3.5×10^{-3} N.

A graph is plotted of $\frac{1}{I}$ against $\sin\theta$.

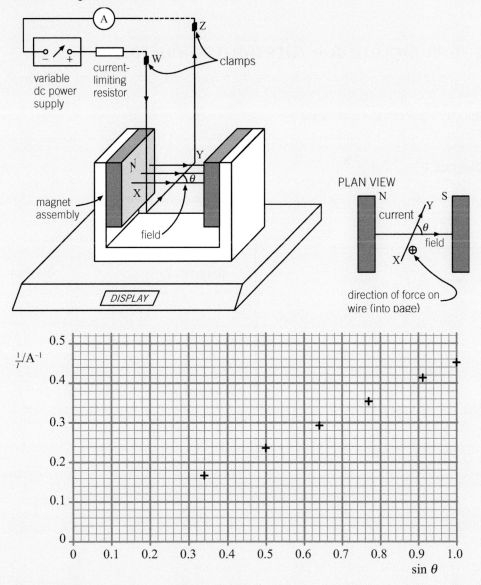

PLAN VIEW

direction of force on wire (into page)

(i) Draw the line of best fit and determine its gradient. [3]

(ii) Discuss briefly to what extent the graph supports the equation for the force on a current-carrying wire in a magnetic field. [4]

(iii) Use the graph gradient and the information given earlier to calculate a value for the magnetic field strength, B, in the region of the wire. [2]

(b) A circular ring of wire has diameter 0.070 m. The wire itself has a diameter of 0.60 mm, and is made of a metal of resistivity $1.7 \times 10^{-8}\ \Omega$ m. There is a magnetic field of strength 0.25 T at right angles to the plane of the ring.

Calculate the mean current in the ring when the field drops to zero in a time of 0.14 s. [6]

Options practice questions

Section B of the Unit 4 examination paper contains four questions worth 20 marks each, one for each option, A, B C and D. You should answer one of these questions only. The following four questions each relates to one of the options. While no attempt has been made to write 'mock exam' questions, the total mark allocation in each question being less than 20 marks and the skills requirements being unbalanced, the level of these questions reflects the expectation of the A level examination.

Question 8: Option A – Alternating currents

(a) The three phasor diagrams are for the pds across the three components in a series RCL circuit, for three different angular frequencies (pulsatances), ω_1, ω_2 and ω_3.

The diagrams are drawn to the **same scale**.

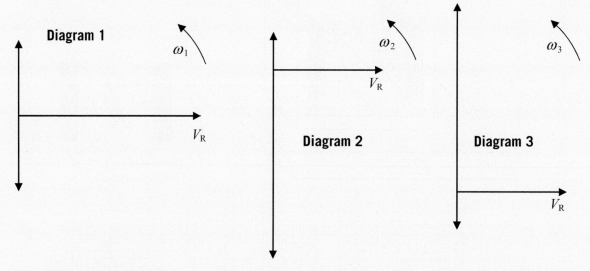

(i) Label the pd phasors for the inductor and capacitor, V_L and V_C, respectively on diagram 1 and explain your answer. [2]

(ii) Complete all three diagrams to show the total pd, V_T, across the combination. Comment on the three values of V_T. [2]

(iii) Comment as fully as possible on the magnitudes of the three values of ω. [3]

(iv) Calculate the Q factor for this circuit and state what change to the value of one or more components would increase it without changing the resonance frequency. [2]

(b) An engineer is designing an inductor for a radio circuit. The inductor needs to have an inductance of $50\ \mu\text{H}$. Because of space requirements the length, l, is to be 1.5 cm and the diameter of the coil 5.0 mm.

(i) Use the formula for the inductance of a solenoid,

$$L = \frac{\mu_0 N^2 A}{l}$$

where A is the cross sectional area of the coil, to calculate the number of turns required. [2]

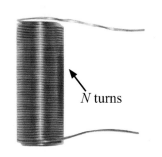

N turns

(ii) The copper wire has a radius of 0.10 mm. Show that the resistance of the coil is $1.50\ \Omega$ and hence calculate the impedance of the solenoid at a frequency of 10 kHz.

[Resistivity of copper $= 1.72 \times 10^{-8}\ \Omega$ m.] [4]

(iii) The graph shows the variation of pd across the inductor over two cycles of a 10 kHz signal.

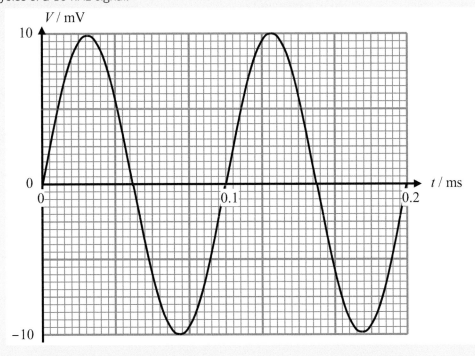

On the same grid, and using the same time axis, sketch a graph of the variation of current with time. Choose a suitable current scale. [4]

Question 9: Option B – Medical physics

(a) The two graphs are different ways of representing the output spectrum of an X-ray tube. It is suggested that they refer to the same tube operated under the same conditions.

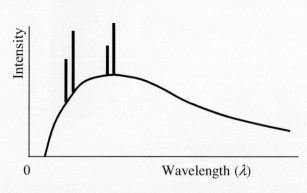 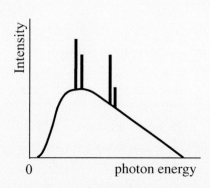

(i) Point out the ways in which the graphs are related and how these relate to the operation of the tube. [4]

(ii) The operating voltage of the tube is halved. Add a sketch of the new expected output spectrum on each of the two axes. [2]

(iii) The attenuation coefficient, μ, for 30 keV X ray photons in muscle is 0.4 cm^{-1}. For bone, the figure is 1.90 cm^{-1}.

    (I) Calculate the fraction of a beam of 30 keV photons which penetrates an 8.0 cm thick arm muscle. [1]

    (II) Explain how 30 keV X ray photons are useful in medical diagnosis. A calculation will help your answer. [3]

(b) An MRI scanner operates with a uniform magnetic field of flux density 1.8 T.

(i) Calculate the Lamor frequency of hydrogen nuclei in this field [$f = 42.6 \text{ MHz T}^{-1}$]. [1]

(ii) Explain the role of non-uniform magnetic fields in MRI scans. [3]

(c) Compare MRI scans with conventional X-ray examinations for medical diagnosis. [4]

Question 10: Option C – The physics of sports

The diagram, which is from a coaching website, shows a hammer thrower in an athletics tournament about to release the hammer. The website identifies the following key variables:

speed of throw, u angle of throw, θ height of release, h

In a calculation tool on the website it makes the following prediction for the throw distance, d:

With $u = 26.5$ m s^{-1}; $\theta = 41°$ and $h = 1.4$ m, $d = 72.46$ m

(a) The diagram below shows the path of the hammer. Use the diagram to help explain the effect of increasing the height of the release on the throw distance.   [2]

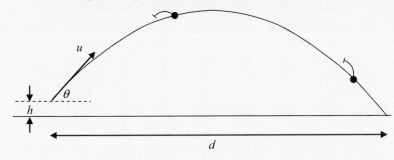

(b) A physics student, comments that the model on the website doesn't appear to take air resistance into account. Show that she is correct.   [4]

(c) A research paper, claims that the effect of air resistance on the sphere is to reduce the throw distance by approximately 4 m.

    (i) Taking the drag coefficient for a sphere to be 1.0, the mass of the hammer to be 7.3 kg and assuming the hammer to be made of steel of density 7900 kg m^{-3}, estimate the mean drag force on the hammer during the flight [air density = 1.3 kg m^{-3}].   [3]

    (ii) Add a suitable line to the diagram below to show the effect on the trajectory of air resistance.   [2]

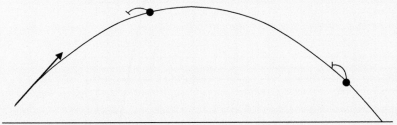

(d) We can model the torque that the athlete, **A**, needs to accelerate the hammer, **H**, up to its throw speed as follows:

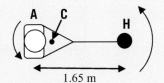

At the release point, the athlete and hammer are rotating about their centre of mass, **C**. A typical separation is 1.65 m. The athlete makes 4 turns before releasing the hammer. Treat both the hammer and athlete as point masses. The athlete has a mass of 80 kg.

    (i) Find the position of the centre of mass.   [1]

    (ii) Show that the total angular momentum of the athlete and hammer at the point of release is approximately 320 kg m^2 s^{-1}.   [3]

    (iii) Assuming a constant angular acceleration, calculate the torque required to produce this angular momentum.   [2]

(e) The athlete's arms must also provide the centripetal force on the hammer. Calculate its value at the point of release.   [1]

Question 11: Option D – Energy and the environment

(a) The temperature of the photosphere of the Sun is 5780 K. The intensity of solar radiation at the Earth's orbit is 1.35 kW m^{-2}.

    (i) By considering the absorption of the Sun's radiation and the subsequent emission by the Earth, show that, if the Earth behaved as a black body, its mean temperature would be approximately 5 °C. [3]

    (ii) Describe how the spectrum of the radiation which is emitted by the Earth is different from that which is absorbed. [2]

    (iii) In fact the Earth reflects 30% of the incident visible and near infrared radiation. Without calculation, explain briefly how this would affect the answer to (i). [1]

    (iv) The effect of the Earth's atmosphere is to raise the mean surface temperature of the Earth to approximately 15 °C. Explain this effect qualitatively. [4]

(b) The graph shows the variation of the output current with terminal pd for a photovoltaic unit under 'ideal conditions'.

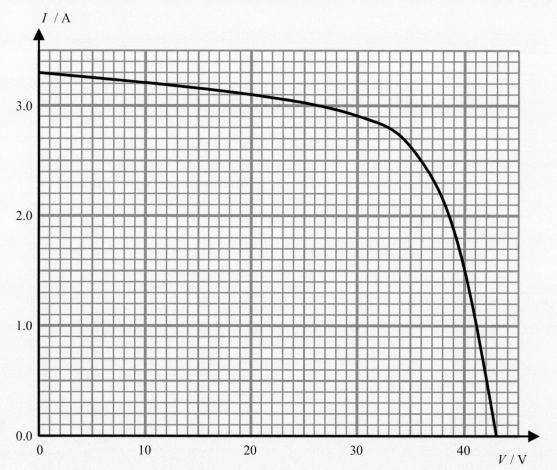

    (i) Suggest what is meant by 'ideal conditions'. [1]

    (ii) Calculate the output power at 20 V. [1]

    (iii) Explain the circuit conditions which would lead to zero output power. [2]

    (iv) The PV unit is marketed as 90 W output. Investigate whether this is a reasonable claim. [3]

Unit 3

Section 3.1

3.1.1 1.3 rad s^{-1}

3.1.2 **(a)** 0.052 cm

(b) Shorter wavelength

3.1.3 **(a)** 1145 N

(b) 877 N

3.1.4 Circumference $= 2\pi r$

$\therefore \text{AB} = \frac{1}{4}2\pi \times 5 = \frac{5\pi}{2}$

$\therefore \text{Time} = \frac{\frac{5\pi}{2}}{20} = \frac{\pi}{8}$ s

3.1.5 For $\pm\frac{\pi}{6}$ rad, $\Delta t = \frac{\pi}{12}$ s

$\Delta v = 2 \times 20\sin\frac{\pi}{6} = 20$ m s^{-1}

$\rightarrow \langle a \rangle = \frac{20}{\frac{\pi}{12}} = 76.4$ m s^{-2}

3.1.6 80 m s^{-2}

3.1.7 $v_{\text{max}} = 6.3$ m s^{-2}

3.1.8 Because the required centripetal force and the maximum grip are both proportional to the mass

3.1.9 1.4 m s^{-1}

3.1.10 $g = 0.0027$ m s^{-2}

3.1.11 $r\omega^2 = 385 \times 10^6 \times \left(\frac{2\pi}{27.3 \times 86400}\right)^2 = 0.0027(3)$ m s^{-2}

Very close!

3.1.12 $r\omega^2 = 6731 \times 10^3 \times \left(\frac{2\pi}{86400}\right)^2 = 0.0356$ m s^{-2}

Section 3.2

3.2.1 **(a)** $\Delta F_A = 4$ W

(b) $\Delta F_B = -6$ N

(c) $F_{\text{res}} = 10$ N (to the left)

3.2.2

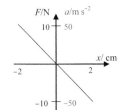

3.2.3 $[k] = \text{N m}^{-1} = \text{kg m s}^{-2} \text{ m}^{-1} = \text{kg s}^{-2}$

$[m] = \text{kg}$

$\left[\sqrt{\frac{k}{m}}\right] = \sqrt{\left[\frac{k}{m}\right]} = \sqrt{\frac{\text{kg s}^{-2}}{\text{kg}}} = \sqrt{\text{s}^{-2}} = \text{s}^{-1}$

3.2.4 **(a)** $0, -2A, 0, 2A$

(b) $20, 0, -20, 0$

3.2.5 **(a)** **(i)** $12, 26, 40$ [all in μs]

(ii) $5, 33$

(iii) $19, 47$

(b) $5, 19, 33\ 47$

(c) $12, 26, 40$

3.2.6 $v_{\text{max}} = 450$ m s^{-1}

$a_{\text{max}} = 1.0 \times 10^8$ m s^{-2}

3.2.7 The 'answers' which can be ruled out are: $\varepsilon = 2.020$ and -4.264

3.2.8 $x\,/\,\text{cm} = 0.2 \cos(220(t\,/\,\text{ms}) - 1.12)$

3.2.9

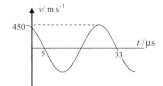

3.2.10 $\omega = 10$ s^{-1}, $T = 0.628$ s, $f = 1.59$ Hz

3.2.11 **(a)** 0.039 m

(b) 8.0 Hz

3.2.12 0.994 m

3.2.13 Tension greatest at bottom because upward centripetal force greatest

$\therefore$ Length greatest at the bottom

3.2.14 Immediately and after each ½ cycle after release.

$F_{\text{max}} = 1.3$ N

3.2.15 **(a)** $E_{\text{total}} = 250$ J

(b) $E_p(0) = 250$ J, $E_k(0) = 0$

(c) $E_p(\frac{\pi}{200}) = 0$, $E_k(\frac{\pi}{200}) = 250$ J

3.2.16 **(a)** $t = 0, \frac{\pi}{\omega}, \frac{2\pi}{\omega} \dots$

(b) $E_{k\,\text{max}} = \frac{1}{2}\frac{m^2g^2}{k}$ at $t = \frac{\pi}{2\omega}, \frac{3\pi}{2\omega}, \dots$

(c)

3.2.17 **(a)** $\omega = \pi$

(b) $\lambda = 0.32$ s^{-1}

3.2.18 **(a)** 0.70

(b) Measure velocity using the gradient where $x = 0$ a whole number of cycles apart. then calculate the ratio of v^2

3.2.19 The amplitude should come down to (every 2.0 s) 7.7 cm, 5.5 cm, 4.3 cm, 3.0 cm, 2.3 cm

3.2.20 Initial gradient = 0

3.2.21 Over-damping – takes a long time to close

Under-damping – continues to swing or hits the stop with a bang

3.2.22 1.3 m s^{-1} and 3.0 m s^{-2}

3.2.23 Oven frequency $\sim 0.000\ 02 \times$ absorption peaks

3.2.24 $T^2 = \dfrac{4\pi^2}{g}l + \dfrac{4\pi^2}{g}r$ where r = radius

Gradient the same, intercept $= \dfrac{4\pi^2}{g}r$

Section 3.3

3.3.1 **(a)** 1.99×10^{-14} J = 124 keV

(b) 1.32×10^{-18} J = 8.2 eV

3.3.2 0.131 kg mol^{-1}

3.3.3 4.0 cm

3.3.4 $A = 235$; $M = 235$ g mol^{-1}; $m = 3.9 \times 10^{-25}$ kg

3.3.5 **(a)** 3.99×10^{-26} m^3

(b) $\therefore$ side of cube $= \sqrt[3]{3.99 \times 10^{-26}}$m $= 3.4$ nm

3.3.6 $R = \dfrac{pV}{nT}$, $\therefore$ for homogeneity,

$[R] = \dfrac{\text{N m}^{-2}\ \text{m}^3}{\text{mol} \times \text{K}}$

$= \text{N m mol}^{-1}\ \text{K}^{-1}$

$= \text{J mol}^{-1}\ \text{K}^{-1}$

3.3.7 **(a)** 1 mol occupies 18 cm^3

$\therefore$ Volume per molecule $= \dfrac{18\ \text{cm}^3\ \text{mol}^{-1}}{6.02 \times 10^{23}\text{mol}^{-1}}$

$= 2.99 \times 10^{-23}$ m^3

$\therefore \dfrac{4}{3}\pi\left(\dfrac{d}{2}\right)^3 = 2.99 \times 10^{-23}$ cm^3

Leading to $d = 390$ pm

(b) ~ 500 nm

3.3.8 **(a)** 3.6 mm

(b) 5.6×10^{-19} m

3.3.9 Whatever the velocity, velocity^2 = speed^2

3.3.10 **(a)** Higher speeds make a proportionately greater contribution to Σv^2

(b)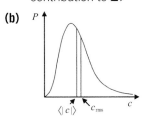

3.3.11 **(a)** 80.8 mol

(b) 4.86×10^{25} molecules

3.3.12 $k = \dfrac{R}{N_A} = \dfrac{8.31\ \text{J mol}^{-1}\ \text{K}^{-1}}{6.02 \times 10^{23}\ \text{mol}^{-1}} = 1.38 \times 10^{-23}$ J K^{-1}

Section 3.4

3.4.1 **(a)** 11.2 kJ

(b) 3.4 MJ

(c) 188 kJ

3.4.2 1.025×10^5 Pa

3.4.3 Equal. Same amount of gas at the same temperature – both monatomic

3.4.4 $U_{\text{He}} = 38U_{\text{Ar}}$

3.4.5 **(a)** $(pV)_C = 4.8 \times 10^5$ J, $(pV)_B = 3.2 \times 10^5$ J, $(pV)_A = 4.0 \times 10^5$ J

$pV \propto T$ hence the result

(b) 0.8

3.4.6 0.32 kg

3.4.7 Electrical work output $= VIt$

Heat output $= I^2Rt$

Change in internal energy $= -EIt$

3.4.8 **(a)** $\Delta U > 0$

(b) $Q > 0$

(c) $W < 0$

(d) $\Delta U < 0$

(e) $\Delta U > 0$

3.4.9 $\Delta U = 2490$ J, $W = 0$, $Q = 2490$ J

3.4.10 -250 ± 26 °C

Section 3.5

3.5.1 $^{235}_{92}\text{U} \rightarrow\ ^{231}_{90}\text{Th} +\ ^{4}_{2}\text{He}$

3.5.2 $^{106}_{52}\text{Te} \rightarrow\ ^{102}_{50}\text{Sn} +\ ^{4}_{2}\text{He}$

3.5.3 **(a)** Baryon number (1)

Charge (0)

Lepton number (0)

(b) Neutrino involvement: only feels the weak interaction

Change of quark flavour: d $\rightarrow$ u

3.5.4 **(a)** 1 MeV/c^2

(b) 1.8×10^{-30} kg

(c) 0.0011 u

3.5.5 **(a)** 0.8 MeV

(b) 5×10^{-12} J

3.5.6 **(a)** A unchanged; $Z \rightarrow Z - 1$

(b) $A \rightarrow A - 1$; $Z \rightarrow Z - 1$

3.5.7 $^{11}_{7}N \rightarrow ^{10}_{6}C + ^{1}_{1}H$

3.5.8 **(a)** 50 000

(b) There are more atoms per µm in a solid than in a gas

3.5.9 **(a)** 0.34

(b) $(0.34)^5 = 0.0045$

3.5.10 Concentration in Al = 0.4 × concentration in Pb

3.5.11 Because it loses more energy in passing through any atom

3.5.12 24.5 (3 s.f.)

3.5.13 1 – 11: 24.3

2 – 12: 24.0

3 – 13: 24.8

4 – 14: 24.9

5 – 15: 24.5

6 – 16: 23.6

7 – 17: 23.6

8 – 18: 23.8

9 – 19: 24.8

10 – 20: 25.1

Similar spread

3.5.14 The β+ tracks would have a similar spread of curvatures to the β− but be oppositely directed (same direction as α)

3.5.15 3.65×10^{10} Bq [36.5 GBq]

Note: Nowadays the curie is *defined* as 37 GBq

3.5.16 5.0×10^{-9} s^{-1}

3.5.17 **(a)** 1.39×10^8 s = 4.39 years

(b) 4.1 GBq [4.1×10^9 Bq]

3.5.18 $T_{400} - T_{200} = 50$ s

From $C_{140} = 100$ s and using $C = C_0 2^{-x}$

$\rightarrow 50.5$ s

From $C_{140} = 100$ s and using $C = C_0 e^{-\lambda t}$

$\rightarrow 50.6$ s

3.5.19 **(a)** $\lambda = 0.086$ day$^{-1} = 9.9 \times 10^{-7}$ s^{-1}

(b) Using 0.086 day^{-1}

$A = 3 \times 10^6 e^{-0.086 \times 21} = 497$ kBq (~500)

3.5.20 $10 \times 10^{-3} = 5 \times 10^6 \times 2^{-n} \rightarrow n = 8.97$

$\rightarrow 225$ min

3.5.21 830, 690, 160

3.5.22 $\left(\dfrac{11}{12}\right)^n = \dfrac{1}{2}$, $\therefore n \ln\left(\dfrac{11}{12}\right) = \ln 2 \rightarrow n = 7.967$

3.5.23 $[\varepsilon] = $ cm, $[k] = $ cm^2 s^{-1}

3.5.24 Gradient $= \dfrac{1}{\sqrt{k}}$, intercept $= \dfrac{\varepsilon}{\sqrt{k}}$

$\therefore \varepsilon = -\dfrac{\text{gradient}}{\text{intercept}}$

Section 3.6

3.6.1 44.01 g

3.6.2 2.5×10^{-12} kg

3.6.3 1.5×10^{-8} u

3.6.4 **(a)** 4.29×10^{-12} J

(b) 27 MeV

~ 2 million times as much

3.6.5 6.5×10^{14} J

3.6.6 Figures give 931.49 MeV

3.6.7 $\Delta m = 1.643 \times 10^{-28}$ kg

$\therefore$ Binding energy $= 1.479 \times 10^{-11}$ J

$= 92.3$ MeV

3.6.8 **(a)** 263 MeV

(b) 8.5 MeV nuc^{-1}

3.6.9 **(a)** $^2_1H, ^3_2He, ^{11}_5B, ^{21}_{10}Ne$

(b) Only one particle in the nucleus

3.6.10 4.2×10^{-13} J, 2.6 MeV

3.6.11 **(a)** $\lambda = 0.00122$ s^{-1},

$A = \dfrac{1 \times 10^{-6}}{27} \times 6.02 \times 10^{23} \times 0.00122$

$= 2.72 \times 10^{16}$ Bq

$\therefore P = 2.72 \times 10^{16} \times 4.2 \times 10^{-13}$ W $= 11.4$ kW

(b) $P = 0.02$ W

3.6.12 **(a)** $^{180}_{74}W \rightarrow ^{176}_{72}W + ^4_2He$

(b) $\Delta m = 179.946\ 70 - 175.941\ 51 - 4.002\ 604$

$= 0.002\ 586$ u

$\equiv 0.002\ 586 \times 931.5$ MeV

$= 2.41$ MeV

$= 3.86 \times 10^{-13}$ J

3.6.13 If $r = kA^{1/3}$, then Mass $= \frac{4}{3}\pi k^3 A \rho$

$\therefore \rho = \dfrac{3 \times 1.66 \times 10^{-27} A}{4\pi \times (1.1 \times 10^{-15})^3 A} = 3.0 \times 10^{17}$ kg m^{-3}

3.6.14 7.6 MeV nuc^{-1}

3.6.15 **(a)** 6.0×10^{26}

(b) 8.6×10^{13} J

3.6.16 4.78 MeV

3.6.17 $r_{H2} = 1.1\sqrt[3]{2} = 1.39$ fm

$r_{H3} = 1.1\sqrt[3]{3} = 1.59$ fm

$\therefore$ Total separation $= 1.39 + 1.59$ fm ~ 3 fm

Unit 4

Section 4.1

4.1.1 Because otherwise the backs of the conductor would be in contact

4.1.2 64 nF = 0.064 μF

4.1.3 30 V. Check the polarity

4.1.4 0.025 m²

4.1.5 $F = kg^{-1} m^{-2} s^4 A^2$

4.1.6 (a) 55 μF, (b) 13.2 μF

4.1.7 (a) $Q = 180$ nC; $V = 8.2$ V

(b) $Q = 320$ nC; $V = 6.8$ V

4.1.8 (a) Parabola

(b) 0.48 μF = 480 nF

(c) Probably not – could be a film capacitor

4.1.9 ~ 4 minutes

4.1.10 (a) 39.6 s

(b) 14 μC

(c) 55 s

4.1.11 (a) 10 V

(b) 6 MΩ

4.1.12 (a) Gradient = −0.85 μC s⁻¹

(b) Current [0.85 μA discharge current]

(c) 0.83 μA

4.1.13 (a) 1.1(2) s

(b) 0.56 s

(c) Charge decay time = 2 × energy decay time

4.1.14 Transferred as internal energy in the connecting wires

4.1.15 (a) 3.3 s

(b) 7.8 V

4.1.16 15 kV m⁻¹ downwards

4.1.17 2.7 μN

4.1.18 The acceleration is constant, so the mean velocity is half the final velocity

4.1.19 3.34×10^{-7} s

4.1.20 (a) 29.5 V

(b) 2.5×10^{-9} s

(c) 6.4×10^{15} m s⁻²

(d) 1640 V

(e) 1.6×10^7 m s⁻¹

(f) 3.6×10^7 m s⁻¹ at 0.464 rad (26.6°)

Section 4.2

4.2.1 F m⁻¹   = C V⁻¹ m⁻¹
         = C J⁻¹ C m⁻¹
         = C (N m)⁻¹ C m⁻¹
         = C² N m⁻²

Other methods are available

4.2.2 (a) 8.2×10^{-8} N

(b) 2.56×10^{-11} m

4.2.3 110 pC

4.2.4 (a) Gradient = − 5 V m⁻¹

(b) 5 V m⁻¹ – in agreement

4.2.5 (a) Assume it behaves as a complete sphere

$Q = -1.1 \times 10^{-6}$ C

(b) 670 kV m⁻¹ towards the centre

4.2.6 0.021 m

4.2.7 0 (zero) – because the potential is 0 at B

4.2.8 (a) $E_h = 0.244$ V m⁻¹ to the left

(b) $E_v = 0.192$ V m⁻¹ downwards

(c) $E = 0.31$ V m⁻¹ at 38.1° downwards to left

4.2.9 $W = 1.2 \times 10^{-13}$ J

Positive as it starts closer to B than A

4.2.10 410 V m⁻¹ horizontally to the right

4.2.11 $\Delta V = 108$ V

4.2.12 Electric field strength is greater

4.2.13 8.3×10^{-8} N

4.2.14 8.3×10^{25}

4.2.15 3.7 N kg⁻¹

4.2.16 17

4.2.17 6.0×10^{30} kg

4.2.18 Gradient = 0.016

Value read off at ~ 0.015 (close)

4.2.19 2270 m s⁻¹

4.2.20 $\frac{x}{y} = 32.36$, $x = 7.55 \times 10^{11}$ m, $y = 2.3 \times 10^{10}$ m

4.2.21 From the ratio of distances to the neutral point, $\frac{M_1}{M_2} = 9$

Section 4.3

4.3.1 $\frac{E_p(P)}{E_p(A)} = \frac{8.2}{0.6} \sim 14$

4.3.2 (a) $E_p = -7.1 \times 10^{21}$ J

(b) $E_k = 5.8 \times 10^{21}$ J

(c) $E_{Tot} = -1.3 \times 10^{21}$ J

4.3.3 $v \propto \frac{1}{\sqrt{r}}$

4.3.4 0.39 AU

4.3.5 Change of velocity in the same direction as the applied force

4.3.6 8.6×10^{25} kg

4.3.7 -1.28×10^{21} J (in agreement)

4.3.8 1.2×10^{10} J

4.3.9 (a) 2130 m s^{-1}

(b) Ratio of period^2 = ratio of radii^3 = 15.6 ✓

4.3.10 (a) gradient = $\sqrt{GM_\odot}$, intercept = 0

(b) gradient = $GM_\odot$, intercept = 0

(c) gradient = $-\frac{1}{2}$, intercept = $\frac{1}{2}\ln GM_\odot$

4.3.11 (a) 28 500 ly

(b) 1 arc second = $\dfrac{1}{3600}^\circ$ = 1.848×10^{-6} rad

$\therefore$ 1 pc = $\dfrac{1 \text{ AU}}{4.848 \times 10^{-6}}$ = 3.09×10^{16} m

$= \dfrac{3.09 \times 10^{16} \text{ m}}{9.47 \times 10^{15} \text{ m ly}^{-1}}$ = 3.27 ly (close!)

(c) 8.7 kpc

4.3.12 Applying $v^2 = \dfrac{GM}{r}$ gives 157 km s^{-1} i.e. 160 km s^{-1} (2 sf)

4.3.13 λ_{meas} = 655.622 ± 0.002 nm

4.3.14 Subst for r_2 in $m_1r_1 = m_2r_2$

$\rightarrow m_1r_1 = m_2(d - r_1)$

Rearranging: $(m_1 + m_2)r_1 = m_2d$

$\therefore r_1 = \dfrac{m_2d}{m_1 + m_2}$ QED

4.3.15 4900 km from centre of the Earth

4.3.16 (a) 2.0 million km

(b) 9.6×10^{29} kg and 6.4×10^{29} km

4.3.17 Ignoring $15° \rightarrow 3.5\%$ error.

Uncertainty in velocity = 7%

4.3.18 (b) v_S = 102m s^{-1}

(c) m_C = 1.36×10^{27}kg = $\frac{1}{74}m_s$

4.3.19 (a) 1.19×10^{10} m

(b) v_s = 102 m s^{-1}

r_s = 16.2×10^6 km

(c) 1.36×10^{27} kg

4.3.20 m_A = 1.09×10^{31} kg

m_B = 1.74×10^{31} kg

4.3.21 Duration = 46 800 s

0.008% dimming

Assumption: exactly in plane of orbit.

4.3.22 67.9 km s^{-1} Mpc^{-1} = $\dfrac{67.9 \times 10^3 \text{ m s}^{-1}}{3.085\ 68 \times 10^{22} \text{ m}}$

$= 2.20 \times 10^{-18}$ s

4.3.23 (a) 2200 Mpc (b) 6.9×10^{25} m

4.2.24 t_H = 4.55×10^{17}s = $\dfrac{4.55 \times 10^{17}\text{s}}{3.16 \times 10^{17}\text{s year}^{-1}}$

$= 1.44 \times 10^{10}$ years

4.2.25 10^{10} light years

Section 4.4

4.4.1 Like poles repel: the plotting compass N pole, which indicates the direction of the field line, points away from the N pole of the magnet

4.4.2 Field lines must get closer together across region 2

4.4.3 Upwards $\uparrow$; to the left $\leftarrow$

4.4.4 0.042 T

4.4.5 (a) 0 (zero)

(b) 2.4 mN upwards

(c) 4.8 mN upwards

(d) 3.4 mN downwards

4.4.6 160 A

4.4.7 $r = \dfrac{mv}{Bq} = \dfrac{2\sqrt{E_k m}}{Be} = \dfrac{\sqrt{2eVm}}{Be} = \sqrt{\dfrac{2Vm}{B^2e}}$

Inserting the data $\rightarrow r = 0.070$ m = 70 mm

This is within the error, so consistent

4.4.8 After n gaps, $E_k = nqV$

$\therefore \frac{1}{2}mv^2 = nqV$

$\therefore v = \sqrt{\dfrac{2nqV}{m}} = w\sqrt{n}$

4.4.9 (a) 4.57 MHz

(b) 0.91 m

4.4.10 $A = 24$

4.4.11 (a) Accelerates along field lines

(b) Remains at rest

4.4.12 1.1×10^6 m s^{-1}

4.4.13 In copper, electrons move much more slowly for the same current, so the magnetic deflection force is much less

4.4.14 e.g. Mains AC circuit or discharging capacitor

4.4.15 Fields due to currents on opposite sides of the coil reinforce at P but oppose at Q

4.4.16 (a) 28µT at 45° (downwards to the left)

(b) 14 µN to right ($\rightarrow$)

4.4.17 5.3 A

4.4.18 17 µWb

4.4.19 Slightly smaller – same flux over a larger area because field lines spread out

4.4.20 1.1 µN downwards ($\downarrow$)

4.4.21 Forces due to the magnetic fields on opposite sides of the tube are attractive, thus pulling the tube inwards

4.4.22 (a) 8.2 mV

(b) Corrected V_H (0.5 A) = 41.6 mV

Corrected V_H(1.0 A) = 82.8 mV

$\therefore V_H(1.0) \sim 2V_H(0.5)$, so proportional

Using 83 mV for 1 A $\rightarrow$ 30 V T^{-1}

4.4.23 gradient $= \dfrac{\mu_0 I}{2\pi}$

4.4.24 They are (a) horizontal and (b) equal and opposite

4.4.25 $B = 21 \pm 2$ mT [accept 21.0 ± 1.6 mT]

Section 4.5

4.5.1 **(a)** PQ and SR – not cutting flux

QR – outside the field

(b) Downwards, $\mathbf{P} \to \mathbf{S}$

(c) Upwards, $\mathbf{S} \to \mathbf{Q}$

4.5.2 From Faraday's law: $[\mathscr{E}] = \left[\dfrac{\Delta N\Phi}{\Delta t}\right] = \dfrac{[B]\,[A]}{[t]}$

$\therefore [B] = \text{V s m}^{-2}$

$\therefore [Blv] = \text{V s m}^{-2}\,\text{m m s}^{-1} = \text{V} = [\mathscr{E}]$ QED

4.5.3 22 mT

4.5.4 **(a)** 0.92 V

(b) There would be an equal and opposite emf induced in the connecting wires

4.5.5 $\langle\mathscr{E}\rangle = 3.7\ \mu\text{V}$

4.5.6 1.03 mV

4.5.7 **(a)** $P = BlvI$

(b) $F_{\text{motor}} = BIl$

$\therefore P = F_{\text{motor}} \times v = BIlv = BlvI$ QED

4.5.8 Out of page

4.5.9 **(a)** $|\mathscr{E}|$ is increasing

(b) Force on side XY is upwards to oppose change. So by FLHR current is out of page

Opposite for WZ

4.5.10 **(a)** 2.9 mWb-turn

(b) 0.37 V

(c) 0.91 mWb-turn

(d) 0.35 V

4.5.11 Eddy currents are induced in the copper pipe which produce a magnetic field which (by Lenz's law) opposes the relative motion of the magnet and pipe

4.5.12 $[\text{RHS}] = [J^2\rho] = (\text{A m}^{-2})^2\,(\text{V A}^{-1}\,\text{m}^2\,\text{m}^{-1})$

$= \text{V A m}^{-3}$

$= \text{W m}^{-3}$

$= [\Pi] = [\text{LHS}]$

$\therefore$ Homogeneous

4.5.13 34 ms

4.5.14 It increases at the rate of 1.5 A s^{-1}

Section 4A

A1 30 Wb-turn

A2 Peak voltage = 470 V

$V = 470 \cos (100\pi t + \varepsilon)$

A3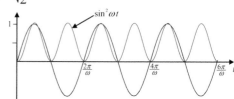

A4 12 ms

A5 Any 3 of $\ldots -4\pi,\ -2\pi,\ 0,\ 2\pi,\ 4\pi \ldots$

A6 $V = 17 \cos (100\pi t + 1.2\pi)$

A7 $V_{\text{rms}} = \dfrac{17}{\sqrt{2}} = 12.0$ V

A8 **(a)**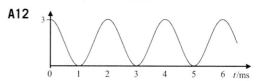

(b) $\cos 2\theta = 1 - 2\sin^2\theta$

$\therefore \sin^2\theta = \dfrac{1}{2} - \dfrac{1}{2}\cos 2\theta$

$\therefore \langle\sin^2\theta\rangle = \dfrac{1}{2} - \dfrac{1}{2}\langle\cos 2\theta\rangle$

But $\langle\cos 2\theta\rangle = 0$

$\therefore \langle\sin^2\theta\rangle = \dfrac{1}{2}$ QED

A9 **(a)** 9.4 V

(b) 1.88 W

A10 **(a)** 75 Ω

(b) $f = 250$ Hz; $\omega = 500\pi\ \text{s}^{-1}$ ($1570\ \text{s}^{-1}$)

(c) $V_{\text{rms}} = 10.6$ V; $I_{\text{rms}} = 0.141$ A

(d) 1.5 W

A11 **(a)** 3 W, **(b)** 3 W, **(c)** 0.

A12

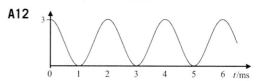

A13 **(a)** 20 V, 1.5 A, 14.1 V, 1.06 A, 100 μs, 10 kHz, $2 \times 10^4 \pi\ \text{s}^{-1}$

(b) 1.2 μF

(c) $\omega C V_{\text{rms}} = 2 \times 10^4 \pi \times 1.2 \times 10^{-6} \times 14.1 = 1.06\ \text{A} = I_{\text{rms}}$

A14 **(a)** 1.0 V, 10.0 mA, 0.707 V, 7.07 mA, 2 ms, 500 Hz, $1000\pi\ \text{s}^{-1}$

(b) 31.8 mH

(c) $I_{\text{rms}}\omega L = 7.07 \times 10^{-3} \times 1000\pi \times 31.8 \times 10^{-3}$

$= 0.706 = V_{\text{rms}}$

A15 $[\frac{1}{2}\varepsilon_0 E^2]$ $\quad = \text{F m}^{-1}\,(\text{V m}^{-1})^2$

$= \text{C V}^{-1}\,\text{m}^{-1}\,\text{V}^2\,\text{m}^{-2}$

$= \text{C V m}^{-3}$

$= \text{J m}^{-3} = [\text{energy density}]$

$$\left[\frac{B^2}{2\mu_0}\right] = (V\ s\ m^{-2})^2 \times (H\ m^{-1})^{-1}$$
$$= V^2\ s^2\ m^{-4}\ (A\ V^{-1}\ s^{-1})\ m$$
$$= (V\ A\ s)^2\ m^{-3}$$
$$= J\ m^{-3} = [\text{energy density}]$$

A16 40 µF

A17 (b) $I < 0$ decreasing; $V > 0$ decreasing

(c) $I < 0$ increasing; $V < 0$ decreasing

(d) $I > 0$ increasing; $V < 0$ increasing

A18 Fig A17(d)

A19

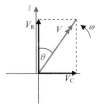

A20 0 (zero)

A21 (a) $V = V_R$

(b) $V = V_C$

A22 141 Ω

A23 $\frac{\pi}{4}$ (45°)

A24 160 nF = 0.16 µF

A25 $V = \sqrt{V_R^2 + V_L^2} = I^2\sqrt{R^2 + \omega^2 L^2}$

$$\tan\theta = \frac{V_L}{V_R} = \frac{I\omega L}{IR} = \frac{\omega L}{R}$$

$$\therefore\ \theta = \tan^{-1}\left(\frac{\omega L}{R}\right)$$

A26 $Z = \sqrt{5^2 + 12^2} = 13\ \Omega$

$$\therefore\ V_0 = 13\ I_0 = 5.2\ V$$

$$\theta = \tan^{-1}\frac{X_L}{R} = \tan^{-1}\frac{12}{5} = 1.18\ \text{rad}$$

A27 $V = 5.2\cos(\omega t + 1.18)$

A28 (a) 40 Ω

(b) 50 Ω

(c) 0.24 A

(d) 0.93 rad

A29

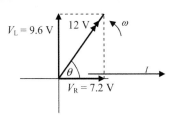

A30 (a) 22.4 Ω

(b) 1.11 rad

A31 $V_0 = \sqrt{2}V_{rms}$ – these are not time variable

But instantaneous values of V are time variable and out of phase

A32 (a) $Q = \dfrac{V_C}{V_R} = \dfrac{I\omega_{res}L}{IR} = \dfrac{\omega_{res}L}{R}$

(b) $\dfrac{\omega_{res}L}{R} = \dfrac{1}{\sqrt{LC}}\dfrac{L}{R} = \dfrac{1}{R}\sqrt{\dfrac{L}{C}}$

A33 $Q = 3.2$

A34 Peak current would be greater (0.6 A); curve much sharper

A35 20 mV div^{-1}

A36 5 µs div^{-1}

A37 $f = 80$ kHz; $V_0 = 48$ mV

A38 (a) $R = 100\ \Omega;\ X_L = 200\ \Omega;\ X_C = 200\ \Omega$

(b) $R = 100\ \Omega;\ X_L = 100\ \Omega;\ X_C = 400\ \Omega$

A39 $V_0 = 1.4$ V; $T = 3.1$ ms

A40 $V_{rms} = 1.0$ V; $f = 320$ Hz

A41 $V_{rms} = 9.2$ V; $I_{rms} = 0.21$ mA, $X_C = 44$ kΩ, $C = 36$ nF

Section 4B

B1 (a) $E = \dfrac{hc}{\lambda},\ \therefore\ \lambda = \dfrac{hc}{E} = \dfrac{6.63 \times 10^{-34} \times 3 \times 10^8}{1.6 \times 10^{-14}}$
$$= 1.24 \times 10^{-11}\ \text{m} \sim 10\ \text{pm}$$

(b) 100 pm, 1 nm

B2 1 kW

B3 (a) 31 pm

(b) Must be greater than 10.2 kV

B4 (a) 36%

(b) 0.26 cm^{-1}

B5 This would need whole body irradiation which would cause too much damage to healthy tissue

B6 Fraction = 1.1×10^{-6} (0.00011%)

B7 1.52 MHz

B8 9.9×10^{-5} s

B9 2.5 cm

B10 Because of the difference in the acoustic impedance between blood fluid and air

B11 130 Hz

B12 (a) 2.5×10^{-26} J

(b) 0.16 µeV (1.6×10^{-7} eV)

B13 250 000 ×

B14 15 MHz

B15 6 tissues × 0.16 = 0.72
1 tissue × 0.08 = 0.08
4 tissues × 0.04 = 0.16
4 tissues × 0.01 = 0.04

Total   1.00

B16 0.76 mSv

B17 (a) Electron anti-neutrino $\overline{v}_e$

(b) $^{99}_{43}\text{Tc}_m \rightarrow {}^{99}_{43}\text{Tc} + \gamma$

(c) 9.4×10^{-5} u

B18 $^{124}_{53}\text{Te} + \text{p} \rightarrow {}^{125}_{53}\text{I} \rightarrow {}^{123}_{53}\text{I} + 2{}^{1}_{0}\text{n}$

$^{123}_{53}\text{I} + {}^{0}_{-1}\text{e} \rightarrow {}^{123}_{52}\text{Te}_m + v_e$

$^{123}_{52}\text{Te}_m \rightarrow {}^{123}_{52}\text{Te} + \gamma$

B19 A photon has momentum $\dfrac{h}{\lambda}$. To conserve momentum, two photons are emitted having equal energy (and hence wavelength) in opposite directions

Section 4C

C1 The C of G is below the midpoint

C2 Horizontal force to left on boat due to water resistance

C3 The anticlockwise moment of the boat's weight will increase

C4 182 N

C5 Because there will be a common factor of $\cos 10°$ in all the terms in the moments equation

C6 2.17 m

C7 Impact speed = 7.06 m s^{-1}

Rebound speed = 5.29 m s^{-1}

C8 $E_{k\,trans} = \frac{1}{2}mv^2$

$E_{k\,rot} = \frac{1}{2}I\omega^2 = \frac{1}{2}(mr^2)\left(\frac{v}{r}\right)^2 = \frac{1}{2}mv^2$

$\therefore E_{k\,tot} = \frac{1}{2}mv^2 + \frac{1}{2}mv^2 = mv^2$ QED

C9 60 J

C10 acceleration = 0.007 m s^{-2}

Tension = 1.9606 N.

This is only a 0.07% change in the tension

C11 kg m^2 s^{-1} (or N m s)

C12 (a) 13.8 kg m^2 s^{-1}

(b) 16.7 rad s^{-1} = 2.7 rev. s^{-1}

C13 Equation gives 16.5 m; so does the figure

C14 (a) 14.3 m

(b) 16.2 m (speed wins over angle)

C15 Drag: [javelin] exerts a forward force on air molecules [N2]

$\therefore$ air exerts an equal backward force on [javelin] [N3]

Lift: [wing] exerts a downward force on air molecules [N2]

$\therefore$ air exerts an equal upward for on [wing] [N3]

C16 'Lift' is backwards for the second half of the flight.

C17 (a) Lift = 1.0 N

(b) For the same lift, the torque will be double so the javelin point rotates downwards too soon

C18 Reasonably so: the drag at 20 m s^{-1} is ~4× that at 10 m s^{-1}

C19 Lift = 39 kN

Section 4D

D1 With 85% energy absorption (albedo = 0.15) the model leads to a temperature of 218 K, which is in good agreement

D2 278 K (5 °C)

D3 Sun: Using 5800 K $\rightarrow \lambda_{max} = 0.5$ μm ✓

Ground: Using 15 °C (288 K) $\rightarrow$ 10 μm ✓

D4 Energy of 7.5 μm photon = 0.17 eV, so vibrational energy levels in N_2O are 0.17 eV apart

D5 0.1%, 0.2% 0.6%

D6 (a) 1.08×10^6 m^3

(b) 0.97×10^6 m^3

D7 Presence of γ photon suggests e-m interaction

D8 Data give: $\Delta m = -0.002\,669\,661$ u

Using 931.5 MeV / u $\rightarrow$ 26.73 MeV

D9 ~ 54 W m^{-2}

D10 (a) $P_{out} = 17.5$ mW

(b) $P_{in} = 50$ mW, $\therefore$ efficiency = 0.35 (35%)

D11 100 parallel branches of ~350 cells in series, $\therefore$ ~35 000 cells

D12 42 W

D13 120 kW

D14 Total area under each graph is 1.0 by definition; location B has a greater max speed so must have a lower maximum probability

D15 (a) 19.8 m s^{-1}

(b) 0.62 m^3 s^{-1}

(c) 620 kg s^{-1}

(d) 120 kW

D16 $[\rho V] = $ kg s^{-1}

$\therefore$ [RHS] = kg s^{-1} × N kg^{-1} × m $-$ (kg s^{-1} m^4 s^{-2})(m^{-2})

= N m s^{-1} $-$ N m s^{-1}

= J s^{-1} $-$ J s^{-1}

= W $-$ W

$\therefore$ RHS homogeneous. And [LHS] = W, so equation homogeneous

D17 (a) ~1.1 GJ

(b) ~ 50 kW

D18 0.9%

D19 (a) 1.4(4), 2.5, $(1.2)^n$

(b) 10.8 (11 to the next whole number)

D20 (a) 2.6×10^{-5} keV

(b) 1.3 keV

D21 0.3 kg

D22 9.8×10^{17} J

D23 $[K] = \dfrac{[\Delta Q][\Delta x]}{[\Delta t][A][\Delta\theta]}$
$= \text{J m (s m}^2\text{ K)}^{-1}$
$= \text{J s}^{-1}\text{ m}^{-1}\text{ K}^{-1}$
$= \text{W m}^{-1}\text{ K}^{-1}$

D24 If K is very small, A needs to be large and Δx small to get a measurable ΔQ

D25 110 W m^{-2}

D26 Comparing the equations $U = \dfrac{K}{\Delta x}$
Using given K values $\rightarrow U_{\text{brick}} = 4.5$ and $U_{\text{ins}} = 0.66$ W m^{-2} K^{-1}

Maths

M1 **(a)** $t = 0, \pm\dfrac{2\pi}{\omega}, \pm\dfrac{4\pi}{\omega}, \ldots$

(b) $t = \pm\dfrac{\pi}{2\omega}, \pm\dfrac{3\pi}{2\omega}, \pm\dfrac{5\pi}{2\omega}, \ldots$

(c) $t = -\dfrac{3\pi}{2\omega}, \dfrac{\pi}{2\omega}, \dfrac{5\pi}{2\omega}, \ldots$

M2 **(a)** 12 cm

(b) 5 Hz

(c) 0.05 s, 0.15 s

M3 **(a)** 1000 s^{-1}

(b) 159 Hz

(c) 62.8 ms

(d) -3.6 ms, 0.5 ms, 2.6 ms, 5.8 ms

M4 $I = 0.1\cos(1000t + 0.5 + \frac{\pi}{2}) = 0.1\cos(1000t + 2.071$ rad)

M5 $\cos\theta = \sin(\theta + \frac{\pi}{2})$
$\therefore \cos(100t - 1.0) = \sin(100t - 1.0 + \frac{\pi}{2})$
$= \sin(100t + 0.57)$
$\therefore$ Yes!

M6

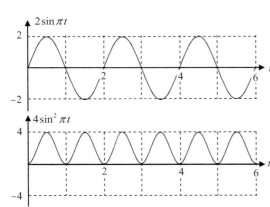

M7 **(a)** 1, 2.7, 4.5, 7.4, 20, 100, 148, 22 000, 2.7×10^{43}

(b) $\sqrt{e^{10}} = e^5$ or $(e^5)^2 = e^{10}$

(c) $(e^{10})^{10} = e^{100}$

M8 **(a)** 0.37, 0.22, 0.15, 0.050, 0.010, 0.0067, 4.5×10^{-5}, 3.7×10^{-44}

(b) $e^{-5} = \dfrac{1}{e^5}$

(c) $e^{-100} = (e^{-10})^{10}$

M9 **(a)** 10^6

(b) 10^7

(c) 14 hours (from the start)

M10 **(a)** $400A$, $160\,000A$, $1.0 \times 10^{26}A$

(b) $403A$, $160\,000A$, $1.1 \times 10^{26}A$

M11 **(a)** $t = \dfrac{0.69}{\lambda}, \dfrac{1.36}{\lambda}, \dfrac{2.07}{\lambda}$ - constant gap

(b) $t = \dfrac{0.51}{\lambda}, \dfrac{1.18}{\lambda}, \dfrac{1.88}{\lambda}$ - constant gap

M12 **(a)** 14.0 days

(b) 0.049 day^{-1}

M13 $\lambda = 0.0472$ s, $C = 2.1$ mF

M14 **(a)** 0.69,

(b) 1.38,

(c) 5.0,

(d) 4.61

(e) 432

M15 **(a)** 10

(b) 48

M16 **(a)** $r = \sqrt{\dfrac{3P}{\rho\pi v^2}}$

(b) $v = \sqrt[3]{\dfrac{3P}{\rho\pi r^2}}$

M17 $\log_{10} 2^x = \log_{10} 35$
$\therefore x = \dfrac{\log_{10} 35}{\log_{10} 2} = 5.13$

M18 Rearranging: $e^{\lambda t} = \dfrac{V_0}{V}$

Taking logs: $\lambda t = \ln\left(\dfrac{V_0}{V}\right)$
$\therefore \lambda = \dfrac{1}{t}\ln\left(\dfrac{V_0}{V}\right)$
Alternative forms: $\lambda = -\dfrac{1}{t}\ln\left(\dfrac{V}{V_0}\right)$, $\lambda = \dfrac{1}{t}\ln(\ln V_0 - \ln V)$

M19 Gradient = 3; intercept = ln 10

Exercises answers

Exercise 3.1

1. **(a)** 754 rad s^{-1} **(b)** 25000 m s^{-2}

2. **(a)** 1.7×10^{-10} m s^{-2} **(b)** 3.4×10^{20} N

3. $a_{\text{geosynch}} = 0.223$ m s^{-2}

   $\dfrac{a_{\text{ISS}}}{a_{\text{geosynch}}} = \dfrac{8.66}{0.223} = 38.8$; $\left(\dfrac{r_{\text{geosynch}}}{r_{\text{ISS}}}\right)^2 = \left(\dfrac{42164}{6783}\right)^2 = 38.6$

4. **(a)** 5000 N **(b)** 26.8 m s^{-1}

5. **(a)** $C \cos\theta + G \sin\theta = mg$

   (b) $C \sin\theta - G \cos\theta = \dfrac{mv^2}{r}$

   (c) $G = -3850$ N

   (d) $58°$

6. 20 rad s^{-1}

7. $4.6°$

8. (At the top) $T_{\min} = 137$ N (3 s.f.)

   (At the bottom) $T_{\max} = 147$ N

9. **(a)** 1.72 m s^{-1}

   (b) $E_k = 0.059$ J; $E_p = 0.235$ J

   (c) 3.84 m s^{-1}

10. **(a)** 0.0333 N **(b)** 4.04 m s^{-1}

Exercise 3.2

1.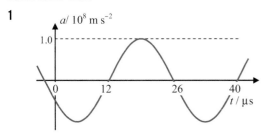

2. $x(0) = 0.87$ mm (2 s.f.)

   $v(0) = 400$ m s^{-1}

   $a(0) = -4.4 \times 10^7$ m s^{-2}

3. ± 300 m s^{-1} (2 s.f.); at any displacement, the motion can be towards or away from the equilibrium point

4. **(a)** 2.2 s, **(b)** 2.5×10^{-4}, **(c)** 21.6 s

5. $x = 2.0 \sin\left(\dfrac{2\pi}{28 \times 10^{-6}}t + 0.45\right)$ mm

6. **(a)** $x / \text{cm} = 2.39 \cos\left(\frac{4}{3}\pi t - \frac{\pi}{2}\right)$

   (b) -2.1 cm, -5.0 cm s^{-1}, 36 cm s^{-2}

7. **(a)** 47 mm s^{-1}, 0.30 m s^{-2}

   (b) $x / \text{mm} = 7.5 \cos(2\pi t + 0.84)$

8. **(a)** $T = \pi(\sqrt{2} + 1)\sqrt{\dfrac{l}{g}}$

   (b)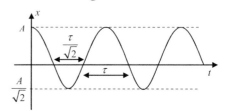

9.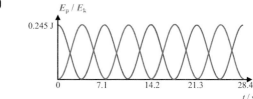

10. **(a)** $x / \text{cm} = 20e^{-0.555t} \cos 4\pi t$

    (b)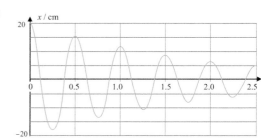

    (c) 3.75 cm

11. Results give $g = 9.83 \pm 0.05$ m s^{-2}

    The range includes 9.81 m s^{-2}, so consistent

12. $g = 9.90$ cm s^{-2}, $h = 4.66$ m

13. Ratio of successive amplitudes are all approximately 0.8 (range, 0.74 – 0.85) so approximately exponential [alternatively, a graph of $\ln A$ against t is a straight line]

    $\lambda \sim 0.0069$ s^{-1}

Exercise 3.3

1. 3.1×10^{24}

2. $n = 840$ mol, $N = 5.1 \times 10^{26}$

3. 96.3 kC

4. 2.23×10^{19} Na$^+$ ions

5. **(a)** $M_r = 352.02$

   (b) $M = 352.02$ g mol^{-1} = 0.352 02 kg mol^{-1}

   (c) 20 mol

6. 280 pm

7. Data give $N_A = 6.023 \times 10^{23}$ mol^{-1}

8 **(a)** 498 m s^{-1}

(b) 42.3 mol

(c) 28.95 g mol^{-1}, c.f. 28.94 g mol^{-1} from composition data

9 $R = 8.30$ J kg^{-1} k^{-1}

10 **(a)** **(i)** 290 m s^{-1} **(ii)** 91 500 m^2 s^{-2} **(iii)** 302 m s^{-1}

(b) 117 K

11 **(a)** Speed ratios – $1 : \dfrac{1}{\sqrt{2}} : \dfrac{1}{2} : \dfrac{1}{4} : \dfrac{1}{4\sqrt{2}}$

(b) Speeds in m s^{-1}, 1110, 787, 557, 278, 197 respectively

12 **(a)** 1×10^{44} molecules

(b) For you to decide!

Exercise 3.4

1 **(a)** 720 kJ

(b) At a higher temperature there are fewer molecules, each with a greater mean KE; these effects cancel

(c) **(i)** 288 K, **(ii)** 1340 m s^{-1}

2 **(a)** 374 kJ

(b) 484 m s^{-1}

(c) The molecules possess 250 kJ of rotational kinetic energy

3 **(a)** $U_A = 3.6$ MJ, $U_B = 2.4$ MJ

(b) $T_A = 578$ K, $T_B = 385$ K

(c) **(i)** 481.5 K

(ii) $p_A = 1.00 \times 10^6$ Pa, $p_B = 5.00 \times 10^6$ Pa

(iii) 608 kJ from A to B

(d) $p = 667$ Pa, $T = 482$ K

4

	ΔU	Q	W
(a)	>0	$=0$	<0
(b)	>0	>0	>0
(c)	>0	>0	~ 0
(d)	$=0$	>0	>0
(e)	>0	>0	~ 0
(f)	$=0$	$=0$	$=0$
(g)	$=0$	$=0$	$=0$

5 **(a)** 170 kJ

(b) 0.075 (7.5%)

(c) $Q = 2.26$ MJ, $\Delta U = 2.09$ MJ, $W = 0.17$ MJ

6 **(a)** 76 kJ

(b) $Q = 76$ kJ, $\Delta U = 76$ kJ, $W = 0$

(c) 34 g

7 **(a)** 0.2, 0.4 and 2.0 respectively

(b) 0.2, 0.4 and 2.0 respectively

(c) & (d)

	ΔU/ kJ	Q/ kJ	W/ kJ
A→B	−12	−20	−8
B→C	3	3	0
C→D	24	40	16
D→A	−15	−15	0
Total	0	8	8

(e) 3.4 mol

8 12 MW

9 502 J kg^{-1} K^{-1}

10 **(a)** 11.2 J [by square counting]

(b) pV is less at C than B. $T_C = 392$ K

(c) 0.0015 mol

(d) 15%

Exercise 3.5

1 **(a)** Gradient = −0.0133, intercept = 6.306

(b) $\lambda = 0.0133$ s^{-1}, $T = 52$ s

2 **(a)** ${}^{192}_{78}\text{Pt} \rightarrow {}^{188}_{76}\text{Os} + {}^{4}_{2}\text{He}$

(b) ${}^{181}_{72}\text{Hf} \rightarrow {}^{181}_{73}\text{Ta} + {}^{0}_{-1}\text{e} + \bar{v}_e$

(c) ${}^{48}_{23}\text{V} \rightarrow {}^{48}_{22}\text{Ti} + {}^{0}_{1}\text{e} + v_e$

(d) ${}^{59}_{28}\text{Ni} + {}^{0}_{-1}\text{e} \rightarrow {}^{59}_{27}\text{Co} + v_e$

(e) ${}^{77}_{32}\text{Ge} \rightarrow {}^{77}_{33}\text{As} + {}^{0}_{-1}\text{e} + \bar{v}_e$

(f) ${}^{65}_{30}\text{Zn} \rightarrow {}^{65}_{29}\text{Cu} + {}^{0}_{1}\text{e} + v_e$

(g) ${}^{65}_{30}\text{Zn} + {}^{0}_{-1}\text{e} \rightarrow {}^{65}_{29}\text{Cu} + v_e$

(h) ${}^{239}_{92}\text{U} \rightarrow {}^{239}_{93}\text{Np} + {}^{0}_{-1}\text{e} + \bar{v}_e$

${}^{239}_{93}\text{Np} \rightarrow {}^{239}_{94}\text{Pu} + {}^{0}_{-1}\text{e} + \bar{v}_e$

(i) ${}^{17}_{7}\text{N} \rightarrow {}^{16}_{8}\text{O} + {}^{1}_{0}\text{n} + {}^{0}_{-1}\text{e} + v_e$

3 For the isotopes with a neutron excess, Ne23 & Ne24, there is a vacant proton state with a low enough energy level for the transformation $n \rightarrow p + e^- + \bar{v}_e$ to occur

For the isotopes with a neutron deficit, Ne18 & Ne19, there is a vacant neutron state with a low enough energy level for the transformation $p \rightarrow n + e^+ + v_e$ to occur

For Ne20, Ne21 and Ne22 neither of these is the case, so the nucleus is stable

4 **(a)** The fractional uncertainty decreases with the total count; a longer count time is used for the lower count rates to increase the total count

(b)

Distance / cm	10	15	20	25	30	50	70
Corrected count, C	729	256	129	89	65.7	22.0	10.5
$1 / \sqrt{C}$ /10^{-3}	37	62.5	88	106	123	213	309

(c) $k = 51700$ cpm cm^2 = 5.2 cpm m^2

$\varepsilon = -1.2$ cm

5 **(a)** $\lambda_{234} = 8.8 \times 10^{-14}$ s^{-1}, $\lambda_{235} = 3.1 \times 10^{-17}$ s^{-1}, $\lambda_{238} = 4.9 \times 10^{-18}$ s^{-1}

(b) U234: 63.5%, U235: 1.6%, U238: 34.9%

(c) 35 MBq

6 Most α particles absorbed by the material; penetration range $\ll 0.1$ mm

7 **(a)** Number of half-lives $\sim 10^4$

$\therefore$ Fraction remaining $= 2^{-10000} = (2^{-10})^{1000} = (10^{-3})^{1000}$
$= 10^{-3000}$

10^{3000} is (vastly) more than the number of baryons in the universe (estimated as 10^{72}), hence there is nothing left

(b) $^{238}_{92}\text{U} \xrightarrow{\alpha} {}^{234}_{90}\text{Th} \xrightarrow{\beta} {}^{234}_{91}\text{Pa} \xrightarrow{\beta} {}^{234}_{92}\text{U}$

8 **(a)** A can only change by 0 or -4; no combination of 0s and 4s can produce a change of -3

(b) U238: 94.4%, U235: 5.6%

(c) 6×10^9 years

9 **(a)** Argon is an unreactive gas – unlikely to bond to other atoms in the rock, hence any argon would have escaped from the original magma

(b) 5.33×10^{-10} year^{-1} = 1.69×10^{-17} s^{-1}

(c) As the rock ages, the number of K40 atoms decreases and the number of Ar40 increases, hence the ratio Ar40:K40 increases. Assumption: all the argon produced remains trapped in the rock.

(d) $^{40}_{19}\text{K} \rightarrow {}^{40}_{18}\text{Ar} + {}^{0}_{1}\text{e} + v_e$

$^{40}_{19}\text{K} + {}^{0}_{-1}\text{e} \rightarrow {}^{40}_{18}\text{Ar} + v_e$

(e)

Time / Myr	0	100	200	300	400	500	600
K40	100	94.8	89.9	85.2	80.8	76.6	72.6
Ar40	0	0.52	1.01	1.48	1.92	2.34	2.74
Ar/K	0	0.006	0.011	0.017	0.024	0.031	0.038

(f) 340 million years, this is the least possible age.

If some argon escaped the true ratio would be higher

10 Using $\pm\sqrt{N}$ as the uncertainty, over 10 minutes:

Background = 250 ± 16 counts

No absorber = 570 ± 24 counts,
so source = 320 ± 40 counts

With absorber = 525 ± 23 counts,
so source now = 275 ± 39 counts

The uncertainty ranges overlap considerably so it is not safe to conclude that the source is an alpha emitter

11 Likely to be β

12 **(a)** 13.5 s^{-1}, **(b)** 10.1 s^{-1}

Exercise 3.6

1 **(a)** 3.0×10^{16} J

(b) 0.33 kg

2 **(a)** 3.6×10^{26} W

(b) 1.6×10^{17} W

3 **(a)** $^{7}_{4}\text{Be} + {}^{0}_{-1}\text{e} \rightarrow {}^{7}_{3}\text{Li} + v_e$

(b) Weak: change of quark flavour (u→d), neutrino involvement

(c) 0.86 MeV; mainly carried by neutrino because it has equal (and opposite) momentum to the daughter nucleus and is much lighter

4 **(a)** $^{241}_{95}\text{Am} \rightarrow {}^{237}_{93}\text{Np} + {}^{4}_{2}\text{He}$

(b) 5.50 MeV = 8.8×10^{-13} J

(c) 2.1×10^6 m s^{-1}

(d) 1.2×10^8 Bq

5 **(a)** β^- gives Ca40: $^{40}_{19}\text{K} \rightarrow {}^{40}_{20}\text{Ca} + {}^{0}_{-1}\text{e} + \overline{v}_e$

β^+ and K capture $\rightarrow$ Ar40 (see Ex 3.5, Q9)

(b) 1.3 MeV

6 **(a)** $^{239}_{92}\text{U} \rightarrow {}^{239}_{93}\text{Np} + {}^{0}_{-1}\text{e} + \overline{v}_e$

$^{239}_{93}\text{Np} \rightarrow {}^{239}_{94}\text{Pu} + {}^{0}_{-1}\text{e} + \overline{v}_e$

(b) U→Np: 1.26 MeV

Np→Pu: 0.72 MeV

(c) Some energy released as KE of the neutrinos, which are not absorbed in the reactor

7 7.3 MeV = 1.2×10^{-12} J

8 **(a)** 168 MeV = 2.7×10^{-11} J

(b) From the Cs: 11 MeV = 1.8×10^{-12} J

From the Rb: 13.7 MeV = 2.2×10^{-12} J

(c) 7.9×10^{13} J (including the subsequent decays)

6.9×10^{13} J from the fission only

9 **(a)** Step 1: 1.44 MeV

Step 2: 5.49 MeV

Step 3: 12.85 MeV

(b) [Note: Total release = 2 × (step 1 + step 2) + step 3]

6.3×10^{14} J

10 **(a)** $Q_{Sn} = 8.0$ aC, $Q_{Mo} = 6.7$ aC

(b) $r_{Sn} = 5.5$ fm, $r_{Mo} = 5.2$ fm

(c) 4.5×10^{-11} J (~280 MeV)

11 Both graphs are useful: the ln graph for the value of the exponent, the linear graph for k. For just the heavier nuclides ($A \sim 120+$) r is proportional to the cube root of A. The relationship departs from this for the lighter nuclides. Again for just the heavier nuclides, the data return a value of 0.9 fm, slightly lower than 1.1 fm.

Exercise 4.1

1. (a) 1.77×10^{-7} F $= 177$ nF

   (b) 8.85×10^{-5} C $= 88.5$ μC

   (c) 0.022 J $= 22$ mJ

   (d) 5×10^{6} V m^{-1} $= 5$ MV m^{-1}

2. 2.8 μF (ignoring dielectric effect)

3. (a) $\frac{2}{3}C$, (b) 3 C

4. (a) $V_R = V$, $V_C = V_{2C} = 0$, $I_0 = \frac{V}{R}$

   (b) (i) $\frac{V}{2R}$, (ii) $\frac{V}{R}$, (iii) (L→R) $\frac{1}{3}CV$, $-\frac{1}{3}CV$, $\frac{1}{3}CV$, $-\frac{1}{3}CV$

   (iv) $V_C = \frac{1}{3}V$, $V_{2C} = \frac{1}{6}V$,

   (v) $U_C = \frac{1}{18}CV^2$, $U_{2C} = \frac{1}{36}CV^2$, $U_{Tot} = \frac{1}{12}CV^2$

   (vi) $P_R = \frac{V^2}{4R}$

   (c) (i) 0, (ii) 0, (iii) (L→R) $\frac{2}{3}CV$, $-\frac{2}{3}CV$, $\frac{2}{3}CV$, $-\frac{2}{3}CV$

   (iv) $V_C = \frac{2}{3}V$, $V_{2C} = \frac{1}{3}V$

   (v) $U_C = \frac{2}{9}CV^2$, $U_{2C} = \frac{1}{9}CV^2$, $U_{Tot} = \frac{1}{3}CV^2$

   (vi) $P_R = 0$

5. (All in μF). Individual capacitors: 10, 22, 33

   Series combinations: 5.7, 6.9, 7.7 and 13.2

   Parallel combinations: 32, 43, 55, 65

   Parallel pair in series with single: 8.5, 14.6, 16.2

   Series pair in parallel with single: 23.2, 29.7, 39.9

6. 11 mF

7. 112 μF

8. (a) 2.2 mC, 11 mJ

   (b) $Q_{220} = 1.51$ mC, $Q_{100} = 0.69$ mC

   (c) 7.6 mJ

9. Capacitance $\rightarrow \frac{1}{2}C$ using $C = \frac{\varepsilon_0 A}{d}$

   pd $\rightarrow 2$ V because Q constant and C halved [or electric intensity, E, and double the separation]

   $Q \rightarrow Q$ because there is nowhere else for the charge to go

   $U \rightarrow 2U$, from any 2 of C, V and Q

10. The pd is the pd across the battery, hence $V \rightarrow V$

    As in Q9, $C \rightarrow \frac{1}{2}C$

    $Q \rightarrow \frac{1}{2}Q$ because C halved and V constant

    $U \rightarrow \frac{1}{2}U$ from any two of C, V and Q

11. (a) 56.4 mC

    (b) I / mA $= 120e^{-0.213t}$

    (c) Area under graph = 56.4 mC

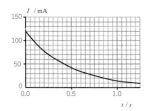

   (d) $U_0 = 0.34$ J

   (e) U / mJ $= 340e^{-4.26t}$

   (f) P / W $= 1.44e^{-4.26t}$

   (g) Area under graph = 0.34 J

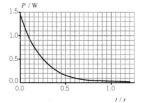

Exercise 4.2

1. Field lines begin and end on the same conductor

   Field line has a sharp corner

   Field lines meet a conductor not at right angles

   Field lines in rings (no beginning and end)

   Field lines change direction

   Field lines in opposite directions between the same two conductors

2.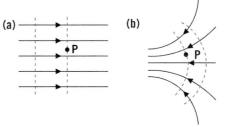

3. (a) X is at the higher potential; work needs to be done to move a positive charge from Y to X

   (b) 15 kV

   (c) 150 mJ

   (d) 1.70×10^{6} m s^{-1}

   (e) 1.23×10^{6} m s^{-1}

4. (a) Uniform distribution on outer surface of the metal

   (b) $V = 108$ kV, $E = 1.08$ MV m^{-1}

   (c) (i) 54 kV, 270 kV m^{-1}

       (ii) 108 kV, 0

5. (a) Repelled by like charge on the sphere

   (b) 1100 m s^{-2}

   (c) 15 m s^{-1}

6. (a) -29.6 J

   (b) -20 μC: 254 N at 45° up to the right

       top 10 μC: 152 N at 12.1° to the left of vertical

       right 10 μC: 152 N at 12.1° below horizontal to the left

7. (a) 254 km s^{-2}

   (b) Accelerates along line of symmetry, reaches maximum speed at midpoint of the two 10 μC charges, decelerates to rest at opposite corner of square; accelerates back and slows down to rest at initial point, etc.

   (b) 173 m s^{-1}

8. (a) 1.50×10^{22} kg, 3580 kg m^{-3}

   (b) 1.00 m s^{-2}

(c) -1.00×10^6 J kg^{-1}

(d) -1.00×10^8 J

9   (a) -2.00×10^5 J kg^{-1}, 0.04 N kg^{-1}

(b) -5.00×10^5 J kg^{-1}, 0.25 N kg^{-1}

10 (a) 1.00×10^6 J

(b) 9.1×10^5 J

(c) 9%

11 (a) 0.100 N kg^{-1}

(b) (i) $-10\,000$ J kg^{-1}

    (ii) $-60\,000$ J kg^{-1}

12 (a) 0.0025 m s^{-2}

(b) 6.5 days

13 (a) 613 km from the centre

(b) The maximum potential energy point

Exercise 4.3

1   0.63 (or 1.59)

2   3.375 (or 0.296)

3   (a) $M_J = 1.90 \times 10^{27}$ kg

(b) $T = 16.9$ days

4   (a) First three points all very close – last one a long way away

(b) gradient $= \frac{3}{2}$, intercept $= \ln\frac{2\pi}{\sqrt{GM}} = \frac{1}{2}\ln\frac{4\pi^2}{GM}$

(c) $m = 2$, $n = 3$; $m = \frac{2}{3}$, $n = 1$; $m = 1$, $n = \frac{3}{2}$ …

(d) Best graph is $T^{\frac{2}{3}}$ against m; data give 8.5×10^{25} kg

5   (a) $a = 1.43 \times 10^{14}$ m

    $M = 7.0 \times 10^{36}$ kg $= 3.5 \times 10^6$ $M_\odot$

(b) $v = 0.023c$

(c) $p = 0.024 \rightarrow \Delta M = \pm 0.8$ $M_\odot$

6   (a) $d = 1.88 \times 10^9$ m; $r_{1.4} = 8.09 \times 10^8$ m; $r_{0.6} = 1.88 \times 10^9$ m

(b) $v_{1.4} = 93.9$ km s^{-1}; $v_{0.6} = 219$ km s^{-1}

(c) $\Delta\lambda_{1.4} = 0.205$ nm; $\Delta\lambda_{0.6} = 0.479$ nm

Exercise 4.4

1   (a) ↓ (b) → (c) ↓ (d) → (e) ↑

2   (a) 0.0125 N 30° above horizontal North

(b) 0.0108 N; 0.006 25 N

3

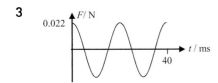

4   (a) 1.5 N m

(b) 0.75 N m

5   (a)

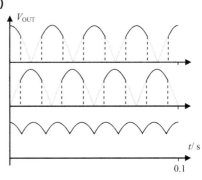

(b) The 'ripples' in the torque would be smaller.

6   (a) 1.71×10^{-23} N s

(b) 0.00213 T

(c) 60 MHz

7   (a) 6.08×10^{-5} T

(b) 930 Hz

8   Radius $= 3.77$ m; pitch $= 41.1$ m

9   (a) 7.62 MHz

(b) 4.0 MeV (6.4×10^{-13} J)

(c) 0.578 m

(d) Energy $> m_e c^2$, so need for relativistic correction

10 (a) 29 000 m s^{-1}

(b) 3.80×10^{-26} kg $= 22.9$ u. $\therefore$ Na$^+$

11 (a) Into the diagram

(b) $v_X = \sqrt{\dfrac{2eV}{m_X}}$ and $v_Y = \sqrt{\dfrac{2eV}{m_Y}}$

(c) $r_X = \sqrt{\dfrac{2Vm_X}{B^2 e}}$ and $r_Y = \sqrt{\dfrac{2Vm_Y}{B^2 e}}$

(d) (i) 120 V (ii) 0.39 cm

12 (a) 3380 Ω

(b) 16.9 V

(c) 0.313 V T^{-1}

(d) 0.34 V

Exercise 4.5

1   The flux linking the coil is decreasing

2   (a) Decreasing

(b) Decreasing

(c) To the right – the direction opposes the decrease in flux

(d) Clockwise viewed from the left

3   (a) The force is to the left – this opposes the increasing separation of the magnet and loop

(b) Towards the viewer at the top (by FLHR)

(c) At the bottom, the current is away from the viewer, so clockwise viewed from the left

4   The flux would be increasing to the left, so the flux due to the induced current is to the right as in 2

The force on the loop must be to the right, so (by FLHR) the current is towards the viewer at the top, as in 3

5   **(a)**   The flux changes slower then more quickly; at the midpoint the rate of change of flux is zero, after this the direction of the change of flux is opposite

(b)   The magnet is travelling more quickly so the rate of change of flux is more rapid

6   **(a)**   0.85 V

(b)   There is no return path for any current

7   **(a)** 1 Wb-turn, **(b)** 0.5 Wb-turn, **(c)** 0

8   **(a)** $N\Phi = \sin 4\pi t$

(b) $\mathscr{E}_{in} = 4\pi \cos 4\pi t$, max voltage = 12.6 V

9

10   $\langle P \rangle = 5.1$ W

11   **(a)** $f = 40$ Hz, $\omega = 251$ rad s^{-1}

(b) $B = 0.24$ T

(c) $T = 12.5$ ms, $V_0 = 12.0$ V

Exercise 4A.1

1   **(a)**   **(i)** 0.90 A,   **(ii)** 8.1 W

(b) $V = 12.7 \cos 100\pi t$ ($12.7 \cos 314t$)

(c) $I = 1.27 \cos 100\pi t$

(d)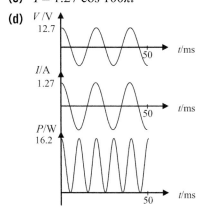

2   1 μF cap: $I_{rms} = 0.075$ A

100 mH ind: $I_{rms} = 0.019$ A

3   $V_{peak}(4.7\ \Omega) = 4.6$ V

4   **(a)** $I_{rms} = 0.70$ A

(b) $I = 0.99 \cos 100\pi t$

5   **(a)** $V_R = 23.5$ V; $V_C = 50$ V

(b) V_R stays the same as its resistance is constant; V_C halves (to 25 V) because its reactance is inversely proportional to the frequency

6   B = resistor (current and voltage in phase). $R = 1.4$ kΩ

The frequency is 31.25 Hz (needed for A and C)

A = capacitor (current leads the voltage). $C = 5.1$ μC

C = inductor (voltage leads the current). $L = 6.1$ F (rather large)

Exercise 4A.2

1   **(a)**   11.2 V

(b)   5.6 Ω

(c)   Resistor: $R = 2.5$ Ω, $X = 0$

Capacitor: $R = 0$, $X = 10$ Ω

Inductor: $R = 0$, $X = 5$ Ω

(d) $R = 2.5$ Ω; $C = 320$ μF; $L = 16$ mH

(e) $\varepsilon = 1.11$ rad (current leads)

(f)   10 W

2   $X_C = 5$ Ω; $X_L = 10$ Ω.

All other quantities remain the same except that the voltage phasor now leads (by 1.11 rad)

3   **(a)** $I_{rms} = 0.707$ A

(b) $L = 0.04$ H = 40 mH

(c) $Z = 23.3$ Ω

(d) $V_0 = 2.33$ V; $V_{rms} = 1.65$ V

(e)   **(i)**   2.0 V (pd across inductor = 0)

(ii)   0.97 V

4   $V = 2.33 \cos (300t + \tan^{-1}0.6)$

5   **(a)** $V_{rms} = 17.0$ V

(b) $C = 55.6$ μF

(c) $Z = 60$ Ω

(d) $I_0 = 0.40$ A

(e) $I_{rms} = 0.283$ A

(f) $I(0) = 0.320$ A

(g) $V_R(0) = 15.4$ V, $V_C(0) = 8.6$ V

Exercise 4A.3

1   **(a)** $T = 92$ μs; $f = 10.9$ kHz

(b) $\Delta T = 12$ μs

(c) $\varepsilon = 0.82$ rad

(d) $V_A(rms) = 7.2$ V; $V_B(rms) = 0.14$ V

(e)   **(i)** $I_{rms} = 0.36$ A ; $\langle P \rangle = 2.6$ W

(ii) $I_{rms} = 1.1$ mA; $\langle P \rangle = 0$

(iii) $I_{rms} = 49$ mA; $\langle P \rangle = 0$

2   **(a)**   A Y gain of 0.5 V div^{-1} will give an amplitude of 3 divisions making food use of the screen

TB doesn't really matter (see ans to Q3); needs at least 20 ms across the screen, which suggests 5 ms div^{-1}

(b) 0.2 V div^{-1}; for TB at least 5 ms across screen so something like 1 ms div^{-1} or more

3 The time-base setting doesn't matter as long as the separate cycles can be clearly seen and counted and that the traces crosses the central (calibrated) line

To measure the peak voltage:

- Increase the Y-gain so the trace occupies as much vertical space as possible for accuracy

- Move trace vertically so the troughs sit on one of the horizontal lines

- Move trace horizontally so that one of the peaks is on the vertical centre line

4 (a) $f = 500$ Hz; the time shift ~ 0.32 ms giving $\varepsilon = 1.02$ rad

This leads to 0.75 H

(b) From peak voltages, $Z \sim 1.8R$ giving $X_C = 2.25$ kΩ and thus $L = 0.72$ H

The two results are similar

5 $I_{rms} = 0.0047$ A; $\langle P \rangle = 0.033$ W

6 Twice as many cycles seen (period = 2.0 div)

Height of V_R trace reduced to 2.28 div (peak to peak)

Time difference = 0.4 div

For accuracy: speed up time base to 0.2 ms/div and increase V_R trace Y-gain to 2 V/div

7 V_{RC} trace after V_R trace with same time delay and the same amplitude.

Exercise 4B

1 (a) $E_{154} = 8.07$ keV; $E_{139} = 8.94$ keV

(b) At least 9 kV

2 (a) Filtering preferentially cuts out low energy photons because they are less penetrating

(b) Low energy photons have carcinogenic properties without being useful for imaging

(c) between 35 kV and 40 kV

(d) $\lambda_{22\,keV} = 57$ pm; $\lambda_{25.5\,keV} = 49$ pm

(e)
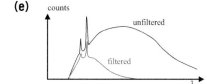

3 (a) (i) 1.0%, (ii) 31.9%, (c) 3.3%

(b) Assuming 8% penetration at 80 keV, $\mu = 0.25$ cm^{-1}

4 (a) Fraction reflected = 0.0013 = 0.13%

(b) 99.7%

5 0.31 m s^{-1}

6 (a) 68.2 MHz

(b) 8520 Hz

(c) Different locations emit different radio frequencies and can therefore be located

(d) 1.7 kHz

7 (a) All the X-rays pass through the tumour; other places receive only a fraction

(b) Most of the energy from a proton beam is discharged at the end of the track. If the correct energy is selected (depending on the depth of the tumour) nearly all the energy is released in the tumour, minimising the effect on tissues before and behind the tumour

8 (a) I123 – useful as a tracer in conjunction with a gamma camera; the γ-rays are not absorbed by the body; its short half-life minimises the long-term effects

I131 – can be used for radiotherapy, the β particles being absorbed by the thyroid (including the tumour); it can also be used as a tracer but is less safe because of the β emissions and the longer half-life

(b) $^{131}_{53}\text{I} \rightarrow {}^{131}_{54}\text{Xe}^* + {}^{0}_{-1}\text{e} + \overline{v}_e$

$^{131}_{54}\text{Xe}^* \rightarrow {}^{131}_{54}\text{Xe} + \gamma$

Exercise 4C

1 A more upright boat presents a larger sail area to the wind. If it heels at angle θ the effective sail area is $A \cos \theta$. The sailor in Fig. C4 should lean further out, thus increasing the anti-clockwise moment of his/her weight.

2 The rudder diverts the water to the starboard (right). The water exerts an opposite force (N3) on the rudder and hence provides a clockwise moment to the boat.

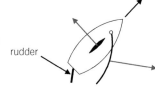

The rudder also exerts a forward force on the water which exerts an equal and opposite force on the rudder.

3 The boat is stable because of the anticlockwise moment exerted by its weight. If the CoG is higher this moment will be less for any particular angle of heel.

4 (a) [Ignoring bounce duration] $t_{max} = 1.095$ s, t_{min} 1.049 s

(b) 1 cm outside range gives a 4 ms difference

5 $e_{max} = 0.84(4)$, $e_{min} = 0.79(5)$

6 29 kN

7 (a) Flight time = 1.91 s; distance = 20.0 m

(b) The flight time is greater (1.92 s) but because the horizontal velocity is less, so is the distance (19.1 m), so this is not advisable

8
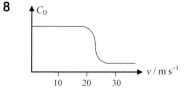

9 Initially, the speed of the ball is high (~ 30 m s^{-1}) giving a low drag force and therefore a small deflection. As the ball slows down (past the 'wall') its drag coefficient becomes much greater, increasing the drag force and therefore the Magnus force. This, in conjunction with the lower speed, gives a greater curvature to the path.

Exercise 4D

1 5880 K

2 (a) 3.847×10^{26} W

   (b) 1.368 kW m^{-2}

3 (a) 40 cm^3

   (b) 43.6 cm^3

   (c) No change; the 43.6 cm^3 of ice melts down into 40 cm^3 of water

   (d) Only melting land-bound ice raises sea level; melting floating ice (e.g. Arctic ice cap) has no direct effect

4 (a) 0.13

   (b) Assuming zero reflection → 167 K; the actual albedo is 0.09, which leads to 163 K

5 106 kW

6 (a) 24 000 kg s^{-1}

   (b) 1.77 MW

   (c) 430 kW

   (d) 1.1 MW

   (e) 60%

7 (a) At 300 K, $kT = 4.12 \times 10^{-21}$ J = 0.026 eV

   (b) $^{233}_{90}\text{Th}^* \rightarrow {}^{233}_{90}\text{Th} + \gamma$
   $^{233}_{90}\text{Th} \rightarrow {}^{233}_{91}\text{Pa} + {}^{0}_{-1}\text{e} + \bar{v}_\text{e}$
   $^{233}_{91}\text{Pa} \rightarrow {}^{233}_{92}\text{U} + {}^{0}_{-1}\text{e} + \bar{v}_\text{e}$

   (c) To provide the source of neutrons for the first step in the reaction chain

   (d) Because its half-life is longer than U235 (7×10^8 years) or U238 (4.2×10^9 years)

8 When U238 absorbs a neutron it decays in two stages to Pu239 which is fissile; the fission of Pu239 provides the additional energy

9 (a) 6.7 kW

   (b) 192 W

   (c) The temperature of the inner surface of the glass is much less than 25 °C and that of the outer surface much greater than 5 °C; a 'skin' of still air on either side of the pane adds to the insulation

   (d) 12.6 W m^{-2} K^{-1}
   [the assumption is not a very good one]

Exercise M

1 (a) $-\cos \omega t$

   (b) $\sin \omega t$

   (c) $\cos \omega t$

   (d) $\cos \omega t$

   (e) $-\sin \omega t$

2 (a) $\cos(\omega t \pm \pi)$

   (b) $\cos(\omega t + \frac{\pi}{2})$

   (c) $\cos(\omega t - \frac{\pi}{2})$

   (d) $\cos(\omega t - \frac{\pi}{2})$

   (e) $\cos \omega t$

3 (a) ± 15 ms, ± 5 ms

   (b) ± 20 ms, 0 ms

   (c) ± 3.3 ms, ±16.7 ms

   (d) ± 6.7 ms, ±13.3 ms

4 (a) (i) −20 ms, −10 ms, 0, 10 ms, 20 ms

      (ii) −12.5 ms, −2.5 ms, 7.5 ms, 17.5 ms

      (iii) −11.7 ms, −1.7 ms, 8.3 ms, 18.3 ms

   (b) (i) −5 ms, 15 ms

      (ii) −7.5 ms, 12.5 ms

      (iii) −16.7 ms, 3.3 ms

   (c) (i) −8.3 ms, −1.7 ms, 11.7 ms, 18.3 ms

      (ii) −10.8 ms, −4.2 ms, 9.2 ms, 15.8 ms

      (iii) −20 ms, −13.3 ms, 0, 6.7 ms, 20 ms

   (d) (i) −18.3 ms, −11.7 ms, 1.7 ms, 8.3 ms

      (ii) −14.2 ms, −0.8 ms, 5.8 ms, 19.2 ms

      (iii) −10 ms, −3.3 ms, 10 ms, 16.7 ms

5 (a) $f = 400$ Hz, $T = 2.5$ ms

   (b) $I = 3.5 \cos (2513t - 0.785)$

   (c) (i) 2.5 A, (ii) 2.5 A, (iii) −2.5 A

   (d) 0.9 ms, 2.2 ms, 3.4 ms and 4.7 ms

   (e) 0.3 ms and 2.8 ms

6 (a) 33, (b) 0.33 (c) 3.2 (d) 0.53 (e) 4.55

7 (a) −1 (b) 15 (c) 6 (d) 216 (e) 24

8 (a) $x = 12.2$

   (b) $x = 2.04$

   (c) $x = 0.777$

   (d) $t = 9.16$

   (e) $t = 1.56 \times 10^4$

9 (a) 2.7 V (b) 497 s

10 The graph of $\ln E_\text{v}$ against $\ln V$ is a good straight line
   n = gradient = 3.34; $k = e^{\text{intercept}} = 0.032$

11 (a) gradient = −0.167

   (b) $C = 1.3$ mF

Exam practice question answers

Question 1: Vibrations and circular motion

(a) (i) The timings are made to the centre of the oscillation rather than the end of the swing. The stopwatch is started and stopped as the vertical string passes in front of a suitable mark. Anticipation both at starting and stopping enables this degree of accuracy.

(ii)

l / m	Time for 10 oscillations / s	Period^2, $(T / s)^2$
0.700	16.85 ± 0.20	2.84 ± 0.07
0.800	18.20 ± 0.20	3.31 ± 0.07

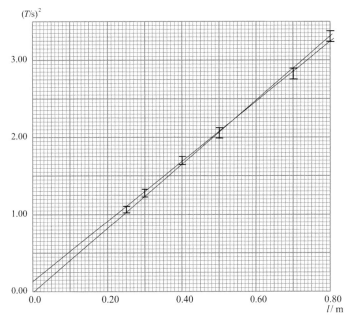

(iii) The gradient, m, of the graph is given by $m = \dfrac{4\pi^2}{g}$

Using the extreme points on the line the maximum gradient, $m_{max} = \dfrac{3.31 - 0.00}{0.800 - 0.002} = 4.148$

And the minimum gradient,

$m_{min} = \dfrac{3.24 - 0.13}{0.800 - 0.000} = 3.888$

∴ Gradient,

$m = \dfrac{4.148 - 3.888}{2} \pm \dfrac{4.148 - 3.888}{2} = 4.018 \pm 0.13$,

i.e. 4.02 ± 0.13 s^2 m^{-1}

% uncertainty in the gradient $= \dfrac{0.13}{4.02} \times 100 = 3.2\%$

∴ $g = \dfrac{4\pi^2}{4.02} \pm 3.2\% = 9.8 \pm 0.3$ m s^{-2}

(b) (i) At the highest point, the level below the pivot
$= 3.0$ m $\times \cos 70° = 1.03$ m

∴ Height of swing, $h = 3.00 - 1.03 = 1.97$ m

∴ Speed, v at lowest point $=$
$\sqrt{2gh} = \sqrt{2 \times 9.81 \times 1.97} = 6.22$ m s^{-1} QED

(ii) Resultant force = centripetal force $= \dfrac{mv^2}{r}$

$= \dfrac{80 \times 6.22^2}{3.0}$

$= 1032$ N

∴ Tension in supporting ropes $= mg + 1032$
$= 1820$ N (3 s.f.)

(iii) Because the velocity is momentarily zero, there is no centripetal acceleration. Hence the acceleration of the man at the extremes is inwards along the tangent to the arc of the swing. Hence the resultant force is inwards along the arc.

Question 2: Kinetic theory and thermal physics

(a) The internal energy = total kinetic energy of the molecules
$= \dfrac{3}{2}nRT = \dfrac{3}{2}pV$

$= \dfrac{3}{2}5.0 \times 10^6\text{Pa} \times 9.00 \text{ m}^3$

$= 6.75 \times 10^7$ J

The number of molecules $=$

$\dfrac{75 \text{ kg}}{4.00 \times 10^{-3} \text{ kg mol}^{-1}} \times 6.02 \times 10^{23} \text{ mol}^{-1} = 1.129 \times 10^{28}$

∴ Mean molecular kinetic energy $=$

$\dfrac{6.75 \times 10^7 \text{ J}}{1.129 \times 10^{28}} = 5.98 \times 10^{-21}$ J

(b) (i) Kinetic energy of O_2 molecules = kinetic energy of H_2 molecules

∴ rms speed of O_2 molecules = ¼ rms speed of H_2 molecules

$= 500$ m s^{-2}

(ii) Momentum change ∝ mass × velocity

∴ Momentum change per collision for
$H_2 = \dfrac{1}{16} \times 4 = \dfrac{1}{4} \times$ that of the O_2 molecules

(iii) Because the hydrogen molecules are moving four times as quickly, the number of collisions (per unit area) per second is four times as great as for the oxygen. The pressure is produced by the product of momentum change per collision and number of collisions per second, which is the same for the two gases.

(c)

Process	ΔU	Q	W
A sample of ideal gas expands at constant of temperature	$= 0$	> 0	> 0
A perfectly insulated sample of ideal gas is compressed	> 0	$= 0$	< 0
A sample of water at 100 °C is boiled into steam at 100 °C	> 0	> 0	> 0
An iron wire at constant temperature has an electric current in it	$= 0$	< 0	< 0

(d) (i) Assuming no heat exchange with the laboratory:

Heat lost by iron filings = heat gained by water + heat gained by the beaker

$$\therefore 200\,c_i\,(100 - 28.5) = 150 \times 4190 \times (28.5 - 18.0)$$
$$+ 50 \times 670 \times (28.5 - 18.0)$$
$$\therefore \qquad 14300\,c_i = 6\,951\,000$$
$$\therefore \qquad c_i = 486\ \text{J kg}^{-1}\,\text{K}^{-1}\ (3\ \text{sf})$$

(ii) A larger mass of iron filings would produce a larger temperature rise which would be easier to measure accurately. However, there is already a rather large mass of iron – more might be difficult to handle quickly, which would give more time for heat to be lost.

If the initial temperature were about 13 °C (i.e. about 5 °C below room temperature) the water would initially be gaining heat from the room. But when it rose by about 10 °C it would lose some heat to the room. The two effects should cancel out. In the original experiment, heat is always being lost to the room, so this should be an improvement.

Question 3: Nuclear physics

(a) (i) Only the lining of the airways and the skin are in contact with radon. Few α particles penetrate beyond dead cells on the surface of skin and so they cause little or no damage to living cells. But lung tissue is not so protected, and is damaged by the highly ionising α particles. Other organs are not (directly) damaged because the α particles from the radon can't penetrate into them.

(ii) (I) For 25% of its initial activity to remain, the initial activity must have halved and halved again, so 2 half-lives must have elapsed.

Thus time taken $= 6.6 \times 10^5$ s.

(II) For 20% of its initial activity, $\dfrac{A}{A_0} = 0.2$. But $\dfrac{A}{A_0} = e^{-\lambda t}$

So $t = -\dfrac{1}{\lambda} \ln\!\left(\dfrac{A}{A_0}\right) = -\dfrac{t_{\text{half}}}{\ln 2} \ln\!\left(\dfrac{A}{A_0}\right)$

$= -\dfrac{3.3 \times 10^5\ \text{s}}{\ln 2} \ln 0.20 = 7.7 \times 10^5$ s

[Alternatively $0.2 = 2^{-n}$,

so $n = -\dfrac{\ln 0.20}{\ln 2} = 2.32$ half-lives, etc.]

(b) (i) $I = I_0 e^{-\mu x} = I_0 e^{-10\ \text{m}^{-1} \times 0.003\ \text{m}} = 0.97\,I_0 = 97\%\,I_0$

So intensity does indeed fall by 3%.

(ii) Alphas would be absorbed by a thin film of aluminium, and betas would be absorbed by 3 mm of aluminium. A small percentage of gammas would be absorbed – so we have gammas in this case. Also the intensity of α and β radiation does not decrease exponentially, unlike γ.

(c) (i) To conserve baryon number and charge, X must be a neutral baryon, so it is a neutron.

All the particles involved are hadrons and there is no change of quark flavour so the strong interaction is responsible.

(ii) Mass loss $= (2.0136 + 3.0155 - 4.0015 - 1.0087)$ u
$= 0.0189$ u

So energy released $= 0.0189 \times 931$ MeV $= 17.6$ MeV

(iii) As mass energy is lost,

total final BE > initial BE

But X, being an individual nucleon, has no binding energy,

So BE of ^4_2He > BE of ^2_1H + BE of ^3_1H.

Question 4: Capacitors

(a) From graph, the initial charge $= 50$ μC.

But initial pd $= 10$ V, so $C = \dfrac{Q}{V} = \dfrac{50\ \mu\text{C}}{10\ \text{V}} = 5.0$ μF.

(b) Since the decay is exponential, Q will have fallen to 37% of its initial value, that is to 18.5 V, at a time equal to RC.

From the graph this is at 15 s, so $RC = 15$ s.

(c) (i)

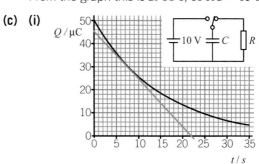

Current at 7.5 s $= \dfrac{\Delta Q}{\Delta t} =$ gradient of tangent at 7.5 s

$= \dfrac{-46\mu\text{C}}{22\ \text{s}} = (-)2.1$ μA

(ii) From graph, at $t = 7.5$ s, $Q = 30$ μC.

So $V = \dfrac{Q}{C} = \dfrac{30\mu\text{C}}{5\mu\text{F}} = 6.0$ V

But, using answers to (a) and (b), $R = \dfrac{RC}{C} = \dfrac{15\ \text{s}}{5.0\mu\text{F}} = 3.0$ MΩ

So, at $t = 7.5$ s, $I = \dfrac{V}{R} = \dfrac{6.0\ \text{V}}{3.0\ \text{M}\Omega} = 2.0$ μA

This is within 5% of the value found in (c)(i), which is reasonable agreement considering that drawing a tangent to a poky little graph is not likely to be very accurate.

(d) Energy dissipated in resistor is energy lost from the capacitor

$E = \dfrac{Q^2_{t=0}}{2C} - \dfrac{Q^2_{t=7.5\text{s}}}{2C}$

$= \dfrac{(50 \times 10^{-6}\text{C})^2}{2 \times 5.0 \times 10^{-6}\text{F}} - \dfrac{(30 \times 10^{-6}\text{C})^2}{2 \times 5.0 \times 10^{-6}\text{F}} = 1.6 \times 10^{-4}$ J

Question 5: Electrostatic fields

(a) (i)

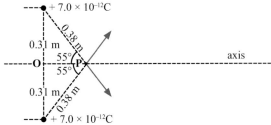

(ii) Field strength at P due to one charge $= \dfrac{Q}{4\pi\varepsilon_0 r^2}$

$$= 9.0 \times 10^9 \text{ F}^{-1} \text{ m} \times \frac{7.0 \times 10^{-12} \text{ C}}{(0.38 \text{ m})^2}$$

$$= 0.436 \text{ V m}^{-1} \text{ [or N C}^{-1}]$$

Field components $\perp$ axis cancel, field components along axis add, so

$$E_{\text{res}} = 2 \times 0.436 \text{ V m}^{-1} \cos 55°$$

$$= 0.50 \text{ V m}^{-1} \text{ [or N C}^{-1}] \text{ to the right } (\rightarrow).$$

(iii) (I) E_{res} is zero at **O** because at this point, field strengths from each charge are equal and opposite.

(II) Fields due to individual charges vary as $\dfrac{1}{r^2}$ and so resultant field approaches zero as r approaches infinity.

(b) (i) $a = \dfrac{F}{m} = \dfrac{Eq}{m} = \dfrac{0.50 \text{ N C}^{-1} \times 4.8 \times 10^{-19} \text{ C}}{4.5 \times 10^{-25} \text{ kg}}$
$= 5.3 \times 10^5 \text{ m s}^{-2}$

(ii) The ion's acceleration is proportional to E_{res}, as in the graph, so the ion's speed increases *at an increasing rate* from 2000 m s^{-1}, its value at O, until the ion reaches the point at which the graph peaks. After that its speed continues to increase, but at a decreasing rate.

(iii) $\text{KE} + \text{PE} = \frac{1}{2}mv^2 + 2\left(\dfrac{Q}{4\pi\varepsilon_0 r}\right)q$

$$= \frac{1}{2}4.5 \times 10^{-26}\text{kg } (2000 \text{ m s}^{-1})^2 +$$

$$\frac{2 \times 9.0 \times 10^9 \text{ F}^{-1} \text{ m} \times 7.0 \times 10^{-12}\text{C} \times 4.8 \times 10^{-19}\text{C}}{0.31 \text{ m}}$$

$$= 2.85 \times 10^{-19} \text{ J}$$

(iv) This energy is conserved, but as the ion becomes very distant from the charges, electrical PE approaches zero, so energy is wholly kinetic,

So $\frac{1}{2}mv^2 = 2.85 \times 10^{-19}$ J, therefore

$$v = \sqrt{\frac{2 \times 2.85 \times 10^{-19} \text{ J}}{4.5 \times 10^{-26} \text{ kg}}}$$

$$= 3.6 \text{ km s}^{-1}$$

Question 6: Gravity and orbits

(a) (i) The square of the orbital period is directly proportional to the cube of the semi-major axis of the orbit.

(ii)

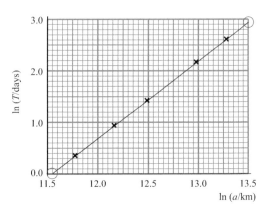

The points lie on a straight line, so the relationship is $T \propto a^n$ where n is the gradient of the log graph.

Using the marked points: $n = \dfrac{2.92 - 0.00}{13.50 - 11.55} = 1.497$

$$= 1.50 \text{ (3 sf)}$$

$\therefore T \propto a^{1.5}$, i.e. $T^2 \propto a^3$ which is in agreement with Kepler's third law.

(iii) Jupiter is more massive than Uranus. Hence, at any particular radial distance, the gravitational force on a moon is greater, so the centripetal acceleration is greater.

The centripetal acceleration, a, is given by

$$a = r\omega^2 = \frac{4\pi^2 r}{T^2}$$

Hence, the value of T, the orbital period is less for Jupiter. The $T \propto a^n$ relationship is still valid, so the graph for Jupiter is parallel to and below the one for Uranus.

(b) (i) The resultant gravitational force from the material outside the orbit of the Sun is zero, assuming that the material is uniformly distributed.

(ii) We can calculate the mass within the orbit of the Sun from the velocity and radius using the centripetal force equation:

$$\frac{mv^2}{r} = \frac{GMm}{r^2}$$

which rearranges to $M = \dfrac{rv^2}{G}$

The mass that astronomers can 'see' is:

$$M = \frac{2.51 \times 10^{20} \times (108 \times 10^3)^2}{6.67 \times 10^{-11}} = 4.4 \times 10^{40} \text{ kg}$$

The observed speed is nearly twice as great which gives a mass 4× this, is ~1.6×10^{41} kg.

Hence, astronomers conclude that most of the mass of the galaxy is in a form which we cannot detect using electromagnetic radiation, which is known as Dark Matter.

(c) The universe is expanding and the Hubble constant gives the rate of the expansion. The value of 2.20×10^{-18} s^{-1}, tells us that two objects are a distance d apart, the rate of increase of that distance is $2.20 \times 10^{-18}d$. If the rate of expansion remains the same then in, say, 1 billion years $(3.2 \times 10^{16}$ s) time, the distance d will have increased to $1.07d$ $(2.2 \times 10^{-18} \times 3.2 \times 10^{16} = 0.07)$.

The time, T, it will take to increase the separation to $2d$ is given by $\dfrac{1}{2.2 \times 10^{-18}\text{ s}^{-1}} = 4.5 \times 10^{17}$ s.

We can also work backwards to say that, at 4.5×10^{17} s in the past, the separation of the two objects was zero. This time is independent of the current separation and so this time in the past marks the start of the universe.

4.5×10^{17} s $= \dfrac{4.5 \times 10^{17}}{86400 \times 365.25}$ years $= 14$ billion years

This is only a rough estimate because it assumes that the expansion is uniform, i.e. that gravity is not slowing down the expansion. If gravity were having an effect then the true age of the universe would be less than 14 billion years. Recent measurements suggest the expansion is actually accelerating after an initial deceleration (due to Dark Energy) so the figure of 14 billion years is quite close.

Question 7: Magnetic fields and em induction

(a) (i)

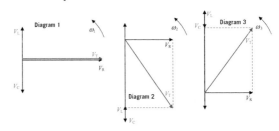

Graph of $\frac{1}{I}$ /A^{-1} (y-axis, 0 to 0.5) against $\sin \theta$ (x-axis, 0 to 1.0).

Gradient $= \dfrac{(0.455 - 0.020)\text{A}^{-1}}{1.00 - 0.00} = 0.435$ A^{-1}

[or, more safely 0.44 A^{-1}]

(ii) $F = BI\ell \sin \theta$, so $\dfrac{1}{I} = \dfrac{B\ell}{F} \sin \theta$

Since F, B and ℓ are constants, the graph should be a straight line through the origin.

The graph is convincingly straight, but there is an unmistakeable small positive intercept (a zero error).

(iii) From (a)(ii) we see that $\dfrac{B\ell}{F} =$ gradient

Therefore $B = \dfrac{F \times \text{gradient}}{\ell}$

$= \dfrac{3.5 \times 10^{-3}\text{ N} \times 0.435\text{ A}^{-1}}{0.025\text{ m}} = 0.061$ T

(b) emf in ring $= \dfrac{\Delta \Phi}{\Delta t} = \dfrac{BA_{\text{ring}} - 0}{\Delta t} = \dfrac{0.25\text{ T} \times \pi(0.035\text{ m})^2}{0.14\text{ s}}$

$= 6.87 \times 10^{-3}$ V

Resistance of ring $= \dfrac{\rho\ell}{A_{\text{wire}}} = \dfrac{1.7 \times 10^{-8}\ \Omega\text{m} \times 2\pi \times 0.035\text{m}}{\pi \times (0.30 \times 10^{-3}\text{m})}$

$= 0.0132\ \Omega$

Therefore current $= \dfrac{6.87 \times 10^{-3}\text{V}}{0.0132\ \Omega} = 0.52$ A

Question 8: Option A

(a)

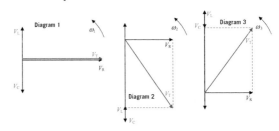

(i) The pd across a resistor is in phase with the current, so the current phasor is horizontal to the right. The pd across an inductor leads the current by $\pi/2$; the pd across a capacitor is $\pi/2$ behind the current.

(ii) The magnitudes of the 3 values of V_{T} are the same.

(iii) ω_1 is at the resonance frequency – we can tell this because $V_C = V_L$. ω_2 is below the resonance frequency $(V_C > V_L)$ and ω_3 above the resonance frequency. The phase angles in diagrams 2 and 3 are equal and opposite which shows that $\dfrac{\omega_3}{\omega_1} = \dfrac{\omega_1}{\omega_2}$.

(iv) $Q = \dfrac{V_L}{V_R}$ (at resonance) $= \dfrac{1.7}{4.2} = 0.4$

It can be increased by decreasing the value of the resistor.

(b) (i) $L = \dfrac{\mu_0 N^2 A}{l}$, $\therefore N = \sqrt{\dfrac{lL}{\mu_0 A}} = \sqrt{\dfrac{1.50 \times 10^{-2} \times 50 \times 10^{-6}}{4\pi \times 10^{-7} \times \pi(2.5 \times 10^{-3})^2}}$

$= 174$ turns

(ii) $R = \dfrac{\rho l}{A} = \dfrac{1.72 \times 10^{-8} \times 174 \times \pi \times 5.0 \times 10^{-3}}{\pi \times (0.1 \times 10^{-3})^2} = 1.50\ \Omega$

$X_L = 2\pi f L = 2\pi \times 10^4 \times 50 \times 10^{-6} = 3.14\ \Omega$

$\therefore Z = \sqrt{1.50^2 + 3.14^2} = 3.48\ \Omega$

(iii) $I_0 = \dfrac{V_0}{Z} = \dfrac{10\text{ mV}}{3.48\ \Omega} = 2.87$ mA

Phase angle $= \tan^{-1}\dfrac{X}{R} = \tan^{-1}\dfrac{3.14}{1.50} = 1.125$ rad

$\therefore$ Current is behind pd by

$\dfrac{1.125}{2\pi} \times 0.1$ ms $= 0.018$ ms

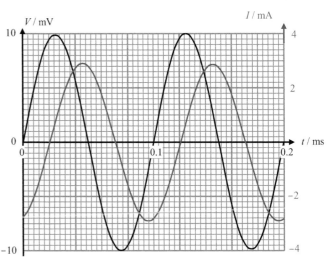

Question 9: Option B

(a) (i)

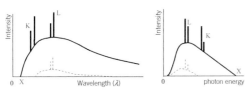

The wavelength and the photon energy are inversely related by $E = \dfrac{hc}{\lambda}$. The point X on the two graphs is the cut off point: the highest energy and the shortest wavelength photons emitted.

The values at X are: $E = eV$ and $\lambda = \dfrac{hc}{eV}$ where V is the operating voltage of the tube. The lines labelled K and L are the line spectra of the tube and are characteristic of the target metal of the tube.

(ii) See dotted lines on the graphs.

(iii) (I) $\quad \dfrac{I}{I_0} = e^{-0.4 \times 8.0} = 0.041$ (4.1%)

(II) If a beam of X-rays penetrates the arm, the intensity of the beam which passes just through the muscle is 4%. Assuming the bone is 2 cm thick, the fraction of the X-rays which pass through the bone also passes through 6 cm of muscle, so the total intensity is reduced to:
$e^{-1.90 \times 2} \times e^{-0.4 \times 6} = 0.002 = 0.2\%$

The ratio of the intensities which has passed through the bone to that which hasn't is about 1:5, so the shadow of the bone can be clearly seen on the detector and breakages seen.

(b) (i) $\quad f = 42.6 \times 1.8 \text{ MHz} = 77 \text{ MHz}$

(ii) The additional non-uniform field adds to the uniform field so that the total flux density varies across the patient. The Lamor frequency is proportional to B. Therefore the hydrogen nuclei in different parts of the patient therefore emit radio waves with frequencies which depend upon their location. For example, a place having a B value 1 mT bigger gives out radiation which is 42.6 kHz greater ($42.6 \times 10^6 \times 1 \times 10^{-3}$ Hz). This difference can be detected by the MRI scanner and used to build up an image.

(c) MRI scans can be used to image soft tissues: conventional X rays are useful for imaging bones, though they can be used for soft tissues with the use of a suitable contrast agent.

X-rays are ionising radiation, which have carcinogenic effects and can also lead to developmental problems in embryos. There are no known hazards associated with MRI scans, though care has to be taken to remove metal objects and they cannot be used with patients who have pacemakers.

X-ray examinations can be undertaken at short notice and are quick and low cost. MRI scans on the other hand require a long time and are expensive because of the capital cost of the equipment.

MRI scans can produce high definition 3D images of soft tissue but require the patient to remain still for a long time. X-rays can produce very high definition images of bones but only 2D.

Question 10: Option C

(a)

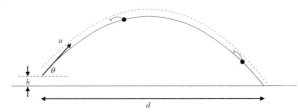

The red dotted line shows the path of the hammer, which is released from a higher point (exaggerated). As the diagram shows, the path higher everywhere than the original and so reaches a further point.

(b) Assuming no air resistance:

The distance $d = u \cos \theta \times$ flight time

$\therefore$ flight time, $t = \dfrac{72.46 \text{ m}}{26.5 \text{ m s}^{-1} \cos 41°} = 3.62 \text{ s}$

The vertical position, y, is given by $y = ut \sin \theta - \frac{1}{2}gt^2$
$$= 26.5 \times 3.62 \sin 41° - 4.905 \times 3.62^2$$
$$= -1.34 \text{ m}$$

The vertical position is thus 1.34 m below the release point which is in good agreement with the initial height of 1.4 m, suggesting that the 'no air resistance' assumption was correct.

(c) (i) Radius of sphere is given by $M = \frac{4}{3}\pi r^3 \rho$,

so $\dfrac{3M}{4\pi\rho} = \dfrac{3 \times 7.3}{4\pi \times 7900}$
$$\rightarrow r = 0.0604 \text{ m}$$

$\therefore$ csa of sphere $= \pi r^2 = 0.0115 \text{ m}^2$

Taking the mean speed of the sphere to be 22 m s^{-1}, the drag force is given by:

$F_D = C_D A v \rho^2 = 1.0 \times 0.115 \times 1.3 \times 22^2 = 7 \text{ N}$

(ii)

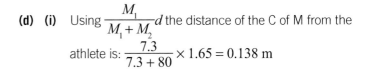

(d) (i) Using $\dfrac{M_1}{M_1 + M_2}d$ the distance of the C of M from the athlete is: $\dfrac{7.3}{7.3 + 80} \times 1.65 = 0.138 \text{ m}$

(ii) Final angular velocity $= \dfrac{v}{r} = \dfrac{26.5}{(1.65 - 0.138)} = \dfrac{26.5}{1.512}$
$= 17.5 \text{ rad s}^{-1}$

Total angular momentum $=$
$I\omega = (80 \times 0.138^2 + 7.3 \times 1.512^2) \times 17.5 = 315 \text{ kg m}^2 \text{ s}^{-1}$

(iii) Mean angular velocity in acceleration $= 8.76 \text{ rad s}^{-1}$.
Angle turned $= 4 \times 2\pi$

$\therefore$ Time taken $= 2.87 \text{ s}$

$\therefore$ Mean torque $= \dfrac{\Delta L}{t} = \dfrac{315}{2.87} = 108 \text{ kg m}^2 \text{ s}^{-2}$

(e) Centripetal force on hammer at release
$= mr\omega^2 = 7.3 \times 1.512 \times 17.5^2 = 3400 \text{ N}$

Question 11: Option D

(a) (i) Power absorbed by the black body Earth $= \pi R_E^2 \times G$, where R_E is the radius of the Earth and G the solar constant.

If the mean temperature of the Earth is T the power emitted by the Earth $= 4\pi R_E^2 \sigma T^4$

$\therefore$ For equilibrium: $\quad 4\pi R_E^2 \sigma T^4 = \pi R_E^2 G$

$\therefore T^4 = \dfrac{G}{4\sigma} = \dfrac{1350 \text{ W m}^{-2}}{4 \times 5.67 \times 10^{-8} \text{ W m}^{-2} \text{ K}^{-4}}$

$\therefore T = 278 \text{ K} = (278 - 273)°\text{C} = 5 °\text{C QED}$

(ii) The solar radiation is a black body spectrum with a maximum at about 500 nm, so most of the power is within the near UV, visible and near IR parts of the spectrum.

The radiation emitted by the Earth is largely in the mid and far IR (i.e. thermal IR) parts of the spectrum with a peak at around 10 µm.

(iii) The power absorbed is only $0.7 \times \pi R_E^2 G$, so the power emitted should be less for equilibrium. This results in a lower temperature than that calculated in part (i).

(iv) The atmosphere is largely transparent to visible and near visible radiation (though 30% is reflected by the atmosphere, clouds and the ground). The ground re-emits radiation in the thermal infrared, much of which is absorbed by polyatomic molecules in the atmosphere, e.g. CO_2, H_2O and CH_4, which have vibrational energy levels with the appropriate energy spacing. The atmosphere is opaque at around 10 µm. These molecules re-emit radiation in the same wavelength range proximately 50% is absorbed by the ground, raising its temperature to a level above its value in the absence of an atmosphere.

(b) (i) The absence of cloud cover and the Sun at right angles to the plane of the unit.

(ii) At 20 V, $P = IV = 3.1 \text{ A} \times 20 \text{ V} = 62 \text{ W}$.

(iii) If the unit were short circuited, the pd would be zero and the power output would hence be zero. If the unit were open circuited, the output current would be zero and hence the power output would be zero.

(iv)

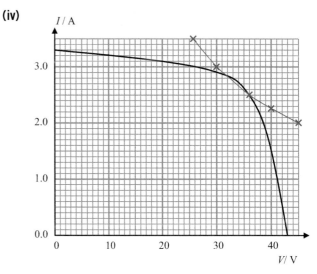

The red line on the graph is the 90 W line, i.e. it has the equation

$$I = \dfrac{90}{V}$$

If the unit's electronic control unit ensures that V_{OUT} is between 32 V and 36 V the power output is above this line, so the claim is justified. The highest output is around 34 V, with a current of 2.7 A giving a power of 92 W.

Under 'non-ideal' conditions the power output would be less!

Index